**EMILY CARR UNIVERSITY OF ART + DESIGN
LIBRARY**
1399 JOHNSTON STREET
VANCOUVER, BC V6H 3R9
TELEPHONE : (604) 844-3840

Culture, Environmental Action and Sustainability

Ricardo García Mira
José M. Sabucedo Cameselle
& José Romay Martínez
(Editors)

Culture, Environmental Action and Sustainability

Ricardo García Mira,
José M. Sabucedo Cameselle &
José Romay Martínez
(Editors)

DISCARDED

**EMILY CARR UNIVERSITY OF ART + DESIGN
LIBRARY**
1399 JOHNSTON STREET
VANCOUVER, BC V6H 3R9
TELEPHONE : (604) 844-3840

Library of Congress Cataloging-in-Publication Data
is available via the Library of Congress Marc Database under the
LC Control Number 2003114186

National Library of Canada Cataloguing in Publication Data

International Association for People-Environment Studies.
Conference (17th : 2002 : A Coruña, Spain)
Culture, environmental action and sustainability / edited by Ricardo García Mira, José Manuel Sabucedo Cameselle, José Romay Martínez.

Proceedings of the 17th International Association for People-Environment Studies Conference, held in A Coruña, Spain, July 23-27, 2002.

Includes bibliographical references.
ISBN 0-88937-282-9

1. Sustainable development--Congresses. 2. Human ecology--Congresses.
3. Culture--Congresses. 4. Globalization--Environmental aspects--Congresses.
5. Globalization--Social aspects--Congresses. 6. Environmental quality--Congresses.
I. García Mira, Ricardo II. Sabucedo Cameselle, José Manuel III. Romay Martínez, José IV. Title.

GF3.I57 2002   333.7   C2003-906311-9

Copyright © 2003 by Hogrefe & Huber Publishers

PUBLISHING OFFICES
USA: Hogrefe & Huber Publishers, 875 Massachusetts Avenue, 7th Floor,
Cambridge, MA 02139
Phone (866) 823-4726, Fax (617) 354-6875, E-mail info@hhpub.com
Europe: Hogrefe & Huber Publishers, Rohnsweg 25, D-37085 Göttingen, Germany,
Phone +49 551 496090, Fax +49 551 49609-88, E-mail hh@hhpub.com

SALES & DISTRIBUTION
USA: Hogrefe & Huber Publishers, Customer Service Department,
30 Amberwood Parkway, Ashland, OH 44805,
Phone (800) 228-3749, Fax (419) 281-6883, E-mail custserv@hhpub.com
Europe: Hogrefe & Huber Publishers, Rohnsweg 25, D-37085 Göttingen, Germany,
Phone +49 551 496090, Fax +49 551 49609-88, E-mail hh@hhpub.com

OTHER OFFICES
Canada: Hogrefe & Huber Publishers, 1543 Bayview Avenue, Toronto, Ontario M4G 3B5
Switzerland: Hogrefe & Huber Publishers, Länggass-Strasse 76, CH-3000 Bern 9

COVER PHOTOGRAPH
Castro de Borneiro (Cabana, A Coruña, Spain). Photographed by A. Romero Masiá.

Hogrefe & Huber Publishers
Incorporated and Registered in the State of Washington, USA, and in Göttingen, Lower Saxony, Germany

No part of this book may be reproduced, stored in a retrieval system, or transmitted, in any form or by any means, electronic, mechanical, photocopying, micro-filming, recording or otherwise, without the written permission from the publisher.

Printed and bound in Germany
ISBN 0-88937-282-9

Table of Contents

I. Introduction

Culture, environmental action and sustainability 3
Ricardo García Mira, José M. Sabucedo Cameselle,
& José Romay Martínez

II. Culture, quality of life and globalisation

Culture, quality of life and globalisation 11
Ricardo García Mira, José M. Sabucedo Cameselle,
& José Romay Martínez

Our uncommon future 21
David Uzzell

Globalisation, commons dilemmas and sustainable quality of life:
What do we need, what can we do, what may we achieve? 41
Charles A. J. Vlek

At home everywhere and nowhere: Making place in the global village 61
Nan Ellin

III. Environmental action and participation

Action competence in environmental education 71
María Dolores Losada Otero & Ricardo García Mira

Participatory action research as a participatory approach to addressing
environmental issues 85
Esther Wiesenfeld, Euclides Sánchez & Karen Cronick

Neighbouring, sense of community and participation:
Research in the city of Genoa 101
Laura Migliorini, Antonella Piermari & Lucia Venini

Environmental values and behavior in traditional Chinese culture 117
Chin-Chin (Gina) Kuo & A. T. Purcell

Some sociodemographic and sociopsychological predictors of
environmentalism 133
Maaris Raudsepp

Sustainable technology and user participation:
Assessing ecological housing concepts by focus group discussions 145
Michael Ornetzeder

IV. Urban sustainability and cultural diversity

Citydwellers' relationship networks:
Patterns of adjustment to urban constraints 161
Gabriel Moser, Alain Legendre & Eugénia Ratiu

Planned gentrification as a means of urban regeneration 171
Miriam Billig & Arza Churchman

Ubiquitous technology, the media age, and the ideal of sustainability 185
Andrew D. Seidel

Applying urban indicators to clarify the urban development of Taipei ... 199
Yung-Jaan Lee

Choosing sustainability:
The persistence of non-motorized transport in Chinese cities 219
John Zacharias

A cultural comparative analysis of two villages in Storm Valley,
Rize, Turkey... 231
Fitnat Cimşit, Erincik Edgü and Alper Ünlü

The dialectics of urban play..................................... 243
Quentin Stevens

Gated communities and urban planning:
Globalisation or national policy 255
Sarah Blandy and David Parsons

Performance and appropriation of residential streets and public
open spaces 267
Maria Cristina Dias Lay and Jussara Basso

V. Children and the environment

Evaluating links intensity in social networks in a school context
through observational designs 287
M. Teresa Anguera Argilaga, Carlos Santoyo Velasco &
M. Celia Espinosa Arámburu

Projects and policies for childhood in Italy 299
Antonella Rissotto

The Body Goes to the City project:
Research on safe routes to school and playgrounds in Ferrara 313
Marcello Balzani & Antonio Borgogni

Environmental comfort and school buildings:
The case of Campinas, SP, Brazil................................. 325
Doris C.C.K. Kowaltowski, Silvia A. Mikami G. Pina, Regina C. Ruschel,
Lucila C. Labaki, Stelamaris R. Bertoli & Francisco Borges Filho

VI. The elderly and the environment

Satisfaction ratings and running costs of nursing homes
Advantages of smaller and more homelike units 337
Karin Høyland

New Urbanism as a factor in the mobility of the elderly............... 351
Michael J. Greenwald

Founding houses for the elderly
On housing needs or dwelling needs? 367
Wim J. M. Heijs

Older Spanish adults' involvement in the education of youngsters 385
Vicente Lázaro & Alfonso Gil

Authors 401

Reviewers.. 411

I. Introduction

Culture, environmental action and sustainability

Ricardo García Mira, José M. Sabucedo Cameselle, & José Romay Martínez

The 17th IAPS Conference held in Corunna (Spain) in 2002 reflected interest in a large number of themes of current importance and relevance in the study of person–built environment interaction in general. On this occasion, the centre of attention focussed on the analysis of the role played by culture in forming quality of life within the framework of a process that we are all witnessing and which goes by the name of globalisation. The study of this singular interaction, under the influence of processes such as those that the various authors included in this book attempt to describe, is of vital importance given its implications to the present and future lives of all citizens.

This book includes research that has taken place into this wide-ranging topic, grouped into the specific subject areas that we will now briefly discuss.

Culture, quality of life and globalisation

It has now been clear for several years that the reciprocal interaction of people with their environment, and the environmental problems that are the result, cannot be the object of thorough study unless we take into account the general dimensions in which people live their lives. In this regard it must be said that over the last few decades we have been faced with three major interrelated dimensions that affect our everyday life on both an individual and group level, and which determine and, at the same time, are determined by the multiple facets of human life. These three super-dimensions are, in turn, shaped by the major changes that have taken place in the world since the end of World War II, and particularly after the 1960s, changes for which an explanation has been sought from a macro-social standpoint such as the post-modern age, whose repercussions affect the way personal values are shaped – as Inglehart, amongst others, has pointed out. Post-modernity, therefore, cannot be understood without reference being made to the new dimensions of culture, the importance of quality of life, and globalisation.

On the other hand, faced with a culture taken almost exclusively to mean Western culture, other cultures such as the Eastern, African, Islamic, Indian, Chinese, etc. have gradually established themselves as cultures in their own right. In short, cultures have bloomed because they have been taken to be what surrounds and is closest to individual lives and actions, and is expressed in the individual's language, traditions, way of life and methods of production. What is more, we are nowadays witnessing the defense of former ways of life that affect not only the landscape, but also the habitat, work, and, in short, what can be called sustainability.

The abandonment in mass of the countryside and farming in many areas of the world has resulted in the loss of many spaces linked to the traditional methods of production and has contributed to a reorganisation of spaces, creating urban conglomerations with all their associated problems of pollution, traffic, noise, precarious employment, often accompanied by a lower level of quality of life. There have undoubtedly been improvements in medicine, hygiene, diet and education, and these have been powerful contributors to the improved quality of life of the population as a whole, but we must not ignore the fact that many marginal city-dwellers have frequently had to suffer the negative consequences of poorly planned or uncontrolled growth. On the other hand, the increase in the older population has led to an increase in applied research into district planning and housing design. A further contribution has been made by the valuing of natural elements such as air and water, precisely because of their scarcity, and of a healthy diet.

The chapter by García-Mira, Sabucedo and Romay analyses these three dimensions of culture, quality of life and globalisation, placing them in their broadest psychosocial context and analysing the tensions generated by a global culture in the smaller-sized spheres of local cultures, whilst at the same time calling for a space for reflection in which the complexity of social organisation in a changing world and the transformation that this works on our lifestyles can be analysed. The study also looks at the role of psychology in interpreting the way in which culture itself constructs the framework for interpreting reality, and puts forward the need to make the existence of a global framework of interaction between cultures compatible with the necessary recognition of the idiosyncratic specificity of each culture.

However, within this cultural specificity, it is only possible to attain a high level of quality of life and harmonious development free of conflicts with other cultures if globalisation policies are developed within a set of fair and balanced parameters. In this context, Uzzell emphasises the concept of "sustainable development" as being one of the most important ideas of the 20th century, since it is around this concept that our common future turns. It demands that all citizens in general, but more particularly those in positions of political power, assume the responsibility for global environmental change, as opposed to local or personal changes. As a result of globalisation, time has conquered space and the mass media play a vital role in the very structuring and definition of reality. For this reason, we have reached a point where people consider global environmental problems to be more serious than those that take place lower down on the spatial scale.

Family and school are the two contexts in which children can become the real promoters of the attitudinal and behavioural change that must spread to all levels of the community, with place identity, manifest through the interrelationship between individuals, and also between these and place, being a major dimension of environmental attitudes.

Vleck presents an expanded commons dilemma model that allows us to understand and handle the tensions between individual and collective interests, paying particular attention to transport problems, for which possible solutions are proposed from a perspective of sustainability.

The close relationship between economic progress, work, income and quality of life is interpreted on a personal level by people who value health, the family and safety as the most important elements in their lives. We can therefore say that the problems of sustainability are multidimensional problems characterised by the impacts, causes and processes of the economic, social and environmental dimensions in which the tensions between the individual and the collective, i.e. between individual freedom and social equity, are involved.

Ellin analyses how it is possible to make a place for oneself in this global village, together with the means that the new technologies provide for doing so. He verifies, however, the failure of postmodernist town planning, which makes it necessary to redefine values, goals and the means with which to reach them. Ellin points out that in fact things have attained crisis proportions in both the natural and the built environment, and draws our attention to the warning contained in the United Nations Environment Programme regarding the very short time left – little more than a decade – within which to achieve environmental sustainability before entering a state of physical and economic decline.

Environmental action and participation

Environmental action and participation is a major component of the person-environment line of study, and several contributions fall within its ambit. The first is a study by Migliorini, Piermari and Venini that deals with the aspects of neighbourhood, the sense of community and participation in Genova, Italy. As its authors point out, the sense of community is a key concept with which people must familiarise themselves in the context of an urban reality, particularly if we consider the massive expansion of cities the world over in the last fifty years. Massive migration to the cities has fractured many of our social networks and the sense of small urban communities, which have to be rebuilt if we want to avoid situations of social anomie. The authors highlight the role of the natural and social environmental context in promoting contacts and shaping feelings and emotions of belonging.

Wiesenfeld, Sánchez and Cronick present a participatory action research model (PAR), which, although not often used in psychology, as the authors themselves mention, may prove extremely useful not only in solving environmental problems but also in the construction of knowledge and the education of the participants.

In this respect, the authors point out that the habitat is a global phenomenon whose study cannot be undertaken as a politically neutral academic exercise, environmental psychology having the possibility to play a major role in redefining the ideological parameters as well as in providing an individual solution to the problems and in the very construction of the matter as an important active agent in the diagnosis and remedying of environmental problems. A significant element in this model is the dialectic resolution that occurs when the various views held by the participants combine to produce new options, which in turn are defined as problems by the participants, who demand practical changes appropriate to their community.

The final goal is to empower the participants so that they can continue to resolve the fundamental environmental problems of their community. In this way, the aim of the authors is not only to help the community overcome its specific problems, but also, and even firstly, to empower and to train the members of the community in methods of solving problems that may arise in the future, through the combination of the contributions of scientific knowledge, common sense, community tradition and the consideration and experience that spring from the practice of communal relationships.

Pro-environmental attitudes and behaviours cannot be understood and sustained on a global level without a close link to the system of social values, which are in turn determined by the cultures in which they are rooted. In this regard, Chin-Chin Kuo and Terry Purcell offer us the results of empirical research into the values and pro-environmental behaviours in traditional Chinese culture, notwithstanding the difficulties inherent in the study of a culture with such a vast history and covering such a variety of ethnic groups. The starting point for these authors is the belief that values are the best predictors of pro-environmental behaviour. When factor analysis is performed on the opinions of the experts, five factors emerge: harmony and balance with nature, simple lifestyle, spiritual communion with nature, altruistic and benevolent norms and *feng shui*. Although these values, when taken together, evoke in us a vision of the oriental way of life, they in fact overlap and interweave with Western cultural values, particularly from the standpoint of the environmental movement.

Raudsepp studies the behaviour of several socio-demographic and socio-psychological variables on environmental attitudes and behaviour in the rural population of Hiinmaa (Estonia). Amongst the socio-demographic variables studied those that appear as significant predictors of environmentalism are age, sex, education and subjective religiousness, whilst values, perceived control over the environment, local identity and childhood experience of nature are the most significant amongst the socio-psychological variables. Other significant predictors are family income and perceived pro-environmental local regulations. The results of this study indicate, however, that socio-psychological factors outweigh socio-demographic ones and point to the existence of factors other than those measured that would be responsible for the unexplained variance in environmental beliefs, attitudes and behaviour within the inherent heterogeneity of environmentalism.

The subject of sustainability, in the form of adopting sustainable technologies in housing construction by means of user participation in decision-making groups, is taken up by Ornetzeder from the perspective of the research recommendations for the "Building of Tomorrow" launched in 1999 by the Australian Ministry of Science. The recommendations made regarding the forming and progress of the group sessions may prove extremely useful within a participatory methodology designed to produce interaction between housing architects and builders and users in the search for a home that will satisfy its inhabitants within a sustainable environmental model.

Losada and García Mira investigate the role of action competence in environmental education, analysing the results of a study involving secondary school students who were presented with an environmental problem upon which they had to demonstrate their competence, identifying its causes and the consequences that derived from it. The problem in question dealt with the degree of cooperation or non-cooperation in waste separation within the home. The students were asked to propose possible actions that would lead to a solution of this problem, and to state the difficulties that a change in behaviour may have to face. The sum of these four dimensions – causes, consequences, actions and difficulties – would result in the concept of *action competence*, as defined by Jensen (1993). The results of this study indicate the advisability of introducing this scheme of action competence into environmental education and educational programmes in general, is a means of contributing to achieving a more sustainable society.

Urban sustainability and cultural diversity

Under this general heading the reader will find various studies that deal with questions which, far from enabling us to see a city as a conglomerate of individuals, present it as a conglomerate of cultures, experiences and actions that may either prejudice or favour its sustainable development. In this context, Moser, Legendre and Ratiu analyse the patterns of adapting the networks of interaction between citizens to the restriction that city life itself imposes. They show the effect that the lack of free time and other restrictions has on relational behaviour. Thus, the world of relationships that have their origin in one's neighbourhood, through neighbourhood associations or at work, is much larger for those who live in Paris and its outlying suburbs than for those who live in smaller towns or cities.

So whilst half of the inhabitants of Paris maintain their social relationships within the above-mentioned ambits, this is only true of one-third of those inhabitants interviewed in a small city (Tours). This means that due to the large distances between places in big cities, our social networks are organised around our neighbourhood and its organisations, or the workplace. However, those inhabitants of large cities who have the opportunity of spending their weekends outside the city (second homes) develop relations of friendship and support similar to those who live in the provinces.

Seidel reflects on the influence of the new technologies, the Internet in particular, on decision-making and how this can affect the distribution of wealth and labour, frequently in opposition to the ideal of sustainability.

Y.J. Lee applies the UIP (Urban Indicators Programme) of the UNCHS (United Nations Center for Human Settlement) to the expansion of Taipei, concluding that sustainability is achievable if attention is paid to technological aspects and efficient governance. In the same order of things, Zacharias highlights the advantages of non-motorised transport for Chinese cities.

Cimsit, Edgü and Ünlu investigate the influences of sociocultural parameters on the physical structure and environmental adaptation of two cities.

"The dialectics of urban play" in the city of Melbourne is the subject of a study by Stevens, since not all aspects of urban social life are rational and predictable, with everyday social practices having a key part to play in its development. Particular attention is paid to spatial design due to its influence on performance, representation and control.

Blandy and Parsons investigate "gated communities" within the ambit of implantation of urban policies in the United Kingdom and the framework of globalisation and national politics. Lay and Basso study the dynamics of social life in the streets and open spaces of three different residential areas of the city of Campo Grande (Brazil).

Children and the environment

This chapter brings together several different themes which cover aspects relating to environmental interaction from the standpoint of childhood. The first of these stresses the methodological aspects, and is a study by Anguera, Santoyo and Espinosa, who by means of observational designs of a diachronic, nomothetic and multidimensional nature measure the intensity of social connections in a school-based social network.

Rissotto presents a project (Città Educativa) that seeks to improve the urban environment from a children's perspective. This project has been operating since 1991, mainly in

Italy but also in some cities in Argentina and Spain. Its principal contributions being those related to the changes in our perception of childhood that has led local governments to introduce innovative choices and to seek the involvement of the community.

Within the scope of this project, Balzani and Borgogni present a research paper on safety in school and leisure routes in Ferrara (Italia), showing how an association, the Faculty of Architecture and the local authorities have managed to cooperate on this matter.

A group formed by Kowaltowski and his colleagues from the State University of Campinas (Brazil) presents the results of a public project to improve environmental comfort in school buildings. To this end the classrooms and playgrounds were observed and measured, and the various agents involved in the educational process were asked to give their opinion and state their wishes, with the aim of improving the degree of satisfaction experienced in the use of the different spaces in schools.

The elderly and the environment

This final section deals with all those matters concerning the elderly, a subject of great current interest given the progressive ageing of the world's population, particularly in the West. Hoyland gives us the results of research into nursing homes in Norway, showing that small homes (6–10 residents) are much more satisfactory than large ones from the point of view of both staff and residents. This model, in our view, could be of great valued when we consider the construction of places in which groups of people cohabit, whether they be old people's homes or residences for other groups (children, adolescents, students, etc.).

Greenwald contributes a study on new urban planning as the key factor in the mobility of the elderly, from the perspective of the role of new urban planning in the design of new instruments that will contribute to the introduction of more manageable environments for the elderly.

The same line is followed by Heijs, whose study explores the preferences, wishes and demands of the elderly with regard to their homes, with the aim of promoting the introduction of appropriate measures in a population in Eindhoven (Netherlands).

Finally, Lázaro and Gil present a study whose purpose is to identify the activities related to environmental education carried out by the elderly in Spain with regard to children and young people in the ambit of the family. The results show a significant influence that complements the environmental education provided by the parents of these children and adolescents.

Last but not least, we would like to express our gratitude to all the authors for their interesting contributions, which we are convinced will help to increase the amount of much-needed cumulative knowledge available to us regarding the study of the nature of the person-environment relationship. We are also deeply grateful to those experts who took part in the scientific review of all the manuscripts received for their time and efforts. It is our hope, then, that this book will prove a useful tool for the future work of researchers and those involved in decision-making processes, and that it will therefore contribute to improving the conditions of quality of life of individual citizens and to a better use of environmental resources in a framework of sustainability, multi-culturality and responsible environmental actions that will undoubtedly act in favour of the overall aim of achieving long-lasting progress and development.

II. Culture, quality of life and globalisation

Culture, quality of life and globalisation

Ricardo García Mira[1], José M. Sabucedo Cameselle[2], & José Romay Martínez[1]

[1]University of Crouña, Spain, [2]University of Santiago de Compostela, Spain

> **Abstract.** This chapter analyses the dimensions of culture, quality of life and globalisation, placing them in their broadest psychosocial context and analysing the tensions generated by a global culture in the smaller-sized spheres of local cultures, whilst at the same time calling for a space for reflection in which the complexity of social organisation in a changing world and the transformation that this works on our lifestyles can be analysed. The study also looks at the role of psychology in interpreting the way in which culture itself constructs the framework for interpreting reality, and puts forward the need to make the existence of a global framework of interaction between cultures compatible with the necessary recognition of the idiosyncratic specificity of each culture.
>
> **Keywords:** culture, quality of life, globalisation, social construction, environment

Culture, quality of life and globalisation are three concepts which, in addition to their generic nature, also provide a clear and concise pointer to basic questions in our academic activity, as well as being in tune with some of the major concerns of our age. At the same time, they represent three interrelated aspects that are not only the subject of academic analysis, but also underlie many of today's political and social debates. It is not in vain that these concepts, in their broadest sense, refer us back to personal and social themes of great significance, such as identity, value systems, intergroup relations, care of the environment, health, etc.

Culture shapes our minds, as various seminal studies such as those by Vygotsky have pointed out, directing and conditioning our actions in the political, social and economic sphere. Culture is also a sign of personal and social identity through which we choose the way we live, our adoption of a particular lifestyle, our intellectual decision-making and our interaction with others. Culture also affects our choice of a place to live, or the selection of physical shapes and contours in all aspects of our lives. This all takes place in a feedback system that in turn contributes to the very creation of culture.

As is the case with many other concepts in the social sciences, culture has been defined in many different ways, which is logical if we take into account the multiple aspects and nuances that can be put forward from different theoretical standpoints. The present study is neither the time nor the place, however, to attempt a definition of culture that brings together the variety of perspectives that surround it, as Kroeber and Kluckhohn (1952) did in their time. This would be a meaningless, as well as complex, task in an article written with the aim of showing that a large number of current intellectual, social, economic and political challenges find their natural place in some of the themes that characterised the 17th IAPS conference. It will therefore be enough to recall the proposal put forward by Tylor (1871), who was above all a generalist, when he defined culture as that complex array of knowledge, beliefs, art, morals, law, customs, habits, etc. that a person acquires when he or she belongs to a particular society. In common with every definition that seeks to be all-encompassing, Tylor's proposal disregards some aspects that are essential if considered from the standpoint of social psychology and a specific idea of what a human being is.

Firstly, we have to recognise that the relation between the *individual* and *society* is more dynamic than has traditionally been supposed. The way in which this relation has been set out from certain theoretical standpoints, linked to specific social problems, has stressed the first element of the pair to the detriment of the second, leading to a totally passive view of the human being, who merely assumes and internalises the dominant beliefs and representations of his or her environment. This becomes apparent in Althusser's concept of ideology, in which individuals find themselves at the mercy of ideological determinations. Similarly, the collective representations proposed by Durkheim reiterate this passive view of the individual, which is why we have questioned the validity of this theory being considered the intellectual predecessor of the work of Moscovici (Billig & Sabucedo, 1994). This line of thinking, in our opinion, implies several serious risks. One of these, which is of particular relevance to the theme under discussion, is that of the reification of social systems and beliefs and of institutions. In opposition to this idea, it should be stated that social systems and everything they represent and mean are by no means alien to human activity. They are not something set apart from people that influence their behaviour. Rather, social systems and all their products are a part of human activity, the result of conflicts and interactions between individuals and groups. Berger and Luckman (1968) produced an excellent analysis of this problem with regard to the social order when they unequivocally stated that the said order is a human product that is constantly being created and modified.

Likewise, culture is also a human product. And if we understand it as such, then there are two aspects which must be emphasised: firstly, the influence that it exerts on individuals who are born into a specific cultural and social niche; and secondly, that culture itself is continuously affected by the action of the same individuals. In other words, contrary to a determinist philosophy regarding context, society or culture, the thesis we defend is that of taking interaction, whether it be between individuals or between individuals and their social and physical environment, as being the most appropriate level at which to analyse, understand and explain human behaviour and its relationship with the environment.

The social construction of interest in the environment

Once the various facets of social reality cease to be seen as objective entities and become just one more human product, the way is left open for the actions of different agents who will attempt to influence the construction of new interpretative frameworks of this reality. The conception of these alternative ways of defining and understanding our environment, physical as well as social, is closely linked to the generation of new values, beliefs, attitudes and behaviours in the face of a large number of matters, amongst which, obviously, is to be found the environment.

We will all agree that interest in the environment has increased significantly in recent years. Proof of this is the fact that all political parties include the environment in their programmes, institutions of further education devote many different types of space to it (subjects, seminars, congresses, journals, etc.), and that governments assign increasingly large sums of money to the study of the multiple aspects that concur in the environment. We will also agree that it is inappropriate to talk of environmental problems, since what we are really referring to are human problems in relation to the environment. The reason for this is that in most cases the environmental problems to which we refer are not the consequence of natural phenomena or accidental events, but rather of the actions, in most cases intentional, of human beings. The so-called environmental problem, therefore, is not significantly different from other problems that owe their origins to certain human practices and behaviours. And precisely for this reason the different theoretical, analytical and methodological approaches to the subject of the environment are shared by all those areas of knowledge that have as their main focus human behaviour in its widest physical and social context.

But the development of a field of study and research dedicated to the environment and its relation with human behaviour is not something that has occurred spontaneously, nor is it the result of a chance discovery by a scientist working alone in his or her laboratory. As the history of science shows, most areas of study arise as the result of a group or collective identifying the subject as one of interest or one which presents problems. In this sense, therefore, the appearance and the subsequent development of environmental psychology is due to the recognition of the fact that the relations between the individual and his or her surroundings are key factors in determining his or her quality of life. In this respect, we should remember Engels' well-known work of 1845, *The Condition of the Working Class in England*, in which he revealed the different psychosocial problems caused by overcrowded living conditions. New technological developments and the use of new sources of energy, such as nuclear power, have also served to increase public awareness of the possible risks they may pose for the environment, and hence for mankind. Once again, however, this concern did not arise spontaneously. Rather, it became necessary to generate a discourse that could oppose that of those groups that supported this kind of energy. This is a good example of how different sectors seek to impose their defining frameworks on reality. Gamson (1988) analysed the struggle between supporters and opponents of nuclear power to impose their interpretative schemata on this question. Faced with the discourse of faith in progress and independence in matters of energy put forward by the defenders of nuclear power, the various different anti-nuclear groups countered with that of the risks inherent to this form of energy and the doubtful economic benefits it brings. The warnings sent out by the antinuclear groups were initially seen as being alarmist, but were later to gain certain credibility, as a result of which they took their place amongst the social representations or common sense of the population at large. The psychology of active minorities (Moscovici, 1981) and studies of social movements (Klandermans et al., 2002) indicate the processes that allow these concerns, initially held by only a small minority of the population, first achieve a position of social visibility and then take their place in the belief systems of the individual.

In this way, through the combined efforts of activists, academics and the media, environmental questions took their place on the public agenda. When this occurs, a subject acquires its own entity in that people begin to talk about it, to discuss it, to adopt a standpoint with regard to it, and, as has been pointed out by Mead (1965), these interactive processes lead to a communion of meanings between the members of a community, and this is precisely what allows a culture to exist, taking this in the sense defined by Tylor but adding the constructive, and therefore ever-changing, dimension that we have defended above.

Culture and quality of life

The environment occupies a leading position in the set of attitudes and sensibilities that characterises present-day culture, principally due to the appearance of a generalised concern for all those aspects related to quality of life. This, at least, is what we can deduce from the numerous opinion polls that seek to determine the degree of concern felt by members of society with regard to a wide range of matters. Quality of life, however, is a complex concept since, in addition to the objective elements involved, it also includes a subjective perception that is the result of a life-time of experience acquired in interaction with the other members of our community, and having a direct effect on the physical environment, whether natural or built. In this sense, we can say that the concept of quality of life is related to the way in which natural resources are managed, on the basis of having to attend to an uncontrolled demand for the consumption of these resources. It is for precisely this reason, due to the extent to which quality of life is affected, that the environment now occupies its current prominent position in social and political life.

However, as the first scholars to study attitudes wisely pointed out, the fact that one manifests a favourable affective posture towards a psychological object does not necessarily mean that one will behave in a manner consistent with this tendency, and this is a crucial point. If, as we have stated above, the problem in this area is basically one of creating behaviours that respect the environment, we need to go from awareness to action and achieve a greater consistency between environmental attitudes and behaviours. If we want to improve our quality of life, if we want to preserve the environment as an economic, cultural and social value, then our culture must incorporate ecologically responsible forms of behaviour. As this book shows, there are many spheres of action which merit an analysis of the interaction between the individual and the environment, and all of them will undoubtedly prove significant for increasing our store of available knowledge in this area, and for contributing to our individual and collective welfare, two concepts that are intrinsically related to that of quality of life. And according to our thesis on the decisive role of human behaviour in creating and, at the same time, solving environmental problems, our knowledge of what determines ecologically responsible behaviour will undoubtedly have to occupy its place at the forefront of the agenda of all those who specialise in the area of environment-behaviour, and in particular that of environmental psychologists.

We are fortunate in that we are well-equipped, theoretically and methodologically, for this purpose. We do not share in vain the formulations and models that have been developed by the social sciences that give our work its scientific status, particularly psychology. It is in the sphere of psychology, by analysing people's social representations of the environment, the process by which responsibilities are attributed, the perceived effectiveness of one's actions, the behavioural costs, our expectations of the behaviour of others, etc., that we can find the keys that allow us to work towards the creation of a new culture that is more respectful of the environment. According to Marx it would never be possible to give different human tasks the same consideration until the idea of human equality acquired the strength and solidity of a human prejudice. In the same way, no solution to environmental problems can be possible until respect for the environment forms part of the set of beliefs that make up the common sense of a community at any given moment. In other words, until the culture of the age incorporates this respect as one of its major elements.

Culture and globalisation

It is obvious that the culture of our times cannot be explained without reference to the movement towards globalisation in which we find ourselves immersed. Globalisation is a relatively recent term that basically describes the increasing uniformity of certain patterns and ways of acting, initially in the economic sphere and then in the sphere of ideology, which in reality represents the hegemony of certain economic and cultural models over others. Thus globalisation, which is not always considered in its true meaning, is related to many other concepts which in turn are linked to the new society of knowledge, information, the world order, our perception of others, the standardisation of lifestyles, genetic engineering, social movements, etc. The fact that we have come much closer to what previously was seen as being only a distant reality has led to social, cultural and political changes, amongst others, and has even rocked some of the solid pillars that previously sustained groups, nations, and even cultures.

The problem we are faced with here has its origins in the attempt to impose, whether explicitly or implicitly, a particular world view. And since different beliefs, attitudes and ways of understanding life are linked to the possessions of different groups – which is another way of saying the social identity of individuals – certain communities and/or collectives may feel threatened by this dominant discourse. This in turn leads, as has often been stated (Tajfel, 1981, 1982; Tajfel & Turner, 1979, 1986), to a variety of responses that range from recurring to strategies of self-affirmation to one of the most perverse psychosocial phenomena imaginable, namely the self-undervaluing performed by one group in the face of others that enjoy greater social status and prestige.

This apart, it should also be pointed out that multiculturality, with the richness and, it must be said, the challenges that it supposes, runs a serious risk of being replaced by behavioural and ideological models that put themselves forward as being universal when in reality, as Martín-Baro quite rightly stated, they are nothing more than particular visions that are the creation of very specific ambits and social interests.

Tension between global and local culture

Globalisation is therefore a process that may well bring with it a series of major advantages, but it also implies major tensions and risks due to the accompanying threats to local culture, to diversity, or to the differentiation between lifestyles that shape the quality of life in each particular place, characterised by its own preferences and cultural manifestations. The second half of the 20th century witnessed these tensions between the forces behind globalisation, exemplified on one hand by the rapid expansion of multinational chains of shops or ubiquitous styles of architecture, and on the other by the presence of a growing attempt to ensure local and regional identities and differences by means of the introduction of planning controls or that of a philosophy of sustainability.

This all poses an enormous challenge that deserves wide-ranging discussion and debate, due to the implications it supposes for quality of life and local culture. In short, the analysis of the impact of global culture on local culture and of the role to be played by the various professional sectors concerned with the study of person-environment relations has now become of the key goals of current research.

The increasing dedication to environmental interaction in IAPS

It is undoubtedly true that attention to and interest in all these matters has increased amongst academics the world over during the last thirty years. Galicia, home to the 17th IAPS Conference in 2002, is no exception. Galician culture, manifest in an idiosyncratic approach to life, a language of its own, or a particular lifestyle of its own, has encouraged research along these lines by different groups in local universities, focussing on the analysis of the environment from the standpoint of personal experience, the personal experience of an individual interacting with his or her community, or with a physical environment endowed with all its special cultural characteristics.

In Spain as a whole, the last twenty years have also seen a noticeable increase in research into person-environment relations. The IAPS conference held in Barcelona in 1982 was the starting point for work by groups that are nowadays carrying out well-consolidated lines of research in many different places in Spain. The texts published in collaboration with IAPS and made available at the 2002 Conference are proof of the wealth of research possibilities now available to us (García-Mira, 2002; García-Mira, Sabucedo & Romay, 2002a, 2002b).

The 17th IAPS Conference received a great number of presentations and participants, evidence of the consolidation and solidity of this scientific society in its academic sphere and of the recognition of the need to tackle environmental problems and changes in the natural and built environments by means of the interaction between individuals and groups. One result of this was the high degree of participation by lecturers, professors and researchers from universities and research institutions the world over. The large number of delegates from European nations served to consolidate the European tradition in this field, whilst the increase in the number of those attending who had come from as far afield as South America and Asia is proof of the expansion of IAPS and the growing interest being shown in the approach of this scientific society.

A multidisciplinary approach to the study of the complex relation between man and his environment

Over four hundred and fifty contributions found their way on to the conference programme after a rigorous selection process, and are evidence of the intensity of the scientific research in this field during the period 2000–2002, following the Paris Conference. This has not just been the work of psychologists, architects or town-planners, but also of professionals from other disciplines such as education, geography, sociology, anthropology, economics, chemistry, law or health sciences, amongst others, who all came together in Corunna in July 2002, to pose new challenges and to share their knowledge and understanding of a problem which affects us all, regardless of the academic discipline we represent.

The Conference was thus proof of the will and the need to tackle these problems from a multidisciplinary standpoint, although when it comes to studying person-environment interaction the emphasis has to be placed on the fact that this is not an environmental question per se, but rather a question of human beings who live and experience relations with their built and natural environments. As a result of these relations, both the individual and the environment are mutually affected, the individual introducing changes and influences on his or her environment whilst simultaneously being influenced and transformed by it.

What singles out the approaches that come together in IAPS from others is that they coincide in analysing all the human processes involved in this singular interaction, which in turn derives from the complexity implicit in all human behaviour. Furthermore, the concentration of the population in urban areas, within this global framework of reference, has favoured the appearance of a set of problems of very diverse nature that have a tremendous effect on our quality of life, on environmental quality and on local culture. Many of these processes are related to the sustainability of environmental resources, pollution, lack of space, noise, decay, lack of safety, and are linked to land management, urban transport or the level of crime, or else to the appearance of interest groups that attempt to obtain maximum profit from this concentration of the population.

The progressive development of urban areas currently taking place all over the word, itself the consequence of the mimetic models favoured by the processes of globalisation, has created the problem of environmental quality, widely linked to the concept of quality of life, to evaluate the effects of human behaviour on the environment, as well as the effect of the environment on behaviour. Urban development, at least in the way that our culture sees it, has given rise to a series of risks to the environment that are at one and the same time risks for the community, such as acid rain, cancer-causing or gene-damaging dioxides, pesticides, pollution of our air and water, all of which are due to only one cause – human behaviour. According to Oskamp (1995) these problems become more serious with every year that passes, making it imperative to take the step from awareness to action, as has been mentioned above.

A multicultural, participative society

In short, even when the evolution of a culture is linked to the progress of a community, room for reflection needs to be found at each stage of this evolution to enable analysis of the complexity of social organisation and the nature of the adaptation that is taking place in such a rapidly changing context. We have witnessed the social acceleration that has occurred over the last forty years, characterised by rapid progress in all areas of science, by a major transformation in our lifestyles and by the shaping of standards in people's quality of life.

Within this process, in which the new information society has undoubtedly played a major role, we have also seen a growing number of tensions arise in opposition to the generalised expansion of the ways in which we use and consume resources, that threaten the survival of locally-based systems of culture and social organisation. These tensions produce important social dilemmas which at times even confuse our common sense by demanding of us that we make the recognition of the need to act together to protect what is common, global and general to mankind compatible with the recognition of the need to preserve what is specific, local, particular and idiosyncratic in our culture, our particular lifestyle, our local customs and habits.

This is no easy task, but neither is it an impossible one if we are capable of building a plural, multicultural society that will protect an asset that is as precious, but at the same time as misused, as the environment. A society that thinks globally in the common interest, but acts locally to promote and improve respect for the environment, this inevitably being linked to both the force of community participation in the management of the environment and to the promotion of the organisation and structuring of society.

This volume contains the contributions of a group of researchers of very different origin regarding culture, their views of what constitutes quality of life, the way in which they have experienced the processes of globalisation and in how these have influenced the multitude of physical and social spaces in which human beings interact with their environment. There is a general consensus of agreement on a good many points, but there are also many differing points of view, as is always the case when questions are approached from a multidisciplinary perspective, differences that are of enormous interest for scientific progress and a greater degree of appropriacy in our interventions in society.

References

Berger, P., & Luckmann, T. (1968). *La construcción social de la realidad [The social construction of reality]*. Buenos Aires: Amorrortu.
Billig, M., & Sabucedo, J. M. (1994). Rhetorical and Ideological Dimensions Of Common Sense. In J. Siegfried. *The status of common sense in psychology*. New Jersey: Ablex.
Gamson, W. A. (1988). Political discourse and collective action. In B. Klandermans, H. Kriesi, & S. Tarrow (Eds.), *Research in social movements, conflicts and change*. London: JAI Press.
García Mira, R. (2002). *Environment – behaviour studies in Spain*. A Coruña: Asociación Galega de Estudios e Investigación Psicosocial.
García Mira, R., Sabucedo, J. M., & Romay, J. (2002a). *Culture, quality of life and globalization. Problems and challenges for the new millennium*. Proceedings of the 17th Conference of IAPS. A Coruña: Asociación Galega de Estudios e Investigación Psicosocial.
García Mira, R., Sabucedo, J. M., & Romay, J. (2002b). *Psicología y Medio Ambiente. Aspectos Psicosociales, Educativos y Metodológicos [Psychology and the environment. Psychosocial, educational and methodological aspects]*. A Coruña: Asociación Galega de Estudios e Investigación Psicosocial.
Klandermans, B., Sabucedo, J. M., Rodríguez, M., & de Weerd, M. (2002). Identity processes in collective action participation: Farmers' identity and farmers' protest in the

Netherlands and Spain. *Journal of Political Psychology,* 23 (2), 235-251.
Kroeber, A. L., & Kluckhohn (1952). Culture: A critical review of concepts and definitions. *Papers of the Peabody Museum of American Archaeology and Ethnology*, 47(1).
Mead, G. H. (1965). *Espíritu, persona y sociedad [Spirit, person and society].* Buenos Aires: Paidós (original edition: 1934).
Moscovici, S. (1981). *Psicología de las minorías activas [The psychology of active minorities].* Madrid: Morata.
Oskamp, S. (1995). Applying social psychology to avoid ecological disaster. *Journal of Social Issues*, 51 (4), 217-239.
Tajfel, H. (1981). *Human group and social categories: studies in social psychology.* Cambridge: Cambridge University Press.
Tajfel, H. (1982). *Social identity and intergroup relations.* Cambridge: Cambridge University Press.
Tajfel, H., & Turner, J. C. (1979). An integrative theory of intergroup conflict. In W. G. Austin & S. Worchel (Eds.), *The social psychology of intergroup relations.* Monterrey, CA: Broks/Cole.
Tajfel, H., & Turner, J. C. (1986). The social identity theory of intergroup behavior. In S. Worchel & W. G. Austin (Eds.), *Psychology of intergroup relations,* pp. 7-24. Chicago: Nelson-Hall.
Tylor, E. B. (1871). *Primitive culture: researches into the development of mythology, philosophy, religion, language, art and custom.* London: J. Murray.

Our uncommon future

David Uzzell
University of Surrey, Guildford, UK

> **Abstract.** Far from being a widely accepted new cultural value the concept of sustainability is neither widely understood, endorsed, followed nor aspired to by large sections of the population. While no end of surveys suggest that people are concerned about the environment, turning that concern into action is not straightforward. Governments have placed considerable emphasis on encouraging individuals to engage in more sustainable behaviours, but the collective problems of waste generation, car use, and electricity consumption are neither caused nor can they be solved by single individuals. Any long-term environmental behaviour strategy has to be located in the relationships which exist between people in the community and the relationship between those people - individually and collectively - and their environment. This research demonstrates that sustainability cannot be considered in isolation from either its social or its place-related context.
>
> **Keywords:** sustainable communities, social cohesion, social exclusion, place identity, architectural education, environmental attitudes

> *According to the Hippocratic tradition, true medicine begins with the knowledge of invisible illnesses, with the facts patients do not give, either because they are not aware of them or because they forget to mention them. The same holds true for social science which is concerned with figuring out and understanding the true causes of the malaise that is expressed only through social signs that are difficult to interpret precisely because they seem so obvious.*
> Pierre Bourdieu, (1999). *The weight of the world*,
> Cambridge: Polity Press, p. 628)

A sustainable future

When the former Prime Minister of Norway Gro Harlan Bruntland introduced the most widely accepted definition of sustainable development as "development that meets the needs of the present without compromising the ability of future generations to meet their own needs" (Bruntland, 1987), she arguably put forward one of the most important ideas of the 20th century. The popular title of the Bruntland Report alone – *Our Common Future* – ought to have caught the imagination of everyone. It was, after all, about our collective future on this planet. One might have thought that if only for self-interest reasons, quality of life issues in the short term and the survival of our species in the longer term would have ensured that sustainable lifestyles and development would have become a personal, community, national and international priority. The case I want to make in this paper is that neither the need for, nor the application of the principles of sustainable development, are widely understood. Even when they are, we cannot assume that the principles of sustainable development are readily and enthusiastically supported. It may be our common future, but the ideas are neither accepted nor owned in common.

Why should sustainable development be a priority for environment-behaviour researchers in general and environmental psychologists in particular? There is now irrefutable evidence that the earth's climate is changing as a result of human activities. It will continue to change with, by and large, detrimental effects on agriculture, water supplies, ecosystems and human health. We can, and of course are going down the "technological fix" route to both mitigate and adapt to the effects of climate change, but technological developments will only take us so far. Environmental psychology has an important contribution to make in terms of raising awareness, enabling attitude and behaviour change and informing policy to bring about change.

Reviewing the research literature on environmental concern over the last few years, several noteworthy features warrant comment. First, there has been an attempt to move away from surveys which seek simply to measure the public's breadth and depth of concern about the environment and to set research within a more theoretically-driven framework. Second, there has been more of an emphasis placed on understanding the determinants and predictors of environmental concern. Third, there has been an increasing emphasis on trying to model environmental concerns and attitudes more effectively, reflecting both the multidimensional nature of environmental concerns and how they may be embedded in broader belief systems and general issues such as quality of life. The research reported in this paper reflects these three trends.

Perceptions, attitudes and attributions of responsibility for global environmental change

Four phenomena have served, intentionally or unintentionally, to focus our attention on the seriousness of global as opposed to local or even national environmental problems.

Through processes of what is now called globalisation (Bauman, 1998; Lechner & Boli, 1999), the economies of nations and communities are being restructured to a degree not witnessed since the industrial revolution. Globalisation is associated with new forms of technology, new forms of management and new forms of media, all of which are oiled by the fluid movement of capital around the globe in a matter of seconds. Time has now conquered space. As a consequence local agendas are being increasingly informed by global perspectives, policies and decision-making. Not surprisingly, it is on the environment that many global processes are acted out.

By selecting, interpreting and emphasising particular events, and by publishing people's reactions to those events, the mass media play a critical role in structuring and defining reality. The visually spectacular and dramatic nature of the effects of phenomena such as El Niño or the emotional impact of the cutting down of the rain forests or the hunting of whales has served to place global environmental problems on the public agenda. Terms such as global warming and the destruction of the ozone layer have entered into everyday language in much the same way as psychoanalytic terminology entered into public discourse in the 1950s (Moscovici, 1976).

Growing ecological awareness and concern amongst the general population has been reflected in the expanding membership of international organisations such as Friends of the Earth and Greenpeace. Although many of these highly conspicuous environmental groups have local membership structures, their public presence and the way in which the mass media interact and report them is often at a national or international level.

Parallel to the popularisation of environmental issues have been high profile scientific and political conferences such as the 1992 Earth Summit in Rio. It is not difficult to believe that global initiatives have lodged in the public's consciousness more firmly than their operationalisation at the local level, such as through LA21 programmes.

While these forces have highlighted global processes, it has also been suggested that people are only able to relate to environmental issues if they are concrete, immediate and local. Consequently, it might be hypothesised that people will be more aware of and consider environmental problems to be more serious at a local rather than global level. In a series of international collaborative studies undertaken in Australia, Ireland, Slovakia and the UK, members of environmental groups, environmental science students, and children were asked about the seriousness of and their sense of responsibility for various environmental problems in terms of their impact at the local, national, continental and global level (Uzzell, Rutland, & Whistance, 1995; Uzzell, 2000)

Using an environmental grid, respondents were asked to assess the seriousness of various environmental problems at different areal scales, to which they responded on a 5-point scale (Uzzell, 2000).

Figure 1. Environmental grid.

	Myself	**Town**	**Country**	**Continent**	**World**
Water Pollution					
Atmospheric Pollution					
Effects of Acid Rain					
Global Warming					
Noise Pollution					
Deforestation					
Ozone Holes					

Without exception in each study, people considered environmental problems at the global level to be more serious than those at lower spatial levels.

In this study we also sought to identify the perceptual areal threshold of attributed personal and institutional responsibility for the environment. We found that perceived

Figure 2. Perceptions from Australia, Slovakia and the UK.

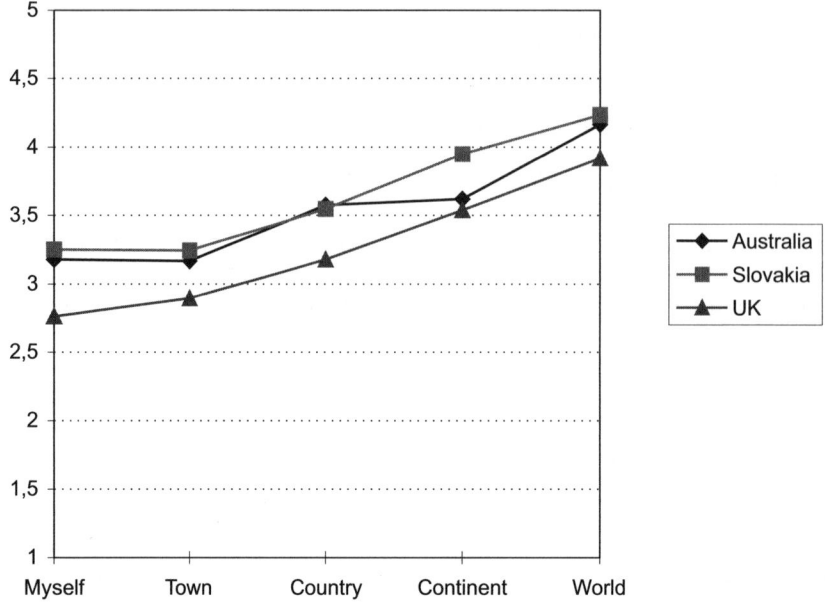

individual responsibility for the environment is greatest at the neighbourhood level and decreases as the areal level becomes more remote. Ironically, although people feel that they are responsible for the environment at the local level this is precisely the level at which they perceive minimal problems. The areal level which they perceive has the most serious environmental problems is the areal level about which they feel least personally responsible and powerless to influence or act. So we are faced with the paradox that statutory and voluntary organisations are trying to raise the public's level of environmental concern and change their behaviours at precisely the level which the public see as unproblematic.

One might look to various theories in psychology to account for what could be termed environmental hyperopia – psychophysiological, social learning, cognitive coping strategies and optimism bias, risk theories through to more social psychological perspectives such as social dilemmas and social representations. For example, if we take a psychophysiological perspective we might argue that however much we are told that global warming is a threat, direct experience of environmental changes at the human psychophysical level is unlikely because the physical signals of global environmental change are way below the thresholds of discernability of human sensory and memory mechanisms (Pawlik, 1991). Temperature changes as a result of global warming are predicted to rise by 1.4°C – 5.8°C over 80 years, or by .05°C per year. In Central England, the diurnal variation in air temperature in the spring is about 9°C, and in the summer can be as much as 17 °C. Pawlik refers to this as the "low signal-to-noise ratio of global change", where the "signal" of global environmental change is small in value and slow in time, and the "noise" of observable changes due to circadian, seasonal and regional temperature changes is large. The environmental crisis can therefore be considered as an abstract concept, because we typically only have a surrogate experience of it since our perceptions and knowledge are derived largely from the media.

One world or two: Trends in environmental concern

But surely, it is assumed, everyone is concerned about the environment – one only has to look at the opinion polls. As can be seen from these figures environmental concerns have been generally rising across both local and global issues for the last 15 years. But these figures warrant closer inspection.

Figure 3. Environmental concerns 1986–2001 (*Social Trends* 32, 2002).

England and Wales				Percentages	
	1986	1989	1993	1996/7	2001
Disposal of hazardous waste	..	..	63	60	66
Pollution in rivers and seas	54	64	63	65	55
Pollution in bathing waters and on beaches	37	59	56	61	52
Traffic exhaust fumes and urban smog	23	33	40	48	52
Loss of plants and animals in the UK	38	45	43	45	50
Ozone layer depletion	..	56	41	46	49
Tropical forest destruction	..	44	45	44	48
Climate change/global warming	..	44	35	35	46
Loss of trees and hedgerows	17	34	36	40	46
Losing Green Belt Land	26	27	35	38	44
Fumes and smoke from factories	26	34	35	41	43
Traffic congestion	..	..	35	42	43
Use of pesticides, fertilisers and chemical sprays	39	46	36	46	43
Using up the UK's natural resources	..	..	27	23	38
Acid Rain	35	40	31	31	34
Household waste disposal	..	..	22	22	33
Decay of inner cities	27	22	26	23	31
Noise	10	13	16	15	22

- Concern over some issues has remained at a stable but high level over that period.
- Concern over some environmental issues has risen substantially – especially in respect of UK environmental .
- But there are also some contradictions – we find lessening concern for acid rain and increased concern about fumes and smoke from factories.
- But perhaps most noteworthy of all is that concern for UK environmental issues has increased, but concern over what might be termed global issues – either where local action has a global impact or in turn, global climate change will impact upon the UK environment – has actually declined through the 1990's and only increased again in 2001 – possibly as a response to the severity of flooding and winter storms over the last few years.

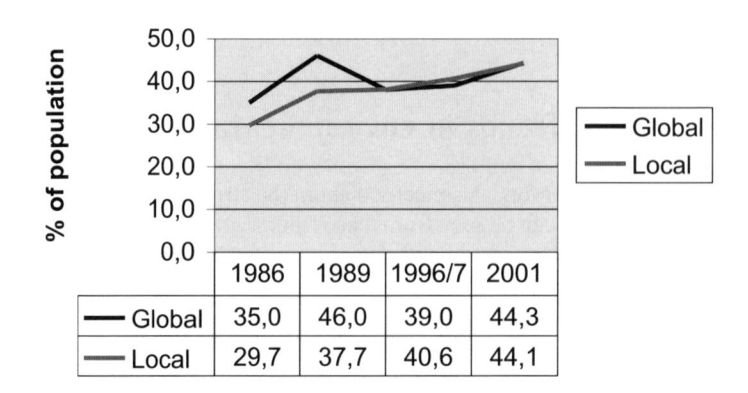

Figure 4. Environmental concerns 1986 – 2001 (*Social Trends* 32, 2002).

	1986	1989	1996/7	2001
— Global	35,0	46,0	39,0	44,3
— Local	29,7	37,7	40,6	44,1

These results show that people do not necessarily see any connection between global and local environmental processes.

The social desirability which surrounds pro-environmental attitudes is considerable. If you ask anyone if they are concerned about the quality of the environment, about air and water quality, about traffic, about urban development, they will, of course, say they are. But if you start to discuss these issues in detail it becomes quickly apparent that these issues are complex and invariably bound up with larger cultural values which influence our society.

As societies become and aspire to become more modern, so it has been assumed they will subscribe to the same set of values and share the same worldview and, as a consequence, the same drivers for development. Although various writers throughout history such as Locke, Rousseau and Marx put forward different and even incompatible views of modernity, they all share certain assumptions in common – that society will become more secular and "the world's divergent cultures meld into a universal humanistic civilisation." (Gray, 2002). One had to look no farther than across our own sitting rooms to the televised millennium eve celebrations broadcast from the capital cities of the world and remote islands in the Pacific Ocean to realise that this was party time in the global village. The fall of the wall and perestroika in the 1990s only seemed to confirm the view that global peace, if not global governance, in some form might be a possibility. The Montreal Protocol, agreed in September 1987, secured the support of governments for the reduction and eventual elimination of the emission of man-made ozone-depleting substances. The significance of the Protocol, apart from the impact that reduced levels of CFCs would have on the ozone layer, was that it was the first international agreement driven by scientific predictions. And so in addition to its scientific value, it represents the embodiment of the principles of the Enlightenment and modernism.

But these feelings of one-worldness and a new global neighbourliness by the West as we entered the 21st century have been tempered by an increasing concern about Third World debt and unfair trade agreements, globalisation and global capital, the intergenerational and inter-species implications of genetic engineering, and international responsibility for ethnic atrocities. And then, of course, there was September 11th 2001. It is quite clear that Al-Qaeda, in addition to demolishing several skyscrapers, destroyed once and for all the idea (or perhaps delusion) that we are living in a consensual world of shared values. In the smoke above Manhattan, it became apparent that the whole world does not subscribe to the Western idea of progress, and there are those who want to follow a different script, with different priorities and which has a different endgame. And what is so at the global scale is equally true at the national or local scale.

Cultural differences and the public

It cannot be taken for granted, despite the rhetoric and expressed concern for the environment, that pro-environmental attitudes and behaviours in any meaningful sense are held in common. Lyons, Uzzell and Storey (2001) recently completed a major study of attitudes towards waste minimisation in Surrey in which over 9,000 people in each of Surrey's 15 Civic Amenity sites were interviewed and questionnaires sent to 16,000 Surrey residents. The research found that people were largely poorly informed and confused about how they could reduce waste. The young are often seen as having the most positive attitudes towards the environment but we found in focus group discussions comprising young people and young parents that recycling and pro-environmental behaviour change was not regarded as a priority because there are perceived to be few immediate, serious and tangible benefits or costs to the individuals concerned. The young also felt that they would only modify their behaviour when the effects started to impact on their lives: whereas drink driving could kill someone and/or result in the loss of a driving licence, the environmental effects of waste generation were too distant to motivate change, and in any case small lifestyle changes by an individual are seen to have "zero effect" on what is regarded as a global problem. Young people were the most strongly opposed to changing their behaviour as they considered being forced to recycle was an infringement of individual freedom. In general, they resented being told what to do and admitted that if they felt under pressure to recycle they were less likely to do it. The young parents' group discussed the possibility of penalties for not recycling goods and generally felt it to be a bad thing - they joked about the "recycling police and a police state," and about having bins with alarms fitted that went off when you threw out a recyclable item.

How do people see those who do recycle? For the focus group discussants, most of the role models associated with recycling were negative. The prototypical recycler identified by the young people was an "old man in his fifties with a beard or a woman in a tie-dyed t-shirts and dungarees". The young parents had various stereotypes of people who recycled: one was the "Swampy" image; other images offered were of Swedes or other Scandinavians, outdoors types, people who bought Ikea furniture, a Blue Peter presenter, or someone who was perfect. The middle-aged group described a recycler as "someone boring."

Therefore, far from being a widely accepted new cultural value, the concept of sustainability is neither widely understood, endorsed, followed nor aspired to, especially by the very groups who are typically regarded as pro-environmental.

Cultural differences in a family context

Whenever adults gather to wring their hands about the environmental crisis, they invariably suggest that we need to capture the hearts and minds of children if we are to deliver on our environmental objectives over the next 20–30 years. Children are regarded as a key audience for environmental messages. They are seen as "tomorrow's opinion leaders and stewards of the earth". But many children and young people resent adults heaping responsibility on them for cleaning up the environment. By laying emphasis on teaching children to become environmentally aware and responsible citizens tomorrow, children interpret this to mean that adults can continue to live their unsustainable lives today, living here on earth as though they were just visiting for the weekend, rather than living as though they were intending to stay for good, as Chris Patten expressed it in his Reith Lecture in 2000 (Patten, 2000). This is not to say that children are not important partners for environmental change, but we cannot just assume that they are either willing partners, interested partners or that they will be effective partners.

In 1992, I directed a four nation study which sought to examine whether children, in conjunction with schools, can act as catalysts of environmental change in the home and the community. The project comprised research teams from Denmark, France, Portugal and the UK and was funded by the European Commission (Uzzell et al., 1994). The idea behind this work was simple: if children could be given environmental education at school and encouraged to disseminate it at home and in the community, this would be an extremely effective way of influencing and educating parents to sustainable environmental behaviours.

Our findings identified very clearly some of the critical barriers to children acting as environmental change agents in the home. These results can be illustrated by four case studies:

1. *Where neither the child nor parent report any inter-generational influence.*
   This was found where there was:

   – low levels of concern about environmental problems by children and parents
   – parents had little knowledge about environmental problems
   – there were negative attitudes towards education
   – low levels of motivation
   – poor self esteem.

2. *The child does not feel they have been influential, but parents report that the child has influenced them:*

   – the parent has a high level of pre-existing knowledge about environmental problems
   – the child is encouraged to talk about the *factual* content of the environmental education they have received.
   – the school is regarded by the parents as an appropriate source of cold facts/hard information about environment/environmental problems.

   However, the child sees themself as uninfluential because:

   – the parent regards themselves as the "expert"
   – the originality and relevance of the educational material to the parent is limited
   – the communication is cold and factual rather than warm and attitudinal.

3. *The child reports having influenced the parent but the parent does not feel the child has influenced them.*

This is the traditional model of influence where the parent adopts the role of expert and the child has minority status. In this case, we found evidence that where the child felt able to tackle small environmental problems, the parent felt a sense of helplessness and exhibited low self-efficacy. This reduced self-efficacy, in turn, prevents information from the child about the environmental being accepted by the parent.

4. *The child reports influencing the parent and the parent reports being influenced.*

Here there is reciprocity in the influence process:

- the child and parent were willing to communicate with each other
- the environment was regarded as an appropriate topic for discussion within the home
- the child's openness and concerns about the environment were valued by the parent resulting in expert status for the child
- the parent was willing to adopt the role of pupil in deference to the expert status of the child.

We concluded that a catalytic effect can occur, but it occurs with a limited number of children/parents, and is conditional upon certain factors:

- the extent to which parents are prepared to acknowledge their children as experts
- the willingness of parents to enter into dialogue with children
- the level of pre-existing concern and knowledge parents have about environmental problems
- the degree to which present teaching and learning methods encourage children to share their learning
- institutional, structural, cultural and interpersonal barriers, e.g., the environment is seen as a suitable and acceptable topic for conversation in the home.

It may seem extraordinary to this audience that these conditions could not be found in most families, but in our research they were the exception.

Place identification, social cohesion and the environment

The social, cultural and political context in which action takes place is an important ingredient in enabling changes in people's attitudes and behaviours. One of the shortcomings of so much psychological research is that it treats the environment simply as a value-free backdrop to human activity, a stage upon which we act out our lives. The environmental context in which action takes place is crucial and for environmental psychologists that means place, the physical setting of human behaviour. As Edward Casey

(1993), the American philosopher writes, "A placeless world is as unthinkable as a bodiless self". Everything we do is influenced by, and in turn, influences our environment. While we as psychologists are interested in the cognitive, affective and behavioural aspects of environmental attitudes and attitude change this has to be situated in a physical setting; the environmental context of community-based social processes may be central to the operation and manifestation of those processes (Moser & Uzzell, 2003).

Over the last twenty years the concept of place identity has taken centre stage in environmental psychology along with concepts such as place identification and place attachment. Place identification refers to the attributes of a place which give it a distinctive identity in the minds of residents (Schneider, 1986). Thus, a person from London may refer to themselves as a Londoner, thereby expressing identification with the city. Place identification reflects membership of a group who are defined by location – it is a place-related social category. It might be investigated by ascertaining those features that enhance identification *with* the place and is measured by ascertaining from the residents the distinctive features of their neighbourhoods.

Whereas the weight of emphasis in place identification is on the distinctiveness of place, the weight of emphasis in place identity is on the psychological construct of social identity in an environmental context. Place identity should thus be regarded as an aspect of social identity.

Places which generate a strong sense of place identification may contribute to a person's sense of place identity. Equally, it may also help to enhance community awareness and bonding, or social cohesion. In turn, social cohesion may contribute to place identity. From this it might be hypothesised that socially cohesive communities which encourage place identity will be more supportive of environmentally sustainable attitudes and behaviours as compared with those communities where cohesiveness and social and place identities are weaker. There is thus an important link between sustainability and social life. This was the premise behind a five-year international collaborative study called the *Cities, Identity and Sustainability Network* (Uzzell, Pol, & Badenes, 2002).

Although the principal hypothesis was that the greater the place identity the greater the probability of sustainable behaviour, there may be different pathways towards sustainable behaviour depending on the different characteristics of the places and the people who reside there. In other words, while place identity may be important, such factors have to be considered along with other social forces such as social cohesion and residential satisfaction. It is the *combination* of these factors - how these processes interact and work together – that we hypothesised would lead to specific environmental attitudinal and behavioural outcomes.

As the UK contribution to this seven nation-study, two neighbourhoods in Guildford – Onslow Village and Stoughton – were selected on the basis of their social histories, housing types and socio-economic composition. Ninety residents in each neighbourhood were interviewed and asked questions about their relationship with both the place and their neighbours. We were interested in whether their neighbourhood generates a strong place identification, how is their own identity enhanced by the place in which they live and what effect does this have on attitudes towards sustainability.

The residents of Onslow Village had a very clear idea of the boundaries of their neighbourhood. Onslow Village residents were also able to point to the architecture of the houses, the high beech hedges which dominate many of the residential roads, and the Cathedral on the boundary all of which provide locational distinctiveness. Clearly the stronger the perceptual boundaries of the neighbourhood in the residents' eyes, the stronger

Our uncommon future   31

Figure 5. Cognitive map – Onslow village

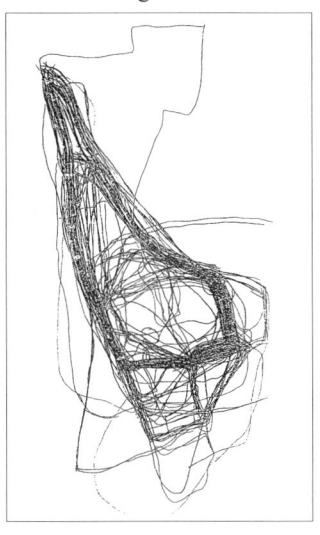

Figure 6. Onslow village – "... hedges and high banks..."

the identification. We asked residents to draw cognitive maps of their neighbourhood. These are freehand maps or representations of the neighbourhood drawn from memory which can tell researchers a great deal about how people structure their neighbourhood in their mind and what places are important to them.

The cognitive maps drawn by the residents of Stoughton of the extent and boundaries of their neighbourhood demonstrated less consensus and were looser in structure.

Stoughton residents identified fewer distinctive environmental features or characteristics. Onslow Village residents were more likely to think that outsiders had a good image

Figure 7. Cognitive map – Stoughton

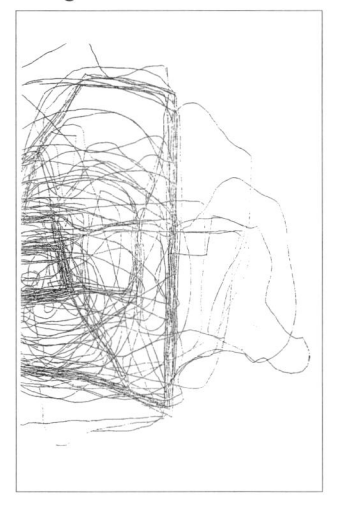

Figure 8. Stoughton – "... fewer distinctive environmental features..."

of their neighbourhood. They also had a greater sense of the village's unique history, which again, perhaps with the exception of the former army barracks, was absent in Stoughton. The distinctive environmental and temporal signature of the area serves to differentiate one area from another and thereby enhance identification.

Feelings of social cohesion in Onslow Village and Stoughton did not differ in a significant way. It was measured through involvement in community action, the perception of similarities-differences between neighbours, social relations, and time spent in the neighbourhood. Although there was no difference in the proportion of residents in each neighbourhood who were involved in actions to improve their neighbourhood (Onslow, 23%; Stoughton, 19%), there were differences though in the kind of activities which residents undertook. Residents in Onslow Village typically were more involved in social issues such as the local Neighbourhood Watch scheme (i.e., crime prevention) or helping at school parties. In contrast, the problems reported with traffic (speed, noise, parking and traffic jams) in Stoughton were more likely to lead residents to focus their energies on environmental issues.

There was a significant difference between the residents of both neighbourhoods in terms of how they would deal with problems: 27% of Stoughton residents said they would go to the local authority for assistance, compared with 13.5% of residents in Onslow Village. On the other hand, 31% of Onslow Village residents were much more likely to use their residents' association compared with only 6% of Stoughton residents who would use one of their community organisations. Stoughton residents were twice as likely to try and solve problems through individual strategies (9.5%).

The clearest differences between Onslow Village and Stoughton were in terms of residential satisfaction. Residents of Onslow Village were more satisfied with their neighbourhood. This was supported by the perception of the environment as being visually and environmentally attractive, less polluted, quiet, secure, and restful. Although the residents of these two neighbourhoods saw the environment differently, there were few differences in terms of environmental attitudes, values and behaviours between the residents of Onslow Village and Stoughton, with both expressing some concern for the environment and sustainability.

Figure 9. Structural equation model – Onslow Village

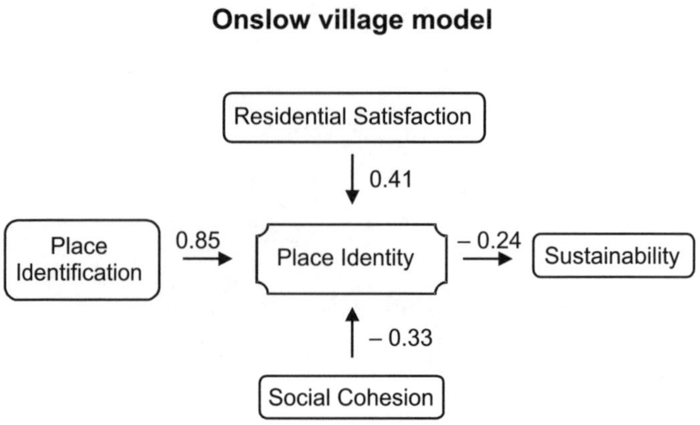

Onslow village model

Figure 10. Structural equation model – Stoughton

Stoughton model

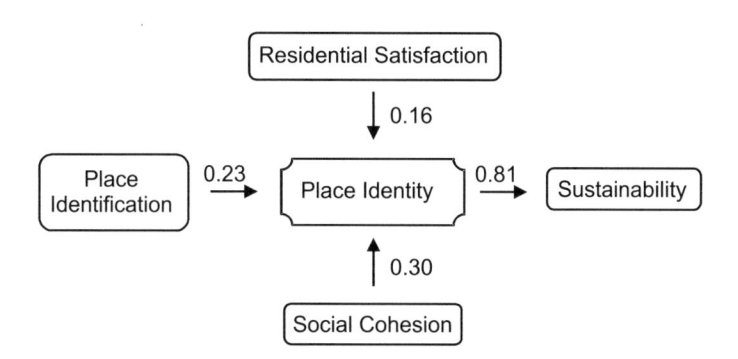

```
                    Residential Satisfaction
                              │ 0.16
                              ▼
   Place        0.23                    0.81
  Identification ──▶  Place Identity  ──▶  Sustainability
                              ▲
                              │ 0.30
                      Social Cohesion
```

Using structural equation modelling we were then able to examine the contribution of place identification, social cohesion and residential satisfaction to place identity, and then in turn the relationship of place identity to attitudes towards environmental sustainability

In the Onslow Village model, the strongest direct effect is from place identification (0.85). This indicates that place identification is an important contributor to the place identity model. Residential satisfaction (0.41) is not as high, but clearly both these factors contribute to the residents' own sense of social identity. What is surprising though is that the relationship between place identity and social cohesion is not just weak but negative (–.33). One explanation of this is that the values tapped into in Onslow Village were highly individualistic resulting in an inward-looking individualism rather than outward community perspective. This conclusion is perhaps reinforced by the findings from the questionnaire survey, and evidence from the physical environment itself. The relationship between identity and environmental sustainability is also in a negative, albeit weak, direction.

In Stoughton, the relationships between place identification, social cohesion, residential satisfaction and place identity is weaker but in a positive direction. There is, however, a strong positive relationship between place identity and environmental sustainability (0.81). Therefore, within Stoughton, social cohesion is more important than in Onslow Village and makes a significant contribution to place identity, which in turn affects environmental sustainability attitudes and behaviour.

Social exclusion

The other side of the coin to social cohesion is social exclusion. Socially excluded groups have typically not been involved in the sustainability debate. The Government's Social Exclusion Unit defines social exclusion as "individuals and areas suffering from a combination of linked problems such as unemployment, poor skills, low incomes, poor housing, high crime environments, bad health and family breakdown." This type of analysis is not new and can be traced back to the work of Charles Booth at the end of the 19th century, whose analysis of the inter-connectedness of urban life and its spatial manifestation laid one of the foundations of social science research. Perhaps what's different now is that our understanding of exclusion has become more inclusive to take on psychological dimensions. For example, some writers will take social exclusion to mean those who are part of disadvantaged communities by virtue of their socio-economic status *or* identity, i.e., how people define themselves and are defined by others (Lucas & Simpson, 1999; Burningham & Thrush, 2001a; 2001b).

Several studies have been undertaken recently by the University of Surrey Environmental Psychology Research Group to examine the attitudes of excluded groups to sustainability, as part of the preparatory work for a sub-regional and district based community strategies (Uzzell and Leach, 2001; Uzzell, Leach and Hunt, 2002). In this research, we talked to members of various socio-economic groups whose voice is rarely heard in the sustainability debate – those in socio-economic groups D/E, young people, ethnic minority groups, and elderly people. One of the most striking, but perhaps not unsurprising, findings common to all groups was the lack of knowledge and understanding of the term sustainable development. The term itself is seen to be a piece of jargon and thus unhelpful and even contradictory. Most people thought that sustainable development was largely concerned with the environment – the social and economic dimensions of sustainability were not referred to at all. Equally, the environment was not thought of as the world within which people live, but as "green" issues or the countryside. Participants mentioned, for example, recycling, maximising resources and creating less waste. One community worker said *"...people with disabilities - they couldn't care less about the environment because they can't get out into it.".* Elderly people thought it had something to do with property development and ensuring that houses are built with consideration for the environment.

A second barrier identified by the groups we spoke to was what might be termed "Surrey culture" where the wealthy "speak" louder than the disadvantaged and the "powers that be" within the councils are not interested in sustainability. As one person said, *"It is very difficult to be sustainable when most people have three or four cars and spend their money on keeping their golf club going...the whole culture is against sustainability."*

Equally this lack of involvement, commitment and motivation to sustainability may reflect the absence of a sense of community amongst Surrey residents: *" ... I go to meetings and everyone assumes that everyone else in Surrey is well off."* and *"If you just listen or read the papers, Surrey don't have any problems."* Surrey is seen to be a fairly homogeneous population dominated by a white, mobile, middle-aged, middle-class culture. Those not falling into these categories are seen to be ignored. It was argued that those who are responsible for getting sustainability messages across need to find a way to address more closely pockets of the community who do not fit into typical target audience, for example, the 33% of Surrey ethnic minorities residing in the Maybury and Sherewater districts of the county.

What can we conclude from this? Not only do people not understand the concept of sustainability principally because we have not been effective in communicating or translating it, but even where there is an understanding we cannot assume that it is seen as uncontentiously desirable. One will not be able to develop sustainable communities if large sections of the population are or feel excluded. Developing social capital by means of strengthening community networks and civic infrastructure, and creating and maintaining a sense of local identity and social cohesion among community members is an important part of sustainability.

Sustainable architecture for sustainable communities

George Miller, (1969) some thirty years ago in a paper entitled "Psychology as a means of promoting human welfare", advocated that we should be giving psychology away. This paper is concluded by briefly describing a joint project between the Environmental Psychology Research Group at the University of Surrey and the Community Design Unit in the Department of Architecture and Building Science at the University of Strathclyde. In the early years of environmental psychology there was an aspiration that psychologists and architects would work together to create buildings and areas that would reflect more accurately and sensitively the needs and wishes of the users. It was not long before disillusionment set in. But twenty years later, with both disciplines wiser and more informed, the sustainability agenda is opening up new opportunities for collaboration and complementary support. Indeed, this will be essential if sustainability programmes are to be effective. Furthermore, this collaboration has the potential to address some of the problems have been identified above in terms of making sustainability meaningful and actionable. In essence, the collaboration between our two departments and disciplines has had the objective of training the next generation of architects, designers and environmental psychologists such that not only will sustainability became part of their taken-for-granted world, but also each of these professions will become more aware of the perspectives of the other and be able to engage in a meaningful and creative dialogue with each other and with their clients.

At the University of Strathclyde one group of fourth year architecture students are required to undertake a year-long community-based project which involves undertaking a complex design task. The process starts by devising a regeneration strategy for a chosen area. This requires the students to interpret and attempt to solve social, economic, physical issues, and then plan and design built structures that will support their strategy. While neither these plans nor buildings will be realised, the work is nevertheless more than simply a simulation. The students are required work with community and youth groups and housing associations in the area to come up with proposals for the design of new facilities and services, which has the potential to provide job opportunities, transport, housing, public spaces and cultural facilities – all with the aim of improving the quality of life in the area leading to an enhanced sense of place and pride.

In 2002 the project area was Govanhill, an area two miles to the south of Glasgow city centre. This is one of the most deprived areas in Britain with male unemployment levels of about 14%, crime, drugs and poor health a major problem. Govanhill has the unenviable distinction of having the lowest life expectancy rate in Europe. It is noteworthy though that in a recent survey in Govanhill, far from being oblivious and apathetic about their surroundings over 90% of residents expressed concern about their local environment.

In order to involve residents effectively in the design process it is essential that students can co-orientate with residents – that is, see the Govanhill area through the residents' eyes, understand what the spaces and places mean to the residents, understand what opportunities they afford, and explore what emotional responses they invite, e.g., relaxation, fear, anxiety, pride, anger. Before any planning or design work can be undertaken, students are required to research the area and its residents. Environmental psychology offers a range of techniques to help accomplish this, and between 2001–2002 the students on the Masters course in Environmental Psychology advised the architecture students on the most appropriate techniques to use. These included cognitive maps, sensory walks, multiple sorting tasks, and interviews. For example, the students walked round the neighbourhood with some of the young people and saw what places are important to them, where they hang out, how they use these places, identify places which instil fear. The architecture students then invited the young people of the area or members of a housing association into the studio to explore further their perceptions and attitudes towards places and spaces.

Figure 11. Understanding residents' perceptions and preferences.

Subsequently the Surrey students acted as consultants to the Strathclyde students. In other words, when the architecture students had particular queries or wanted advice on how to collect or analyse certain types of perceptual, attitudinal or behavioural data they contacted our students and used them as consultants. To this end, an interactive discussion room was set up on the web where questions were raised and answered, and suggestions made. The Strathclyde students spent four days in London with the Surrey students looking at community regeneration and architectural projects that might contribute ideas to the Govanhill project. They also engaged in crit sessions at which each student presented their provisional plans and models and the Surrey students discussed their strengths and weaknesses from an environmental psychological perspective. Some of the Surrey students then went to Glasgow to discuss the Govanhill designs further in the context of their own appraisal of the Govanhill environment.

This was a genuinely collaborative process. For the Surrey environmental psychology students they were exposed to the socialisation and culture of design professionals and how their view of the world leads to particular ways of dealing with and responding to user needs. They also had a chance to advise trainee architects in the use of environmental psychology techniques and see how they work in practice. For the architecture students it was an opportunity to explore different forms of participatory design based on research in environmental experience, to link environmental psychology and architecture, devise planning and design solutions which are grounded in the needs of local communities which are the product of a participatory process appropriate to the community's needs and lifestyles. This surely is what sustainable development is all about.

Figure 12. Web-based discussion room.

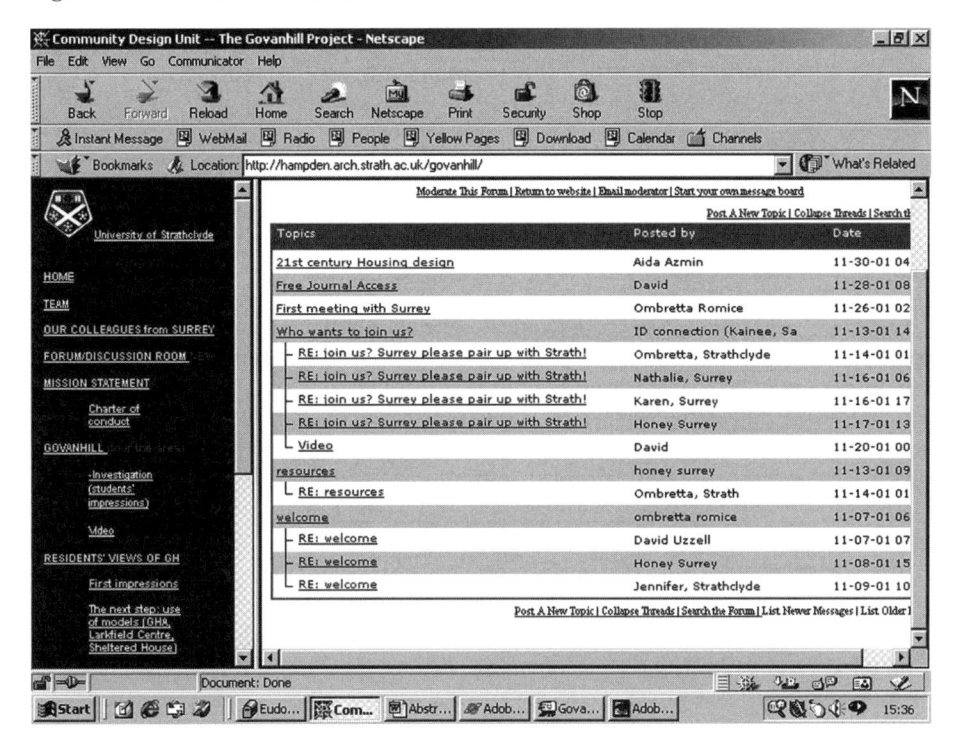

Conclusion

It is quite clear that while no end of surveys suggest that people are concerned about the environment, turning that concern into action is not easy. The prospective sustainability of the planet may be our Common Future, but it is quite clear that it is not commonly regarded so. One can have no end of Government reports and initiatives which are called *Our Common Future* (Bruntland, 1987), *This Common Inheritance* (Department of the Environment, 1990), or the *Common Agenda* (Surrey County Council, 2000) – but if these initiatives are neither truly understood, nor held in common then we will not witness a significant change in people's environmental attitudes and behaviours.

The paper began by discussing some early research which focussed on individual attitudes and behaviours – how seriously do people rate environmental problems, and can they see a link between the global and the local. The research undertaken for the EU was set much more firmly within a social framework – the family. It looked at how the social and cultural context of the family, in relation to the school and the wider community, can enable or prevent children taking environmental action and persuading others to do the same. The research on social exclusion, and more particularly the research on social cohesion and place identity have taken these ideas much further. These results suggest that place identity as expressed through both collective social relations and individual and collective relationships with place may form an important dimension of environmental attitudes. While one can address the problem of sustainability at an individual level, it would seem that any long-term environmental behaviour strategy has to be located in the relationships which exist between people in the community, and the relationship between those people - individually and collectively - and their environment. If we are to argue that change can only come about through social and collective action that is grounded in identity processes and people's identification with place, then we need to devise social and political strategies that recognise these processes. This research demonstrates that sustainability cannot be considered in isolation from either its social or its place-related context.

Acknowledgement

I would like to express my sincere gratitude to colleagues in the Environmental Psychology Research Group at the University of Surrey, and in particular to Rachel Leach. I would also like to acknowledge the enthusiasm and support of my colleagues at the University of Strathclyde, Dr. Ombretta Romice, Mr. Colin McNeish and Dr. Hildebrand Frey, without whom the Surrey-Strathclyde collaborative project would not have been possible.

References

Bauman, Z. (1998). *Globalization*. Cambridge: Polity Press.
Bourdieu, P. et al. (1999). *The weight of the world*. Cambridge: Polity Press.
Bruntland, G. H. (1987). *Our common future: Report of the world Commission on Environment and Development*. Oxford: Oxford University Press.
Burningham, K. & Thrush, D. (2001a). *Environmental Concerns, the Perspectives of Disadvantaged Communities*. London: Rowntree Trust
Burningham, K. & Thrush, D. (2001b). *Rainforests are a long way from here: the environmental concerns of disadvantaged groups*. London: Rowntree Trust
Casey, E. S. (1993). *Getting Back into Place: Towards a renewed understanding of the place-world*. Bloomington: Indiana University Press.
Department of the Environment (1990). *This Common Inheritance*. London: The Stationery Office.
Gray, J. (2002). Why terrorism is unbeatable. *New Statesman*, 25 February 2002, p. 50.
Lechner, F., & Boli, J. (1999). *The Globalization Reader*. Oxford: Blackwell.
Lucas, K., & Simpson, R. (1999). *Transport and Accessibility: The Perspectives of Disadvantaged Communities*. Working Paper No. 1 for JRF project. Transport Studies Group, University of Westminster.
Lyons, E., Uzzell, D.L., & Storey, L. (2001). *Surrey Waste Attitudes and Actions Study*. Report to Surrey County Council and SITA Environmental Trust.
Moscovici, S. (1976). *La Psychanalyse: son image et son public* (2nd Edition). Paris: Presses Universitaires de France.

Moser, G., & Uzzell, D. L. (2003). Environmental Psychology, in T. Millon & M.J. Lerner (Eds.), *Comprehensive Handbook of Psychology, Volume 5: Personality and Social Psychology,* pp. 419–445. New York: John Wiley & Sons.

Office for National Statistics (2002). Environmental concerns, 1986–2001: Social Trends 32: Social Trends 31.

Pawlik, K. (1991). The psychology of global environmental change: some basic data and an agenda for co-operative international research. *International Journal of Psychology,* 26(5), 547–563.

Patten, C. (2000). Respect for the Earth: Governance, Reith Lecture, 12 April, 2000. (http://news.bbc.co.uk/hi/english/static/events/reith_2000/lecture1.stm).

Surrey County Council (2000). *The Common Agenda.* Kingston Upon Thames, UK (see http://www.sustainable-surrey.org.uk/).

Uzzell, D. L., Davallon, J., Bruun Jensen, B., Gottesdiener, H., Fontes, J., Kofoed, J., Uhrenholdt, G., & Vognsen, C., (1994). *Children as Catalysts of Environmental Change.* Report to DGXII/D-5 Research on Economic and Social Aspects of the Environment (SEER), European Commission, Brussels. Final Report, Contract No. EV5V-CT92-0157.

Uzzell, D. L., Rutland, A. & Whistance, D. (1995). Questioning Values in Environmental Education, in Y. Guerrier, N. Alexander, J. Chase, & M. O'Brien (Eds.), *Values and the Environment,* pp.171–182. Chichester, UK: Wiley.

Uzzell, D. L. (1999). Education for Environmental Action in the Community: New Roles and Relationships *Cambridge Journal of Education,* 29(3), 397–413

Uzzell, D., & Leach, R. (2001). *Involving Groups in a Sustainable Future.* Kingston-Upon-Thames, UK: Report to Surrey County Council.

Uzzell, D., Leach, R., & Hunt, L. (2002). *Woking Community Safety Strategy: Minority Group Perceptions of Crime in Woking.* Report to Woking Crime and Disorder Partnership.

Uzzell, D. L. (2000). The psycho-spatial dimension to global environmental problems, *Journal of Environmental Psychology,* 20(3), 307–318.

Uzzell, D. L., Pol, E., & Badenes, D. (2002). Place identification, social cohesion and environmental sustainability, *Environment and Behavior,* 34(1), 26–53.

Globalisation, commons dilemmas and sustainable quality of life:
What do we need, what can we do, what may we achieve?[1]

Charles A. J. Vlek
University of Groningen, The Netherlands[2]

Abstract. A behavioural science view is unfolded on sustainable development of society from a European perspective. Free-market expansion implies significant pressures on environmental and social qualities in many countries. An extended commons dilemma (ExCD) model is presented to capture the tension between individual and collective interests. A variety of people-environment studies can be well aligned with the successive ExCD steps. The model is applied to the expanding domain of motorised transport. Here, market-stimulated individual freedom is insufficiently checked by government policies aimed at securing vital collective goods. Current (Dutch and EU) policy principles are criticised and a set of guidelines for sustainable transportation is advanced. In view of these and on the basis of Dutch survey research, we may question economic growth, work and income as overriding factors of future human well-being. Conclusions are drawn about comprehensive sustainability research, the structuring of people-environment studies, and the issue of "practicing theory versus theorising practice".

Keywords: sustainable development, commons dilemma, behaviour change, motorised transport, quality of life.

---

[1] This is a shortened and revised version of an invited lecture for the 17th Conference of the International Association for People-Environment Studies (IAPS), A Coruna, Spain, 23-27 July 2002.
[2] The author is Emeritus Professor of Environmental Psychology and Behavioural Decision Research.

I. Globalisation and environmental decline

Society has always fought against ignorance, poverty, diseases, threats from nature, and competitors from abroad. It has always aspired for knowledge, freedom, health, power, security, comfort and pleasure. Important means to get from Stone Age conditions to the security and opportunities of Modern Times are: hard work and social organisation, education, science, technology, and capital formation. This is schematised in Figure 1.

Figure 1. Development from Stone Age conditions (top oval) to Modern Times (bottom).

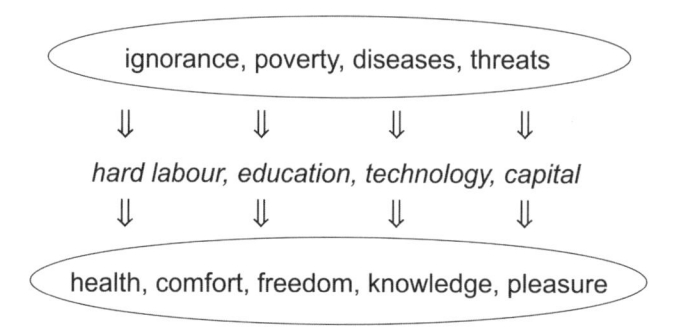

```
      ⟨  ignorance, poverty, diseases, threats  ⟩
         ⇓        ⇓        ⇓        ⇓
      hard labour, education, technology, capital
         ⇓        ⇓        ⇓        ⇓
      ⟨  health, comfort, freedom, knowledge, pleasure  ⟩
```

So far, in various places this natural process has been leading to overpopulation, overconsumption and an overwhelming technology. Material wealth and technical power are especially strong in the western-industrialised countries, while population growth and poverty are abundant in the less industrialised countries, especially in Africa (e.g., Goodland, Daly & Kellenberg, 1994). The worldwide inequities involved are widely exposed via modern communication media. Thus the "underdeveloped" peoples are strongly motivated (if not: seduced) to grow economically and reduce their backlog in security and opportunity.

Overpopulation, overconsumption and overtechnology are sustained and stimulated by cultural beliefs and values, many of which are crystallised in the institutions on which society rests. Figure 2 (from Steg and Vlek, 2003) summarises important driving factors of economic, social and environmental Impacts: Population, Affluence and Technology (well known from Ehrlich and Holdren's $I = P \times A \times T$ formula, 1971; see Schmuck and Vlek, 2003 for a psychological discussion). Each of these is under the influence of Culture and Institutions. In the lower oval of Figure 2, economic wealth, social well-being and environmental quality represent three components of the sustainability concept (e.g., Munasinghe ,1993). Each of these may be decomposed into specific quality variables, in relation to various human needs and desires, beliefs and attitudes.

The modernisation of society since the First Industrial Revolution (abt. 1850) and particularly since the Second World War (abt. 1950) has gradually led us into an expanding and noisy fossil-fuel economy. Because of increasing differentiation and power of economic actors (households, business companies, branch organisations), this seems ever more hard to control. Its most impactful manifestations are motorised transport, building

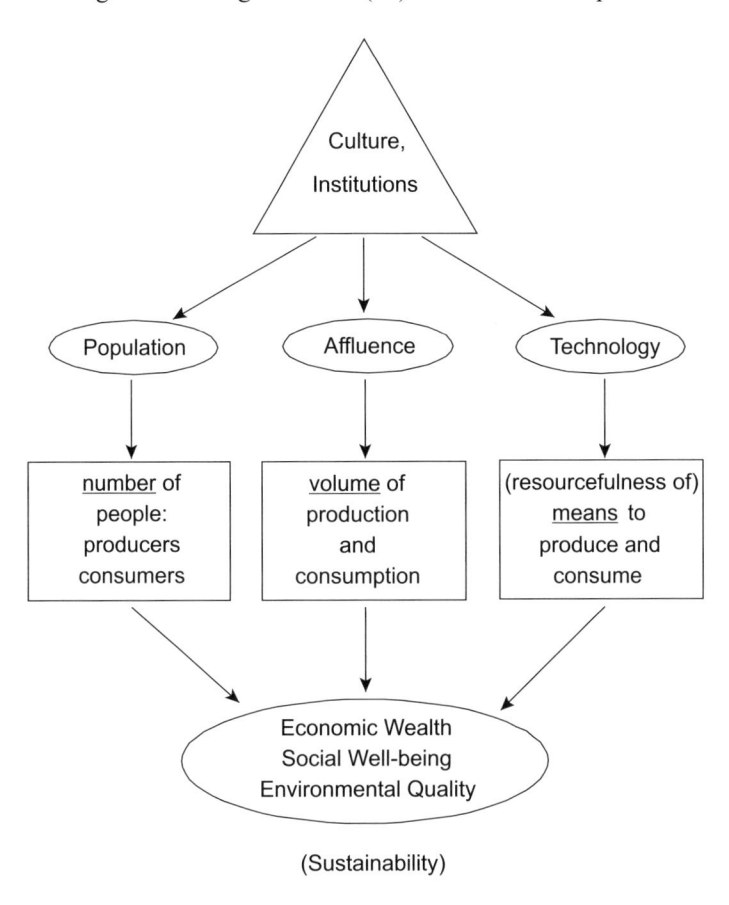

Figure 2. Five general driving forces for (un)sustainable development of society.

industry, agriculture and livestock business, tourism and recreation, and of course, family householding. In many domains, the classical conflict between individual and collective optimality has acquired alarming proportions. The Belgian philosopher Vermeersch (1988, p. 29) dramatically wrote (translated from the Dutch):

> "...the whole forms a system which rushes on autonomously, and nobody can guarantee that somewhere at the end of the route there is a goal waiting which is still meaningful for people. (...) The aimlessness, the irrationality of the total system is being obfuscated by the utter rationality of the system's separate components."

Phrased differently, although individual actors here-and-now operate very efficiently to increase their own quality of life, the society of all actors together moves in a direction which seems unsustainable in the long run, whereby all individuals' quality of life will eventually deteriorate.

Our current concerns are focused on the negative effects of globalisation, individualisation and commercialisation. We wonder why economic growth should degrade natural landscapes and our own living environment to the detriment of human health and well-being, and even at the risk of long-term economic depression. Our proliferating "human biomass" seems both intelligent and selfish enough to exploit ever more of the earth's resources whilst literally crowding out other living species. Recently, Wackernagel and colleagues (2002) have attempted to measure "the ecological impact of humanity (...) as the area of biologically productive land and water required to produce the resources consumed and to assimilate the wastes generated" (p. 9266). These authors conclude:

> "Our accounts indicate that human demand may well have exceeded the biosphere's regenerative capacity since the 1980s. (...) humanity's load corresponded to 70% of the capacity of the global biopsphere in 1961, and grew to 120% in 1999."

Wackernagel et al. point out that much of the overdemand for bioproductive space is related to the need for sequestering the huge amounts of CO2 from fossil fuels used in human activities.

Modernisation, inequalities and human limitations

As environmental planners, psychologists, architects, urban and landscape designers concerned about sustainable development we may ask: which problems do we see, what do we want, what could we do, how may this best be done, what may be achieved and who would use our results? In finding answers to these questions we should not forget three things:
   (a) that modern industrial society with its security and comfort has emerged from times of great hardship extending until less than 100 years ago; the availability of cheap fossil fuel has been strongly instrumental in this;
   (b) that the majority of people worldwide are still suffering from great hardships, mostly in relation to colonisation and sudden decolonisation, unfair trade, lack of democratic government and a less favourable climate;
   (c) that we are all living our own life under more or less restrictive physical and social conditions here-and-now, which makes any "grand design for sustainability" rather hard to prepare, organise and realise.

From needs to impacts. As people-environment specialists we must understand that human needs, values and beliefs are collectively translated into positive, but also negative economic, social and environmental impacts. The translation vehicles are various plans, designs, decisions and behaviours that we find necessary, comfortable or attractive, because they provide for increasing quality of life – as we see it; see Figure 3.

Certain kinds of environmental pollution, e.g., extreme noise, exhaust emissions or nature destruction, are directly related to people's desire to take it easy, to impress others and/or to confirm their own autonomy. Changing collective impacts from negative to positive ones requires changing human behaviour patterns, which may require changes in human needs, values and beliefs. This is not easy to do and often requires concerted efforts of different parties.

There is a comprehensive model for interdisciplinary work aimed at diagnosing, decision-making, intervention planning and evaluation about collective (environmental,

Figure 3. Translation from human needs (etc.) to economic, social and environmental impacts.

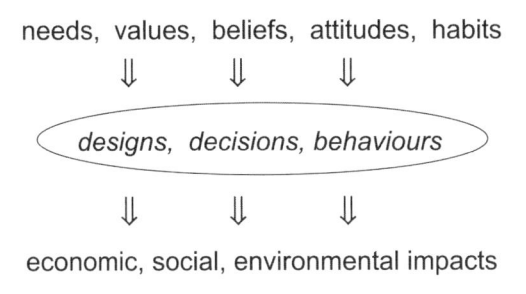

needs, values, beliefs, attitudes, habits

⇓ ⇓ ⇓

(designs, decisions, behaviours)

⇓ ⇓ ⇓

economic, social, environmental impacts

social and economic) sustainability problems. This so-called extended commons dilemma paradigm, obviously inspired by Hardin's (1968) "tragedy of the commons", leads the way along several issues and tasks for different kinds of experts as well as policy makers.

II. Commons dilemmas as private–public antagonisms

Are we living in a time where the private is becoming the enemy of the public and perhaps vice versa? Individual households, businesses, travellers and transporters, local chambers of commerce and municipal governments: are they all overexploiting their freedom of enterprise at the expense of long-term collective social and environmental goods and values? Have government offices lost public trust and popularity to the extent that individual freedom of enterprise became the dominant existential philosophy? Might we even fear that sustainability problems will get worse because citizens in democratic societies nowadays will simply not elect governments serving them further than what citizens' short-term benefit seeking requires?[3]

The tension between individual and collective interests is well captured in the term commons dilemma (Dawes, 1980; Hardin, 1968; Liebrand, Messick, & Wilke, 1992; Vlek, 1996; Vlek & Steg, 2002).[4] Vivid examples of commons dilemmas are the exploitation of fishing grounds by various companies, metropolitan air pollution through massive use of motor vehicles, and large-scale damage to natural ecosystems by expanding agriculture and livestock industry. Many specific problems in environmental and social research and even in economic research, can be better placed when both the micro and the macro perspective, or the private and the public perspective are taken into account.

A commons dilemma is a situation of conflict between an aggregate collective interest and numerous individual interests. In pursuing their own personal interests, many individuals may easily shift off limited negative effects on their common environment.

---

3 This, by the way, may be called the democratic dilemma in managing long-term common goods for society, which is based on the assumption that most citizens are short-sighted and normally discount things that are psychologically further away from them.
4 The term "commons dilemma" goes back to Hardin's (1968) inspiring analysis of the disastrous overgrazing of common pasture land (the "commons") by ever more cattle owned by individual farmers trying to improve their income position

By accumulation of the numerous small contributions the overall effect may be alarming and collective environmental qualities may significantly deteriorate. In many cases, collective risks increase through increases in material production and consumption patterns. Risks may also increase through the sheer growth in the number of separate actors such as inhabitants, households and commercial enterprises. Thus factors like population, consumption (or affluence) and technology are driving sustainability risks (cf. Figure 2 above).

For a political authority at the macro level of society, a commons dilemma – if recognised as such – presents itself as a permanent contrast between a (changing) collective risk, such as air pollution, and a large collection of (changing) individual benefits, from car driving, for example. Minimisation of risk and maximisation of benefits are incompatible social goals between which a trade-off must be made. Or, more concretely, one cannot have clean air *and* unlimited motorised traffic, produce butter *and* guns, or maximise individual freedom *and* social equality. In many domains a balance must be struck. However, the political authority may be tempted to do little or nothing about the collective risk, being elected to office for only a limited term, possibly by voters from the very population of individuals whose behavioural freedom is to be checked. In addition, the political authority may itself have an interest in continuing the collectively risky activity, due to considerations of tax revenue, employment or ownership of public infrastructure, for example. The dominant system of motorised transport may serve as an example.

In contrast, the individual actor's perspective at the micro level of society is naturally focused on his or her own benefits "here and now". There are several reasons why individuals may not recognise a commons dilemma as such:
- because they are unaware of any collective damage,
- because they do not appreciate their own responsibility for collective problems,
- because they do not feel the long-term collective risk is serious enough in relation to the numerous short-term benefits, and/or
- because they know there is a collective risk but feel little can be done about it (because of a lack of feasible alternatives and/or a lack of trust in others' cooperation).

The combination of personal benefit, "short-sightedness" and perceived lack of control over the situation, in particular, may readily lead to a denial or belittling of the risk and thus to a refusal to participate in risk control efforts. CO_2 emissions, global warming and the risks of climate change constitute examples.

An extended commons dilemma (ExCD) model

Environmental risk management is a matter of decision-making about risk acceptance ("does it weigh up against the distributed benefits?") and of practical intervention for controlling collective risk via behaviour change of contributing individuals. Risk management may be most effective if it links up with the diagnosis made about the social processes by which the risk is being generated or enhanced. And it needs monitoring and evaluation, so that its effects and side-effects may be fed back towards the contributing actors involved and policies may be revised.

Thus, understanding commons dilemmas and managing collective (environmental and social) risks revolves around problem diagnosis, policy decision-making, practical intervention and effectiveness evaluation. Together, these four components may be elaborated into twelve points of attention, as listed in Table 1.

Table 1. Twelve key issues for research and policy-making on commons dilemmas (from Steg and Vlek 2003).

I. Problem diagnosis
1. Analysis and assessment of collective risk, annoyance and stress
2. Analysis and understanding of socio-behavioural factors and processes underlying risk generation
3. Assessing problem awareness, risk appraisal and actors' individual values and benefits

II. Policy decision-making
4. Weighing of collective risk against total individual benefits ("need for change?")
5. If "risk unacceptable": setting objectives for reducing environmental and/or social risks
6. Translation of risk reduction objectives into individual behavioural goals

III. Practical intervention
7. Focusing on individual target groups and considering essential conditions for policy effectiveness
8. Specifying feasible behavioural alternatives and selecting effective policy instruments
9. Target group-oriented application of strategic programme of behavioural change

IV. Evaluation of effectiveness
10. Designing a monitoring and evaluation programme to determine policy effectiveness
11. Systematic, comprehensive evaluation of observable effects and side-effects
12. Intermittent and post hoc policy feedback, and possible revision of policies

The four divisions of Table 1 indicate the key problems in understanding and managing common resource dilemmas: appreciating the collective risk in light of associated benefits, weighing the risk against (total) individual benefits and deciding about action, promoting social behaviour change (or restraint), and evaluating the actual effects of policy interventions.

General strategies for behaviour change

Commons dilemmas reflect persistent conflicts between many individual (producer and consumer) interests on the one hand, and a small number of (large-scale) collective interests on the other. As dilemmas they may be resolved only by the achievement of a safer, sustainable *balance* of individual and collective benefits and risks. The nature and the effectiveness of various solution approaches have been investigated in a great number of laboratory and some field experiments (see, e.g., Dawes, 1980; Messick & Brewer, 1983). Most of these approaches may be categorised under seven general strategies for social behaviour change (Vlek, 1996; see also DeYoung, 1993; Gardner & Stern, 1996, Ch. 7; Gifford, 1997, Ch. 14). These are given in Table 2 – along with exemplary specifications.

Table 2. General strategies for behaviour change in managing commons dilemmas.

1. **Provision of physical alternatives, (re)arrangements**
   [adding/deleting/changing behaviour options, enhancing efficacy]
2. **Regulation-and-enforcement**
   [enacting laws, rules; setting/enforcing standards, norms]
3. **Financial-economic stimulation**
   [rewards/fines, taxes, subsidies, posting bonds]
4. **Provision of information, education, communication**
   [about risk generation, types and levels of risk, others' intentions, risk reduction strategies]
5. **Social modelling and support**
   [demonstrating cooperative behaviour, others' efficacy]
6. **Organisational change**
   [resource privatisation, sanctioning system, leadership institution, organisation for self-regulation]
7. **Changing values and morality**
   [appeal to conscience, enhancing altruism towards others and future generations]
[8. (by default) **Wait and see**: "do nothing, the quay will turn the ship"]

Strategies 1, 2 and 3, and certain (physical) forms of strategy 6 would initiate so-called *structural* (or: "hard") solutions to a commons dilemma, whose basic nature or type would thereby be altered. Strategies 4, 5, certain other ("mental") forms of strategy 6, and strategy 7 would imply *cognitive-motivational* (or: "soft") solutions. Through the latter, individual actors would be induced to behave in a cooperative (i.e., collectively optimal) manner, while the basic nature and payoff structure of the commons dilemma would be maintained. Structural solution strategies are generally more effective, but they are often not available or not easily implemented. Specific cognitive-motivational solution strategies are more easy to design and apply, but their effectiveness is generally lower. "Changing values and morality" (no. 7) stands relatively by itself as a cultural solution on which much behaviour change might come to rest, so that explicit regulation is no longer needed.

One solution-by-default is nicely expressed in a Dutch saying translated as "the quay will turn the ship". That is, strategy 8 in Table 2: "Wait and See" will automatically elicit behavioural (catastrophe) responses from causal agents, but these behaviour changes then cannot be but inappropriate, too little and too late. Making such unsustainability scenarios palpable – e.g., via scenario studies – would mean that individual actors are (better) enabled to "live under the shadow of the (common) future" and duly undertake preventive action. Disaster scenarios may thus initiate "self-destroying prophecies", as society tries to steer away from them.

Each strategy in Table 2 has its own strengths and weaknesses. For example, new physical infrastructure may *not* be effective because subjects turn away from it or start overusing it. Lawful regulation may just as well *not* work because rules are unknown or the needed enforcement is perceived as violating acquired civil rights. Pricing policies may have unexpectedly weak effects because subjects do not feel the price increases hard enough in their purses, while hard-enough prices would be politically unacceptable. In view of these considerations, various behavioural-change strategies, hard and soft ones alike, may best be combined such that their common effects may be optimal, while negative side effects are largely avoided.

The commons dilemma model and people-environment studies

From the perspective of sustainable development, the extended commons dilemma model in Table 1 may help us in surveying and categorising the field of people-environment (P-E) studies; a basic reference here is the recent volume edited by Moser et al. (2003). For each of the twelve focal points of the ExCD model we may work out the kind of P-E studies needed to clarify relevant issues and questions. Table 3 illustrates the various topics of study as we go along the model divisions of problem diagnosis, policy decision-making, practical intervention and effectiveness evaluation.

Now the list of P-E study topics in the right half of Table 3 comes to life only to the extent that each topic is linked to a practical issue or problem such as a housing project, the planning of road infrastructure, airport expansion or farm management policies. In each case, the particular problem under consideration may elicit a wider (sustainability) context such that all twelve focal points in Table 3 can be meaningfully addressed.

The ExCD model as explained above may be fruitfully applied to well-known sustainability problems such as sea-fishery, motorised transport and the use of drinking water: what is the problem, which policies are to be chosen, and how could one optimally

Table 3. People-environment study topics (right) as aligned with the ExCD model foci of Table 1 (left).

ExCD model focus (cf. table 1)	People-environment study topic
Problem diagnosis 1. Analysing risk, annoyance, stress 2. Understanding risk generation 3. Analyzing problem awareness, benefits and values	**[Problem diagnosis]** 1. Behavioural and social impacts of and responses to environmental risk 2. Behavioural and social causes and mechanisms of risk 3. Assessing problem awareness of different groups, their benefits involved and people's needs, values and expected quality of life
Policy decision-making 4. Risk-benefit weighing 5. Setting risk-reduction objectives 6. Setting of behaviour goals	**[Policy decision-making]** 4. Trade-offs, public values, temporal/geographic discounting 5. Social clarification of sustainability, deriving and specifying risk-reduction objectives 6. Translating environmental goals into behavioural goals, scenario design and ex ante evaluation
Practical intervention 7. Involving target groups 8. Selecting policy instruments 9. Actual intervention	**[Practical intervention]** 7. Selection of relevant target groups, social tuning of policy instruments, target group involvement 8. Specification of behaviour alternatives, effectiveness of behaviour-change strategies, campaign design and organisation 9. Actual implementation of policy measures and instruments, focusing on target groups while supporting their cooperation

Table 3 continued

Effectiveness evaluation	**[Effectiveness evaluation]**
10. Designing evaluation programme	10. Designing, planning and organisation of systematic evaluation of behavioural and attitudinal effects for different involved target groups
11. Systematic evaluation	11. Actual methodical evaluation of effects and side-effects for different groups; analysis of results and formulation of summary conclusions
12. Feedback and policy revision	12. Design and provision of effective feedback for target groups and possible adaptation of policies under way

Note. The term *risk* is also meant to indicate a broader sustainability problem. In addition, *risk* is conceived as a process rather than a static condition (as in gambling).

intervene to safeguard the sustainability of the relevant domain? This requires a careful intertwining of scientific ideas and policy issues, such that the actual problem may be dealt with in a comprehensive way. For a demonstration, let us consider more substantively how the four divisions of Table 1 may aid in the understanding and management of collective problems involved in motorised transport.

III. Motorised transport as a commons dilemma

Thus far, the area of human mobility and transport is largely escaping from effective policy-making in view of sustainable development. During the past 100 years increasing motorisation has brought about an enormous revolution in everyday life, in human interaction and in the spatial configuration of social and economic activities. Currently, there are widespread and growing concerns about congestion, (other) environmental impacts and reduced quality of life in relation to unsustainable transport developments (i.e., a general over-motorisation).[5]

Transport problem diagnosis

Currently in OECD countries, the private motorcar is dominating human settlements, consumers consider their car a daily necessity, the car industry keeps cherishing its sales' growth, and governments willingly cash in tax money and supply infrastructural facilities. Meanwhile, most parties involved tend to look away from short-term as well as long-

---

[5] Motorised transport is, of course, part of a more-encompassing development of mass motorisation of numerous human activities. Divergent examples are building construction, street cleaning, lawn mowing and water scooting. For any of the relevant tools, implements or vehicles used, pertinent 'sustainability' questions are: why is this used; what are its social and environmental costs; to what extent is the user aware of the costs; what would be feasible alternatives; how could the user be encouraged to use a more environment-friendly alternative? There is rather little systematic analysis and environmental policy making on the various motorised vehicles and tools.

term sustainability problems related to accessibility, liveability and environmental quality, especially in metropolitan areas. At least for the past 50 years, car ownership has been strongly facilitated and gradually promoted into an acquired civil right. Car ownership has led to structural changes in economic, social and government practices. Therefore, sustainable changes require comprehensive, multi-level innovations. To get effective policies going, society wants to know:
– What are the main reasons for change?
– What are possible substantive transitions?
– How will needed transitions be managed?

National, regional and local governments are reluctant to come up with convincing answers. At some distance, however, international organisations are fairly concerned about developments in motorised transport. For example, at the end of the international OECD-conference *Towards sustainable transportation* in Vancouver, 1996, it was concluded (OECD, 1997, p. 56):

> "Sustainable transportation is achieved when needs for access to people, services, and goods are met without producing permanent harm to the global environment, damage to local environments, and social inequity. (...) Systems of transportation used in OECD and some other countries are unsustainable."

In its second *Environmental Assessment Report* the European Environment Agency (EEA, 1999, p. 30, 62 en 416) writes:

> "Transport and mobility is jeopardising the EU's ability to achieve many of its environmental policy targets. (...) improved eco-efficiency is not a sufficient condition for sustainable development. (...) In the past, economic growth and lowering transport prices have raised demand for transport. Where congestion occurred, new roads, airports and other infrastructures were constructed. (...) This closes the vicious circle of ever expanding transport volumes. (...) While many instruments are being applied to reduce transport damage, these are being overwhelmed by the rapid rise in demand for transport."

And in its broad-ranging *Global Environment Outlook*: GEO 2000, the United Nations Environment Program (UNEP, 1999, p. xxix and 13) notes about traffic and transportation:

> "Means must be found to tackle the root causes of environmental problems, many of which are unaffected by strictly environmental policies. (...) If current rates of expansion continue, there will be more than 1000 million vehicles on the road by 2025. Transport now accounts for one-quarter of world energy use, and about one-half of the world's oil production; motor vehicles account for nearly 80 per cent of all transport-related energy. (...) The transport sector has, so far, proved highly resistant to attempted policy reforms."

From a commons-dilemma perspective there is an impressive contrast between the individual benefits and the collective costs and risks of motorised transport, as illustrated in Table 4. This hardly needs detailed explanation. Collective (system) benefits of motorised transport, such as government income, business profits and employment, are not accounted for in Table 4. We must realise, however, that in motorised transport conflicts of interest occur at all system levels.

Table 4. Individual benefits versus collective costs/risks of motorised transport.

Individual benefits	Collective costs/risks
– Availability (continuous)	– Space occupation (roads, parking, street life)
– Payability (low costs)	– Traffic jams, congestion, delays
– Speed	– Landscape fragmentation (biodiversity)
– Comfort (passengers, baggage)	– Traffic accidents (prevention and handling costs)
– Flexibility (from door to door)	
– Reliability	– Energy consumption
– Safety-en-route	– Use of raw materials
– Privacy	– Solid wastes
– Freedom, autonomy	– Harmful emissions, air pollution (local-global)
– Pleasure, sensation	– Environmental noise
– Social embeddedness	– Costs of infrastructure and maintenance
– Social status, distinction	– Costs of lawful regulation and enforcement
	– Decline of transport alternatives (train, bus, bike)

Market versus government operation. The oppressing problem reflected in Table 4 is that the supply, achievement and improvement of individual benefits are in the hands of the market: vehicle producers, sales companies, customers, specialised media and sports organisations. In contrast, the monitoring and control of the collective costs and risks is considered to be a government responsibility. What complicates matters is that the government is in a double position: on the one hand it should limit the collective costs and risks; on the other hand, it eagerly profits from tax and excise revenues. Thus it is not entirely incomprehensible that – under the dominant free-market ideology – governments tend to view the traffic and transportation domain as rather uncontrollable.

Freedom versus equality. As in every commons dilemma, in transportation too, two classical ideals struggle for priority: freedom for the individual (household, business) and equality for all. The dilemma implies that too much individual freedom of (private) transport leads to unacceptable violations of equal rights to mobility, liveability and environmental quality for all. In traffic and transportation, individual freedom and its risks of social inequality play a prominent role. The public space of roads, streets, squares and parks withdraws itself more than other locations (homes, offices, restaurants, theatres) from social control of behavioural rules and norms. Thus in private transportation people are tempted to manifest extensive freebooting, at the expense of traffic safety, the mobility of vulnerable groups, environmental quality and ultimately the equal accessibility of chosen destinations.

In combination with the increasing scarcity of desired high-speed mobility (at least on the roads), the dominant principle of individual freedom brings along widespread feelings of social deprivation, envy and competition, three factors that guarantee a strong upward drive for "more, bigger, stronger, faster and further". In contrast, collective transport systems (bus, train, airplane) bring along greater social equality, less individual freedom and therefore less competition among travellers.

Transport policy decision-making

In densely populated countries like The Netherlands, government policy on traffic and transportation is a topic of national debate, political crisis and social controversy. It is perhaps illustrative to briefly consider ten traditional policy foundations for this domain and (in brackets) the critique they provoke, as follows.
1. *(Motorised) Mobility = prosperity.* [But individual prosperity is being reduced by the deterioration of environmental and social goods and qualities.]
2. *Let there be freedom of (motorised) mobility.* [But this dogma of free-market thinking neglects the importance of equal rights for all citizens, including future generations.]
3. *Reducing the volume of traffic is impossible.* [But behavioural and social scientists surely can come up with effective measures for traffic volume reduction.]
4. *Government should meet people's transport needs.* [But transport needs are not inevitably given; needs/values can be changed and shaped to meet collective interests.]
5. *Local problems are unrelated to national problems.* [But local policy options and limitations are strongly dependent on national policies (e.g., taxes and fuel price).]
6. *Both car use and transport alternatives should be facilitated.* [But effective promotion of one transport mode requires simultaneous discouragement of the other(s).[6]]
7. *The supply side of car use does not need more attention.* [But many excuses of transport users lie in the force of system properties; the supply side actually plays a dominant role.]
8. *Technology will eventually offer decisive solutions.* [But technology is only one way of eliciting behaviour change; several other strategies exist (cf. Table 3).]
9. *Factor 4: prosperity can be doubled while environmental impact is halved.* [But Factor 4 is proving to be an illusion, due to steady volume growth and to rebound effects of environmental technology.]
10. *Society cannot be shaped following a preferred policy scenario.* [But society is continually shaped through infrastructure, technology, economic policy and socio-cultural events.]

What seems to be missing in (Dutch) policy decision-making is a clear future perspective on sustainable transportation, and a consistent set of long-term policy goals. By consequence, motorised transport seems to be a domain of unbridled developments where governments cannot or will not (yet) introduce effective sustainability policies.

Practical intervention for sustainable transportation

What then might be sustainable transportation? Following its definition of sustainable transportation (see above), the OECD conference (1997, p. 56) also concluded:

> "Achievement of sustainable transportation will likely involve improvements in vehicles, fuels, and infrastructure (…), and reductions in personal mobility and in the movement of goods, (...)."

---

6 This is precisely what happened in favour of road transport through the long-term neglect of the railway system.

More recently, OECD ministers of transport have agreed to a more elaborate definition, as summarised in Table 5.

Table 5. OECD ministers' (2001) requirements for sustainable transport.

- Meeting basic access needs of individuals, companies and societies ...
- Safe, healthy and environmentally friendly ...
- Equitable within and between generations ...
- Affordable, fairly and efficiently operated ...
- Supportive of a competitive economy ...
- Allowing balanced regional development ...
- Optimally using renewable energy, materials ...
- Controlled use of non-renewables ...
- Minimising land use ...
- Minimising noise generation.

While the original 1997 definition (see above) is strict and clear, the revised (ministers') definition is much more encompassing and considerably more vague. For one thing it involves trade-offs which have not been explicated. Another question is whether this definition is not inherently conflictual, depending on how liberally one defines the term *basic access needs*.

Eight policy guidelines. Elaborating on the idea of sustainable transport in light of Dutch mobility and transport problems, we may consider the eight guidelines for achieving a sustainable transportation system as displayed in Table 6.

Table 6. Eight policy guidelines for sustainable traffic and transportation.

1. Freedom of transport is inversely proportional to the environmental impact of the chosen transport mode.
2. Transport policy will be tuned to the fundamental causes and driving forces of motorised transport.
3. Policies will be aimed at strategic changes in transport needs, system, infrastructure and technology.
4. The user pays for of his/her part in all collective costs and risks of traffic and transportation.
5. An expansive, freedom-loving market will be checked by a responsible, equality-guarding government.
6. The environmental responsibility of transport users will be significantly improved.
7. Social acceptance of sustainable transportation must be continually increased.
8. Authorities (cabinet ministers, politicians, the royal family) will set good examples.

As for Principle 1, the simple observation that environmental decline is gradually boomeranging on individual living environments and thus actually is reducing individual freedom, makes it inevitable that limits be set to individual behaviour expansion and that individual freedom be subordinated to the safeguarding of collective goods and qualities.

Under Principle no. 3 it must be stressed that *hard* physical and social *system* changes are likely to bring about greater behaviour change than direct *soft* campaigns appealing to transport users who are dependent on given system characteristics. However, the usual *and and* policies (e.g., stimulating public transport *and* facilitating car traffic) would have to be left behind: you can't have you cake and eat it, too.

Lifting sustainability barriers. More generally, psychological conditions for social behaviour change can be derived from the various strategies given in Table 2 above. Policy measures should lift the barriers for sustainable transport behaviour and facilitate and stimulate change. For example, problem awareness should be increased, people's own responsibility should be stressed, policy goals should be clarified, behaviour alternatives must be available, and recommended sustainable behaviours should be attractive (or at least acceptable) from both a personal and a social point of view.

Policy effectiveness evaluation

Much policy making about transportation and traffic is focussed on the resolution of immediate problems (which may have aggravated for a while) and the introduction of policy measures that seem to have some promise of alleviating the problem noticeably, however temporarily. Actual effects and side-effects of measures taken are rarely observed and analysed systematically, and this especially applies to long-term and indirect effects and side-effects of policy measures.

Policy-relevant effectiveness evaluation is a professional affair, which should be carefully designed and conducted according to plan. To the extent that practical interventions should somehow link up with the problem diagnosis performed, effectiveness evaluation should be related to the nature of interventions, their application conditions and the target groups involved. This involves repeated monitoring of the variables for which policy goals (environmental, social, behavioural) had been set, differentiation among target groups and of physical locations and time periods, and a valid description of what has and what has not been achieved and what will likely be the future trends in observed effects and side-effects of policies.

Patterns of behaviour change set in motion via hard and soft policy measures are usually not limited to transport mode choice, travel speed and/or parking location themselves. Behaviour change in relation to mobility and transport may also involve changes in daily activities, in spending patterns and in the nature and intensity of social interactions. It is in these and other behaviour domains that one may find various reasons for the acceptance or rejection of innovative transport policies aimed at strengthening system sustainability.

Conclusion on motorised transport

Considering problem diagnosis, policy decision-making, practical intervention and effectiveness evaluation together, we see that applying the extended commons dilemma model to motorised transport on the one hand reveals rich possibilities for research and policy-making. On the other hand, however, in the actual practice of policy there is a poor reflection of systematic analysis. We should, of course, realise that motorised transport is an enormous domain of popular human activity, where sustainability questions are being asked only recently while collective problems are pressing already for some time.

IV. Sustainable quality of life

Let us, finally, consider the crucial concept of (future) quality of life. Shouldn't human life at any level be sustainable in terms of economic security, social well-being and environmental quality? To what extent does this Brundtland conception (WCED, 1987) fit into the psychology of human motivation?

In order to explain people's feelings and judgments about prospective changes in their country's environmental and economic conditions, Vlek, Skolnik and Gatersleben (1998) used a shortlist of twenty-two different Quality-of-Life (QoL) variables. The list had resulted from a survey of various theories and taxonomies of human needs and values. A slightly revised list was used in a more extensive study by Poortinga, Steg and Vlek (2003), aimed at studying expected QoL effects from significant energy savings through various sustainable household scenarios. Table 7 gives the average importance ratings of 22 QoL variables as evaluated by 455 subjects in Poortinga's field study.

Table 7. Twenty-two QoL-attributes ordered from most to least important following mean judgements given by 455 respondents in field study by Poortinga et al. (2003)

QoL attribute	Mean rating	QoL attribute	Mean rating
Health	4.87	Security	4.13
Partner/family	4.68	Nature/biodiversity	4.12
Social justice*	4.65	Leisure time	4.00
Freedom	4.53	Money/income	3.63
Safety	4.51	Comfort	3.54
Education	4.30	Aesthetic beauty	3.49
Personal identity	4.22	Variation	3.33
Privacy	4.21	Challenge	3.18
Environmental quality	4.20	Status/recognition	3.03
Social relations	4.19	Spirituality	2.94
Work	4.15	Material beauty	2.62

Note. The judgement scale ranged from 1: unimportant to 5: very important, with midpoint 3: reasonably important. *Social justice (italicised) did not fit into the factor analysis reported below.

It appeared that the average (Dutch) subject found health, partner/family and social justice to be most important, while status/recognition, spirituality and material beauty were found least important. Note that, on the average, almost all QoL attributes were judged to be at least reasonably important. After factor analysis on the importance ratings it appeared that, with one exception (social justice), the twenty-two variables could be compactly summarised into seven factors, together explaining 60% of the variance in people's importance judgments. These factors, together with the variables they include, are given in Table 8, in top-bottom order of averaged importance. Note that "maturity", "openness-to-change" and "material wealth" were judged to be less important than "family, health and safety", "personal freedom" and "achievement".

Poortinga et al. (2003) also conducted multivariate analyses of variance to examine whether various subgroups of respondents differed in their importance judgements of the various QoL attributes. Since it was possible to condense respondents' aspectwise judgements into seven QoL factors, average factor scores were used as dependent variables. It

Table 8. Seven quality-of-life (QoL) factors summarising 21 out of 22 specific QoL variables (in brackets); from Poortinga et al. 2003 (in press).

Quality-of-life factor (and specific aspects bundled)	Mean importance factor	% explained variance
health, family and safety (health, partner/family, safety)	4.69	7.0
self direction (freedom, privacy, leisure time)	4.35	9.8
achievement (education, work)	4.22	6.4
environmental quality (environmental quality, biodiversity, aesthetic beauty)	3.93	9.9
maturity (identity/self-respect, security, spirituality/religion)	3.77	7.6
openness to change (social relations, change/variation, excitement)	3.57	8.5
self enhancement (money/income, comfort, status, material beauty)	3.20	11.2

Note. From top to bottom, the factors bundle QoL variables of decreasing importance. Ratings could range from 1 to 5 (= most important). The singular aspect "social justice" (mean rating 4.65) fell outside the above factor structure.

appeared that the latter varied significantly across subgroups of respondents. Significant multivariate effects were found for age, sex, household type, income, level of education and environmental concern. With reference to Table 8, some results on subgroup differences are:
- *young people* found *openness to change* more important,
- *women* found *personal freedom* and *maturity* more important,
- *the higher-educated* found *achievement* more important,
- and those *higher in environmental concern* found *environmental quality* and *personal freed*om more important; they found *material wealth* less important.

V. General conclusions

Sustainability problems are multidimensional problems characterised by economic, social and environmental impacts, causes and processes. A comprehensive approach to such problems means working in a multidisciplinary team maintaining a common focus on a particular policy problem. Key questions may be derived from an extended commons dilemma model (cf. Table 1) capturing the tension between individual and collective interests, and between individual freedom and social equality. Main questions are:
- What is the nature and seriousness of the problem; what are its behavioural causes?
- What should be done about the collective risk; which sustainable-development scenarios may be envisaged; which behaviour alternatives are available or should be designed?
- Which policy interventions can be effectively carried out to reduce or eliminate the problem and get onto a sustainable-development path?

Under a general concern about unsustainable developments, the area of people-environment studies may well be aligned with an extended commons dilemma model. On the one hand, a thorough commons dilemma analysis of a given sustainability problem may be based on a systematic inventory of existing research articles and reports in the professional literature (we do know a lot already). On the other hand, new comprehensive sustainability studies may start from a commons dilemma perspective and then proceed in a fruitful science-policy interaction. An overwhelming example is the area of motorised transport, where policy making could profit from a comprehensive and systematic assembly of relevant studies and multidisciplinary expertise.

Those of us who have experienced multidisciplinary work about concrete policy problems may have few illusions about the effective implementation of what seems really needed. As an environmental psychologist myself I might say that psychology itself has not been very effective in convincing environmental policy makers to practically apply psychological ideas and findings. Some reasons for this are:
- the historical lead of the natural sciences in supporting environmental policy making,
- policy makers' conservative preference for hard data,
- governments' enduring technological optimism ("little need for behavioural science"),
- policy makers' ignorance of pertinent psychological theory and methods,
- psychologists' love (and incentive structure) for small-scale theory-driven experiments,
- society's dominant free-market ideology implying a general neglect of commons dilemmas,
- lack of comprehensive multidisciplinary problem solving wherein psychology fits.

Theory and practice. Concerning the nagging theme of practicing theory versus theorising practice the following may be relevant. It is the business of science to efficiently describe, explain and predict puzzling phenomena, so that these may be well understood and possibly managed. In contrast, it is the business of practice to make concrete plans and designs to meet specific goals for environments that should satisfy users. Much scientific work does not get operational enough to predict how users would actually respond to given plans or designs. On the other hand, practical planners and designers seem to keep scientific explanation and prediction all too implicit. Social psychologist Kurt Lewin (about 1950) has been frequently quoted as writing: "There is nothing so practical as a good theory." Following another saying, however, "the proof of the pudding is in the eating", meaning that for things that work we may not need elaborate recipes. Let us hope that future people-environment theorists become good enough to get their feet on the ground, however dirty their toes might get. Let us also hope that people-environment practitioners will increasingly realise that theorising is indispensable for getting beyond the clinical experience of trial and error, however creative their thinking has been.

VI. References

Dawes, R. M. (1980). Social dilemmas. *Annual Review of Psychology*, 31, 169-193.
DeYoung, R. (1993). Changing behavior and making it stick. The conceptualization and management of conservation behavior. *Environment and Behavior*, 25, 485-505.

EEA: European Environment Agency (1999). *Environment in the European Union at the turn of the century*. Copenhagen. Luxemburg: Office for Official Publications of the European Communities.
Ehrlich, P. R. & Holdren, J. P. (1971). Impact of population growth. *Science*, 171, 1212-1217.
Gardner, G. T. & Stern, P. C. (1996). *Environmental problems and human behavior*. Boston: Allyn and Bacon.
Gifford, R. (1997). Environmental psychology; principles and practice. 2nd edition. Boston: Allyn and Bacon.
Goodland, R., Daly, H. & Kellenberg, J. (1994). Burden sharing in the transition to environmental sustainability. *Futures*, 26, 146-155.
Hardin, G. (1968). The tragedy of the commons. *Science*, 162, 1243-1248.
Liebrand, W., Messick, D. & Wilke, H. (1992). *Social dilemmas; theoretical issues and research findings*. Oxford/New York: Pergamon.
Messick, D. M. & Brewer, M. B. (1983). Solving social dilemmas: a review. In L. Wheeler & P. Shaver (Eds.), *Review of personality and social psychology*. Volume 4. Beverly Hills, CA: Sage.
Moser, G., Pol, E., Bernard, Y., Bonnes, M., Corraliza, J. A. & Giuliani, M. V. (Eds., 2003). *People, places and sustainability*. Seattle/Toronto/Bern/Göttingen: Hogrefe & Huber.
Munasinghe, M. (1993). *Environmental economics and sustainable development*. Washington, D.C.: World Bank.
OECD: Organisation for Economic Cooperation and Development (1997). *Towards sustainable transportation*. Summary proceedings of an international conference held in Vancouver, March 1996. Hull (Quebec): Environment Canada.
Poortinga, W., Steg, L. & Vlek, C. (2003, in press). Values, environmental concern, and environmental behavior. *Environment and Behavior*, 35.
Schmuck, P. & Vlek, C. (2003). Psychologists can do much to support sustainable development. *European Psychologist*, 8 (2, June).
Steg, L. & Vlek, C. (2003, in press). Understanding and managing environmental resource use: a behavioural science perspective. In J.J. Boersema & L. Reijnders (Eds.), *Principles of environmental science*. Dordrecht/Boston: Kluwer Academic Publishers.
UNEP: UN Environment Program (1999): *GEO 2000: Global Environment Outlook*. London: Earthscan.
Vermeersch, E. (1988). *De ogen van de panda; een milieufilosofisch essay* [The eyes of the Panda; an environmental-philosophical essay]. Brugge: Marc van de Wiele.
Vlek, C. (1996). Collective risk generation and risk management; the unexploited potential of the social dilemmas paradigm. In W. B. G. Liebrand & D. M. Messick (Eds.), *Frontiers in social dilemmas research*, pp. 11-38. Berlin/Heidelberg/New York: Springer Verlag.
Vlek, C., Skolnik, M. & Gatersleben, B. (1998). Sustainable development and quality of life: expected effects of prospective changes in economic and environmental conditions. *Zeitschrift für experimentelle Psychologie*, 45 (4), 319-333 (with German summary).
Vlek, C. & Steg, L. (2002). The commons dilemma as a practical model for research and policy-making about environmental risks. In G. Bartels & W. Nelissen (Eds.), *Mar-

keting for sustainability; towards transactional policy making. Amsterdam/Burke (VA)/Leipzig/Oxford/Tokyo: IOS Press, 285-303.

Wackernagel, M. et al. (2002). Tracking the ecological overshoot of the human economy. In *Proceedings of the National Academy of Sciences*, PNAS, July 9, no. 14, 9266-9271.

WECD: World Commission on Environment and Development (1987). *Our Common Future*. Oxford/New York: Oxford University Press.

At home everywhere and nowhere:
Making place in the global village

Nan Ellin
Arizona State University, Tempe, AZ, USA

> **Abstract.** A "quiet" revolution in architecture and urban planning has been taking place over the last decade, aiming to heal the wounds inflicted upon the landscape during the modern era. It is quiet because its practitioners are not united under a single banner and because their sensitivity to people and the environment often translates into design that does not call attention to itself. Nonetheless, its impact can not be overstated. Surfacing from all corners of the globe, this transnational revolution is dramatically reshaping our physical environment, the social life that occurs in it, and the practice of urban design. In this essay, Professor Nan Ellin describes this transformation focussing on five principle qualities – hybridity, connectivity, porosity, authenticity, and vulnerability. She also examines the larger social changes in which these qualities are emerging featuring a shift towards slowness, simplicity, sincerity, spirituality, and sustainability.
>
> **Keywords:** globalization, postmodernism, ecological design, landscape urbanism

Since moving to the desert of the American southwest several years ago, I have been thinking a lot about the breakdown of traditional boundary markers. Boundaries between countries and ethnic groups; between central cities, their suburbs, and the countryside; between major world cities and the periphery; among professions, areas of knowledge, and much more, have been breaking down thanks to the end of the Cold War, the expansion of the European Community, the possibilities allowed by computer technologies, and the overall acceleration of global flows. All of these have made the world seem much smaller – a kind of global village.

Reaction

The speed and novelty of these changes have also generated a sense of emptiness, anxiety, and insecurity. Indeed, the words most commonly used to describe places today suggest an absence or aftermath such as placelessness, deterritorialization, translocality, and postcolonial. Some of the the adjectives most commonly used to describe space include: abandoned, blank, vacant, abstract, and anonymous.

Not surprisingly, we have felt nostalgic for the erstwhile clarity of the traditional boundary markers. During the 1970s and 1980s, this was apparent in the desire to retribalize or to assert cultural distinctions. It was apparent in the search for "roots" through the tracing of family lineages, resurrecting old customs, and even inventing "new" traditions. And it has been apparent in the retreat to ones' own kind: "We want to be with people like us" is the common refrain. Although American popular culture celebrates multiculturalism, certain trends suggest we are in fact becoming a more phobic and segregated society. Paradoxically, it is as we are achieving the modernist ideal of universalism that it has become unacceptable.

In anthropology, the study of cultural diversity, the modernist quest for an objective truth was manifest as an effort to scientifically study society, without imposing one's own viewpoint, to apply controls, develop hypotheses and come up with answers. The statistical average was sought – the norm – and exceptions were dismissed as irrelevant, uninteresting, or abnormal. People became factors in equations – the human factor – devoid of flesh, blood, and emotional complexity. By the 1970s, it became painfully obvious that this attempt to be an absent author only revealed an absent subject. One attitude held that we can not ever really know others, so don't even try. Retreat to your own backyard (and study your own kind) or inward (and study yourself). This attitude has been described as reflexivity, or more pejoratively, as navel-gazing.

Another reaction was to fictionalize one's accounts. What used to be a search for truth became an acknowledgment that there are many or no truths and that autobiography and fiction are all we can aspire towards. These tendencies in anthropology to retreat or escape are, needless to say, counter to its pursuit of better understanding those different from ourselves and thereby decreasing cultural conflicts, granted a modernist one.

In city building, the retreat and escapist reflexes have been apparent in segregated and defensive urbanism as well as theme-ing. The expansion of segregated communities is most blatant in the growth of retirement communities, like Sun City outside of Phoenix, but there are also many neighborhoods segregated by ethnicity and social class.

The impulse to retreat is epitomized by the growth of gated communities for all age and income levels, despite recent findings that gating communities has little effect on crime, either within the gates or outside them. This atavistic desire to mark one's turf with walls, gates, and prohibitions lends a new and eerie resonance to Max Weber's "iron cage" metaphor.

Outside of gated communities, security signage is ubiquitous. When designing new homes or renovating, safety features are of paramount importance. Sometimes, a client asks for an appearance that conveys a "don't-mess-with-me" attitude or which appears inconspicuous to conceal the residents' wealth. These have been described as stealth houses (Mike Davis's term). In the house he designed for actor Dennis Hopper in Venice, Brian Murphy set a bunker-like structure with a windowless corrugated metal facade behind a white picket fence mimicking those in the neighborhood. In a house around the corner (the Dixon house on 5th Avenue, Venice), Murphy simply left the shell of the existing dilapidated house, built a new house inside it, and pre-graffitied the facade to fit into the surroundings.

Other houses take the opposite tack and elaborately appoint their entryway, perhaps in a show of intimidation. These houses assure protection through a variety of means such as sophisticated security systems, the posting of signs which warn trespassers not to enter or indicate "armed response," and so-called "security gardens," which group shrubs beneath windows and around yards specifically for the purpose of obstructing intruders. Architects whose clients live in Beverly Hills and Bel-Air report that one of the most often requested features is the "safe room," a "terrorist-proof security room" concealed in the houseplan and accessed by sliding panels and secret doors, reminiscent of a James Bond movie.

The mentality of fear among homeowners of all kinds has led to a pronounced anti-growth movement. People who do not want development to occur near them have been referred to as NIMBYs (not in my back yard) or as BANANAs (build absolutely nothing anywhere near anything). Such separatism offers a certain sense of security, but it also allows for more ignorance of others and less tolerance for difference. It feeds an us against them mentality and a tendency to defend one's borders, family, and self with gates as well as with guns (vigilantism). There are currently more than 200 million guns in private hands in the US [total population: 243 million] and over the last decade, the number of women with guns has doubled. Gun control versus the freedom to bear arms. (Amendment 2 of the US Constitution) is hotly debated around the country.

The popularity of the 4-wheel drive sports utility vehicle, especially in cities, also expresses a desire to defend oneself. Although equipped for off-road driving, very few actually ever leave the roads. The appeal of this sort of vehicle is epitomized by the current vogue for the Humvee (human military vehicle or high-mobility vehicle) which was recently released in a civilian edition called the Hummer, available for $65,000 and up. Actor Arnold Schwarzenegger purchased the very first one (Rugoff). While the Hummer may be "the ultimate in body armor" (Rugoff), the safety of all cars today is a major selling point, including a wide range of options from alarms to car phones, built-in car seats for children, air bags, nonbreakable glass, and more. There are now microwave-activated security systems which, sensing that a body is near the car, emit a rough man's voice saying "Get away from this car or an alarm will go off in 5 seconds."

Our public spaces have been disappearing and those that remain often relay the message, Go away, or, Don't linger long, since they have been stripped of public rest rooms, telephones, and even water fountains. This barrel-vaulted bench was created for the city of Los Angeles to discourage people without homes from sleeping on them. It is called the "bum-proof bench." Not incidentally, since the 1980s, commissions for corporate buildings have been declining (because of an office glut in most parts of the country), while commissions for prisons, police substations, and homeless shelters have been growing. Also on the rise – and demonstrating the escapism into fantasy worlds – is the building of theme parks and of megastructures devoted to leisure and recreational activities, particularly sports stadiums and convention centers.

In the architecture and planning professions, the escapist reflex has been apparent in the call, since the 60s, to refer to cities of the past, described as regionalism, historicism, neo-traditionalism, or the New Urbanism. The threat to previously clear boundaries has incited an anxious effort to produce places that mask what is going on behind the facades and that look as though they grew spontaneously over time without planning.

The escapist nature of all these undertakings – behind gates or prison bars, away from our downtowns, into the past, other places, or fantasy worlds – may emit signals that the present is indeed unsavory. The rising tide of fear has led people to stay at home more. Activities that once occurred outside the home are increasingly satisfied now inside the home with the television or computer. Or if we go out, in the strictly controlled settings of the shopping mall, theme park, or sports arena. We no longer go out to mingle with the anonymous urban crowd in the hope of some new unexpected experience or encounter, a characterizing feature of earlier urban life. Unexpected experiences and encounters are precisely what we do not want. We go out for specific purposes, with specific destinations in mind, and with a knowledge of where we will park and whom shall meet.

But these retreat and escape reflexes provide only temporary relief from the crises in society, in the urban design professions, and in the built environment. And not part of the solution, they contribute to the problem.

The predominant urban design reflexes of the 70s and 80s – historicism, regionalism, theme-ing, and defensive urbanism – were attempts to rekindle a sense of community and with it a sense of security and meaning as well as interest. Although this postmodern urbanism has offered certain correctives to the modern urbanism that preceded it, it failed to sufficiently satisfy these persistent longings, and in many instances, deeply intensified them.

Proaction

The failure of postmodern urbanism has led to a reconsideration of values, goals, and means of achieving them, especially marked over the last decade. In contrast to the fast-paced more is more mentality, the appeals of simplicity, slowness, spirituality, and sincerity are clearly on the rise. Instead of responding reactively to rapid change through escapist and distilling strategies, there have been efforts to embrace, steward, or partner with (rather than control or manage) it.

The modernist quest for objective truth compelled creators and their interpreters to model themselves after scientists and engineers. In architecture and urban planning, this was manifest in the emulation of the machine, the quest for efficiency, and the desire to produce universal plans. But the disappointing outcome there, as well as in parallel efforts in the social sciences and humanities (described above), incited attempts to bring the author back and thereby breathe life back into all these endeavors. In the process, the guiding metaphor shifted from the machine to the collage or the text (story, narrative).

In architecture and planning, the goal became that of creating legible, meaningful, and sacred places (placemaking), in contrast to the anonymous rational spaces that were a product of the post-war period. These metaphors dominated the decades of the 1970s and 80s.

More recently, however, the metaphors for city and culture have been shifting once again. Like the collage and text, these emerging metaphors also suggest an inclusivity. But this time, it is no longer for the sake of inclusivity itself. Rather, there is attention paid towards whole-ness, a more calculated beauty, a smoothness, a lightness (lack of heaviness), and a strong sense of connectedness. In this spirit, these emerging metaphors are also more than metaphors, carrying literal and place-derived meanings as well.

The most overarching of the current metaphors is *ecology*. Architects and planners are increasingly seeking inspiration for their work in "nature's deep interconnections" (Van der Ryn & Cowan, 1996)[1] while anthropologists and cultural theorists are increasingly regarding culture as a part of nature rather than in opposition to it. Cultural theorist Catherine Roach, for example, argues "against the idea that nature and culture are dualistic and opposing concepts," suggesting that this idea is "environmentally unsound and [needs] to be biodegraded, or rendered less harmful to the environment" (1996, p. 53).

While these understandings of connected-ness have many precedents, there is something qualitatively different this time around in the emphasis on change as a constant and on the reconfiguration of space and time due to digitalization.

The other prevalent metaphor for city and culture is the *border* or *edge*. Current buzz-phrases amongst anthropologists, cultural theorists, architects, and urban planners include border cultures, borderlands, edge conditions, edge cities, and cities on the edge. Clearly, the current fascination with borders and edges is another response to the dissolution of traditional markers discussed earlier, but now it is proactive not reactive. It is along these borders that our greatest dilemmas reside as well as our greatest opportunities for resolving them.

Amongst architects and planners, there is a great deal of attention paid towards the borders and edges in both their literal and figurative manifestations. Theory and practice have been focussing on places which are betwixt and between, places which are perceived as somehow liminal in space and/or time. This is apparent in the obsession with spaces considered interstitial, "terrain vagues," "no man's lands," or "ghost wards" (Mitchell Schwarzer). It is also apparent in the attention towards designing along national borders and between ecologically-differentiated areas such as along waterfronts and coastlines and towards preserving or creating edges between city and countryside (e.g., Daniel Libeskind's 1987 City Edge project for Berlin, Steven Holl's 1991 proposal for creating edges on urban peripheries in an effort to counter sprawl, the Banlieues '89 project in France).

The notion that the talents and energies of architects and urban planners should contribute to mending seams, not tearing them asunder, to healing the world, not to salting its wounds, has grown much more widespread in acceptance.

Another – although very different – expression of interest in borders/edges is the obsession with the "fold" (via Gilles Deleuze) amongst the contemporary Eisenman School. "Unlike the space of classical vision," Peter Eisenman contends, "the idea of folded space denies framing in favor of a temporal modulation" (Eisenman, 1992). Given "the exhaustion of collage as the prevailing paradigm of architectural heterogeneity," architectural theorist Jeffrey Kipnis suggests that "folding holds out the possibility of generating field organisations that negotiate between the infinite homogeneity of the grid and the hierarchical heterogeneity of finite geometric patterns" (Kipnis). In cooking, Greg Lynn explains, a "folded mixture is neither homogenous, like whipped cream, nor fragmented, like chopped nuts, but smooth and heterogenous." Likewise, he sees "pliant systems" in architecture as an opportunity to "neither repress the complex relations of differences with fixed points of resolution nor arrest them in contradictions, but

---

[1] James Wines, John Todd, and others share this view which has evolved from the earlier discussions of Aldo Leopold (1949), Ian McHarg (1968), Gregory Bateson (Ecology of Mind), Charles and Ray Eames (powers of 10), E.F. Schumacher (1973), Ivan Illich, Murray Bookchin, and others. It is also an extension of Jane Jacob's 1961 understanding of the city as a "problem of organized complexity" and Robert Venturi's 1966 discussion of complexity.

sustain them through flexible, unpredicted, local connections" (Lynn), practices which are "capable of bending rather than breaking." (ibid.). Charles Jencks describes this process as "enfolding," connecting that which is different by smooth transitions to reach a reconciliation, not a resolution.

In anthropology and cultural studies, the border has become significant as a place (again geographic as well as conceptual) where people engage in defining and re-defining themselves and others. As global flows have accelerated, there has been a perceived need to negotiate one's identity on a virtually continual basis, and perhaps in a chameleon-like fashion, with different identities surfacing depending on the circumstances. While posing a potential threat to individual and group identity, this condition also presents an opportunity for less prescriptive groupings. The anthropologist Renato Rosaldo speaks of "border crossings" as the "sites of creative cultural production" where interconnections take place (1989, p. 208).

Part of the appeal of *ecology* and of *borders/edges* is their ability to adapt creatively to change, their inherent flexibility. As these new metaphors suggest, the celebration of diversity persists but no longer for its own sake. Rather, there is an emphasis on what happens when diverse regions, peoples, styles, technologies, and so forth, collide or merge. And on what should happen. The timidity characterizing much postmodern commentary is being gradually eclipsed by bolder personal positions and polemics which recognize that excessive striving for evenhandedness and thoroughness ultimately allows the market to hold sway.

After centuries of increasingly dividing labor; cataloguing things and knowledge; segregating the landscape according to function as well as social class, age, and ethnicity; objectifying nature and people and fetishizing objects; we are now witnessing concerted efforts to de-alienate by bringing it all back together, albeit in a new way. The question is no longer whether or not to grow or to apply new technologies but how best to accomplish these. Some of the manifold ways in which this re-integration is apparent are a shift back from monoculture to polyculture and from functional zoning to mixed use, massive restructurings of the labor force (initiated from above as well as below); re-envisioning the purpose and structure of museums, schools, libraries, and zoos; an increased public role in local politics, in urban development, and in what we consume from food, to goods, advertising, and information; and in new collaborations amongst professions and between professions and academia.

While simplicity is sought, it is not the pared down "form follows function" of modernism. From "less is more," the goal might now be described as "more from less,"[2] after scenic detours through "less is a bore" and "more is more." The difference is in the inspiration (not platonic forms and geometry, but nature, the vernacular, the mundane, the "everyday") and the goal (not universality or nostalgia, but a critical regionalism or appropriate modernity). The resultant product is therefore also different, not a generic machine for living, nor an escape from the present into the past or from reality into fiction or virtual reality, nor a surrender to market forces. Rather, it is a place which sustains the environment including the people who use it. From the modern "form follows function" to the postmodern "form follows fiction, fear, finesse, and finance," perhaps now "form follows desire,"[3] symbolic and sensual as well as programmatic.

---

[2] Ian Ritchie 1994 (Well)Connected Architecture pp.70-3, London: Academy Editions, excerpted in Jencks and Kropf.

Rather than respond to specific problems with piecemeal solutions that only exacerbate the problems or push them elsewhere (reactive solutions), the emphasis on holism and seeing or forging connections at a higher and more complex level is leading to some more proactive responses. As our connections to the environment and other people grow increasingly tenuous – a condition commonly described as the breakdown in community and the family as well as the ecological crisis – efforts to re-think urban design have been seeking to resurrect such connections or to provide spaces which allow them to occur and to flourish. Some examples include the emphasis on bioregions and on world congresses to protect the environment [ecological crisis can be incentive for peace], the growth in metropolitan governments on a regional scale, increased consideration of culture in discussions about contextualism and (critical) regionalism, public discussions about "smart growth" and the creation of quality public spaces and public transit systems, urban infill projects and adaptive reuse, the revitalization of central cities and of public housing projects, the building of transit-oriented developments, and the exponential growth of neighborhood associations and community gardens along with the important establishment of community land trusts.

Perhaps we have reached a place where the question of whether to continue or abandon the modern project has become moot. Our hyper-rational embrace of computer technologies along with the simultaneous revalorization of simplicity, slowness, spirituality, and sincerity may be conspiring to eradicate the either/or proposition. This is because now we are doing both simultaneously, each providing feedback for and adjusting the other accordingly. We know we will never return to a pre-industrial integration, but the possibility of integration at another level now appears within our reach.

Conclusion

The sense that we are poised on a threshold is widespread. Over the last decade, economist Francis Fukuyama intoned the end of history (1989, 1992), philosopher Rorty the end of philosophy, Arthur Danto the end of art, Lukacs the end of the modern age (1993), Jean Baudrillard[4] and Homi Bhabha the end of modernity, Peter Eisenman the end of humanism (the classical), Peter Blake the end of cities (1982), Richard Ingersoll the end of suburbia (1992), and Michael Sorkin the end of public space (1992). Others have described the end of utopia, of taboos, affluence, intelligent writing, religion, comedy, libraries, law, art theory, beauty, conversation, sex (due to sexually transmitted viruses or to cybersex), and desire (all cited by Murphy, 1992, p. 79). Some of these declarations are clearly despairing while others are hopeful; others simply mark a departure, the destination of which is as yet uncertain.

---

3 This phrase is suggested by Ritchie, 1994. Perhaps lending confirmation was the ACSA/AIA Teachers' seminar of June 1998 at Cranbrook Academy of Art which scheduled six workshops: The Desire to Imagine, The Desire to Democratize, The Desire to Connect, The Desire to Amplify the Body/The Body Electronic [cyborgs], The Desire to Proliferate Information, The Desire to Control Chaos vs. Promotion of Heterotopia.

4 According to Baudrillard (1982), modernity was an "aesthetic of rupture," of the "destruction of traditional forms" and of the authority and legitimacy of previous models of fashion, sexuality and social behavior. But because of this, modernity lost "little by little all its substantial value, all moral and philosophical ideology of progress which sustained it at the beginning, and becomes an aesthetic of change and for change .... ultimately, becoming purely and simply fashion, which means the end of modernity" (1982).

We are at a crossroads. Not because of some arbitrary year number, but because of a general awareness that things have reached crisis proportions in our natural and built environments. The United Nations Environmental Program reports that we have about a ten-year window to turn the tide and achieve environmental sustainability, or we will descend into economic and physical decline. We now face the task of intervening in a way that nurtures the communities and the environment which ultimately sustain us. I am feeling optimistic these days since promising examples are much easier to find today than they were just one decade ago. With continued vision, diligence, and a bit of luck and goodwill, what is now a ray of hope just beginning to pierce the cloud cover may expand into a veritable sunburst.

References

Davis, M. (1990). *City of Quartz: Excavating the Future in Los Angeles.* New York: Verso.
Eisenman, P. (1984). The End of the Classical: The End of the End, the End of the Beginning. *Perspecta: The Yale Architectural Journal, 21.*
Eisenman, P. (1992). Visions' Unfolding: Architecture in the Age of Electronic Media. *Domus, no.734*, January.
Jencks, C. (1995). *The Architecture of the Jumping Universe: A Polemic: How Complexity Science is Changing Architecture and Culture.* London: London Academy Editions.
Kipnis, J. (1993). Towards a New Architecture: Folding. *Folding in Architecture, Architectural Design, 63,* No.s. 3-4.
Lynn, G. (1993) Architectural Curvilineariy: The Folded, the Pliant and the Supple. In R. Rosaldo, *Culture and Truth: The Remaking of Social Analysis.* Boston: Beacon.
Rugoff, R. (1995). L.A.'s New Car-tography. *LA Weekly*, October 6.
Schwarzer, M. (1998). Ghostwards: The Flight of Capital from History. *Thresholds, 16,* 10-19.

III. Environmental action and participation

Action competence in environmental education

María Dolores Losada Otero & Ricardo García Mira

University of Coruña, Spain

Abstract. Action competence is a social process which refers to the ability to assess and look for solutions to current environmental problems and carry them out in practice, within a democratic system of cooperation and participation (Uzzell, 1997).
Environmental action competence has several components: skills at a general level, knowledge of and insight into environmental problems and possibilities for solving them, environmental commitment, visions about our future lives, and environmental action experiences (Jensen et al., 1997).
The objective of this paper focuses on the analysis of the school context, where we try to explore the adequacy of a responsible ecological behaviour model towards domestic waste based on the environmental action competence model. We also attempt to improve our previous study about the relationship between the attitudes towards the environment and the appearance of responsible ecological behaviours (Losada & García-Mira, 2001), incorporating action competence within the general framework of the theory of planned behaviour (Ajzen, 1985, 1991), as well as comparing the effect of action competence both on behavioural intention and environmental behaviour.

Keywords: action competence, critical thinking, environmental education, school context, environmental problems

Introduction

The solution to current environmental problems requires us to identify and analyse the conflict of interests in the use of natural resources, both between groups and/or individuals and the social framework, to know how they affect our future (Breiting et al., 1999). Education and, more specifically, school education appears as the most effective way for present day students, "tomorrow's opinion leaders and stewards of the Earth" (Uzzell, 1999, p. 397), to improve this relationship. To achieve this, it is necessary for students to perform both theoretical analysis and educational practice on their immediate surroundings, and for the teaching-learning processes to be able to socialize and to democratize them, so that students can be helped not only with the acquisition of knowledge, but also with the development of abilities, skills, attitudes of critical thought and action competence.

Action competence is a social process which refers to the ability to assess and look for solutions to current environmental problems and carry them out in practice, within a democratic system of cooperation and participation (Uzzell, 1997). This involves not only that young should know understand, and analyse their environment (Giordano & Souchon, 1995), but also that they must play a democratic critic, reflective and active role which encourages them to create new responsible patterns of environmental behaviour. Critical thinking, which is considered to be essential to students' development of action competence, is based on two aspects: reasoning and judgement (Mogensen, 1997).

Steele (1980) has considered three aspects in the development of environmental competence. One part of environmental competence is awareness of one's own environmentally important abilities, skills, needs and values. A second part is the individual's knowledge about his or her surroundings, and finally it involves practical skills related to the environment.

Environmental action competence has several components: skills at a general level, knowledge of and insight into environmental problems and possibilities for solving them, environmental commitment, visions about our future lives, and environmental action experiences (Jensen et al., 1997). These elements should form part of the educational process in environmental education.

Environmental Education, whose basic framework was developed in the Belgrade Conference on Environmental Education (1975) and specified in Tbilisi Intergovernmental Conference on Environmental Education (1977), proposes to help every person create new patterns of respectful behaviour towards the environment through the acquisition of knowledge, attitudes, skills, awareness and environmental concern. Its main goal should be that students learn to be active citizens in a democratic society (Jensen et al., 1997). Knowing current environmental problems, their causes and the barriers that can appear, they will be able to consciously and intentionally look for possible action strategies in such a way that they will choose the most suitable actions and behaviour in their everyday life, assuming that the human being-nature relationship is an inseparable one and that it is necessary to find a balance between development and environment, between current needs and those of future generations.

We think that environmental competence may be developed through educational programs both within and outside school contexts. The information a schoolchild possesses about ecosystems and the effects of their own actions on them is not enough for them to act in an ecologically responsible manner. The construction of ecologically responsible behaviours requires a change in values and attitudes towards the environment and natural processes (Benayas del Álamo, 1992).

The development of environmental attitudes, defined as *"an individual's concern for the physical environment as something that is worthy of protection, understanding, or enhancement"* (Gifford, 1997, p. 47), is one of the main objectives in environmental educational programs.

Attitudes are considered as a predictor of behaviours, and knowing the psychological processes that mediate between attitudes and behaviour can help us to change environmental behaviours, since there are theories in the field of social psychology that suggest that *"attitudes could explain human actions"* (Ajzen & Fishbein, 1980, p.13). In consequence, we applied the theory of planned behaviour (Ajzen, 1985, 1991), which provides an explanation of how attitudes influence human behaviour (to act in a particular way) by their influence on the intentions to carry out that behaviour. According to the theory, behaviour intention is the most important determinant of a person's behaviour. Environmental behaviour is, in a certain way, controlled by the intention of carrying out that behaviour. The determinants of an individual's intention to perform a behaviour are the attitude toward the behaviour, as well as the subjective norm and the perceived behavioural control. The attitude towards the behaviour depends on beliefs that the behaviour leads to certain outcomes and the evaluation of the outcomes. The subjective norm depends on the beliefs that specific referents think "I should or should not perform the behaviour" and the motivation to comply with the specific referents. Perceived behavioural control is one's perception of how easy or difficult it is to perform the behaviour. Intention is considered to be the immediate antecedent of behaviour but many behaviours have difficulties of execution that may limit volitional control and therefore it is useful to consider perceived behavioural control and a degree of actual behavioural control in addition to intention (Ajzen, 2002).

Therefore, we look for an explicatory model for developing environmental attitudes that involves the students in the acquisition of action competence about environmental problems.

On the other hand, the multidisciplinary nature of Environmental Education means that in oder to study environmental problems and to obtain sustainable development, it is necessary to carry out a critical analysis of present-day social, economic, politic, ethical and ecological dimensions and to know to what extent the interdependence between urban and rural areas or between global and local determines the current unsustainable tendency. It also requires the encouragement of the necessary abilities in students in order to transform them (International Conference on Environmental and Society, Salónica, 1997), through both solidarity and an equitable environmental lifestyle.

Objectives

The objective of this paper focuses on the analysis of the school context, in which we try to explore the adequacy of a responsible ecological behaviour model towards domestic waste, based on the environmental action competence model. We also attempt to improve our previous study about the relationship between the attitudes towards the environment and the appearance of responsible ecological behaviours (Losada & García-Mira, 2001), incorporating action competence within the general framework of the theory of planned behaviour (Ajzen, 1985, 1991), as well as comparing the effect of action competence both on behavioural intention and environmental behaviour.

Method

Sample

In order to collect the relevant information, a sample of 190 secondary school students, aged from 12 to 18 years, was interviewed. 43.4% were female and 56.6% were male.

Procedure

We used a questionnaire methodology in order to obtain data on the relevant variables, including the following scales: Responsible Ecological Behaviour (Hess, Suárez & Martínez-Torvisco, 1997); Environmental Concern Scale (Weigel & Weigel, 1978; Amerigo & González, 1996); Behavioural Intention; Subjective Norm; Perceived Behavioural Control; and Action Competence (Jensen, 1994; Uzzell, 1997).

In order to evaluate action competence, we asked the individuals for the solution to several questions related to domestic waste management. We defined the problem through the following sentence: *"Some people do not cooperate with domestic waste management at home"*. Subjects had to identify the *causes* and *consequences* of the problem and look for possible *actions* for its solution, stating what *difficulties* they saw for producing a change. These four dimensions (causes, consequences, actions, and difficulties) formed our global concept of action competence, according to the theoretical framework proposed by Jensen (1993). Responses were evaluated by 10 judges, who gave a score from 1 to 5 to each (causes, consequences, actions and difficulties) according to the environmentally adequate/inadequate response of the individuals (see Appendix 1).

The questionnaire was filled out by students from different classes and years in a Corunna secondary school in February 2001.

Data analysis

Reliability alpha coefficients were computed for each scale. In order to obtain the latent variables defining environmental concern a factor analysis was used (varimax rotation; method: principal components). With respect to the subjective norm, two items related to waste management, taken from our previous study, were used. We also used one item from our previous work in order to measure both behavioural intention and perceived behavioural control. Action competence variables were generated from the mean scores of the 10 judges. AMOS was used software (Arbuckle, 1997) to test the fit of the models.

Results

Reliability analysis

Cronbach Alpha coefficients are shown in Table 1. These coefficients were in general good for all the variables in the model.

Table 1. Cronbach Alpha Coefficients

Scale	Cronbach Alpha
Responsible Ecological Behaviour	.7433
Environmental Concern	.8009
Behavioural Intention	.6350
Perceived Behavioural Control	.5016
Subjective Norm	.6945
Action Competence Towards Waste	.8410

Table 2. Rotated component matrix with 5 components (see Weigel & Weigel, 1978, Amérigo & González, 1996)

Item		I Attitudes nature 19.00%	II Attitudes pollution 15.84%	III Consumption transport 12.41%	IV Personal efforts 12.35%	V External attribution 11.86%
V32	Birds and animals of prey (falcons, ravens, foxes, wolves, etc.) which live from grain crops and farmyard animals should be exterminated.	.852				
V26	We needn't worry about hunting too many animals as in the long run everything will balance out.	.840				
V30	We should avoid the extinction of any animals even though it means us making a sacrifice to do so.	.666	.443			
V33	Ecological groups are more interested in maintaining an opposite point of view than in avoiding pollution.	.562				
V28	Nature is so wise that everything will return to its natural state in the end, although there is pollution in the air, rives and lakes.	.561	.512			.469
V29	Those who pollute should be banned by the Government as people are unable to avoid pollution by themselves.		.861			
V27	Pollution is something which doesn't affect me at all.		.750			
V31	The government should give everybody a list of places where we can go to complain about pollution.		.652			
V21	Although public transport produces less pollution, I prefer using my motorbike or car.			.868		
V23	The advantages of the current products we consume are greater than the pollution produced by their manufacture and use.			.827		

Table 2. continued

Item	I Attitudes nature 19.00%	II Attitudes pollution 15.843%	III Consumption transport 12.406%	IV Personal efforts 12.345%	V External attribution 11.855%
There should be obligatory lessons on environmental conservation at high school.	V19				.835	
I would be willing to donate part of my monthly allowance to try to improve the use of natural resources.	V22				.641	
I would be willing to make personal sacrifices to reduce environmental pollution, although the immediate results may seem unimportant.	V18				.587	
I would be willing to spend some of my money and time on associations like ADENA, which work to improve environmental quality.	V20				.569	
The government has got good inspectors and it is therefore difficult for pollution from the power stations and industriy to become excessive.	V24					.861
Industries do their best to develop anti-pollution technology.	V25					.791

Factor analysis

A solution of 5 components was derived accounting for 71.45% of the variance: 1) Attitudes Towards Nature; 2) Attitudes towards pollution; 3) Consumption and Transport; 4) Personal Efforts for the Environment; 5) External Attribution (see Table 2).

Figure 1. A model of action competence within the frame of the Ajzen's planned behaviour theory.

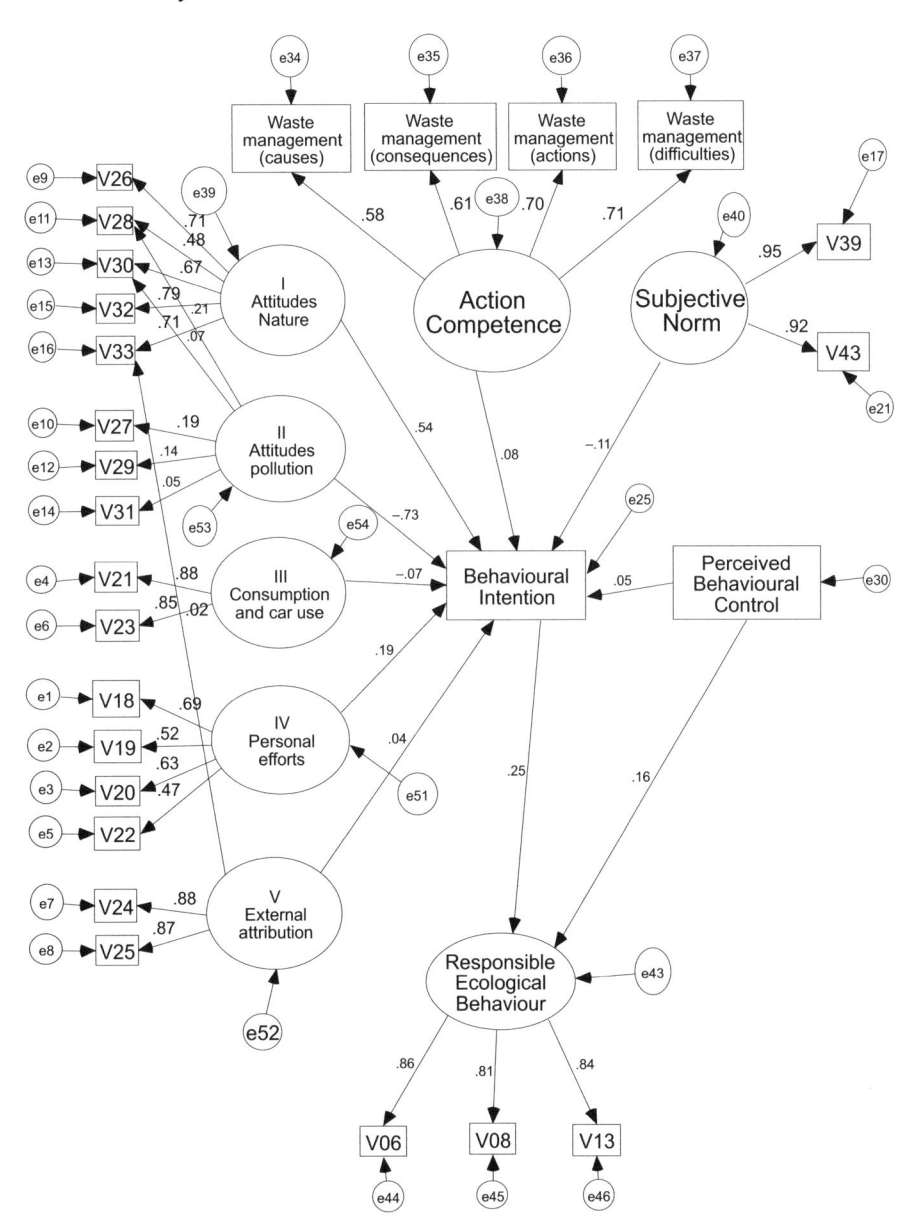

Figure 2. A model for predicting responsible ecological behaviour from action competence variables.

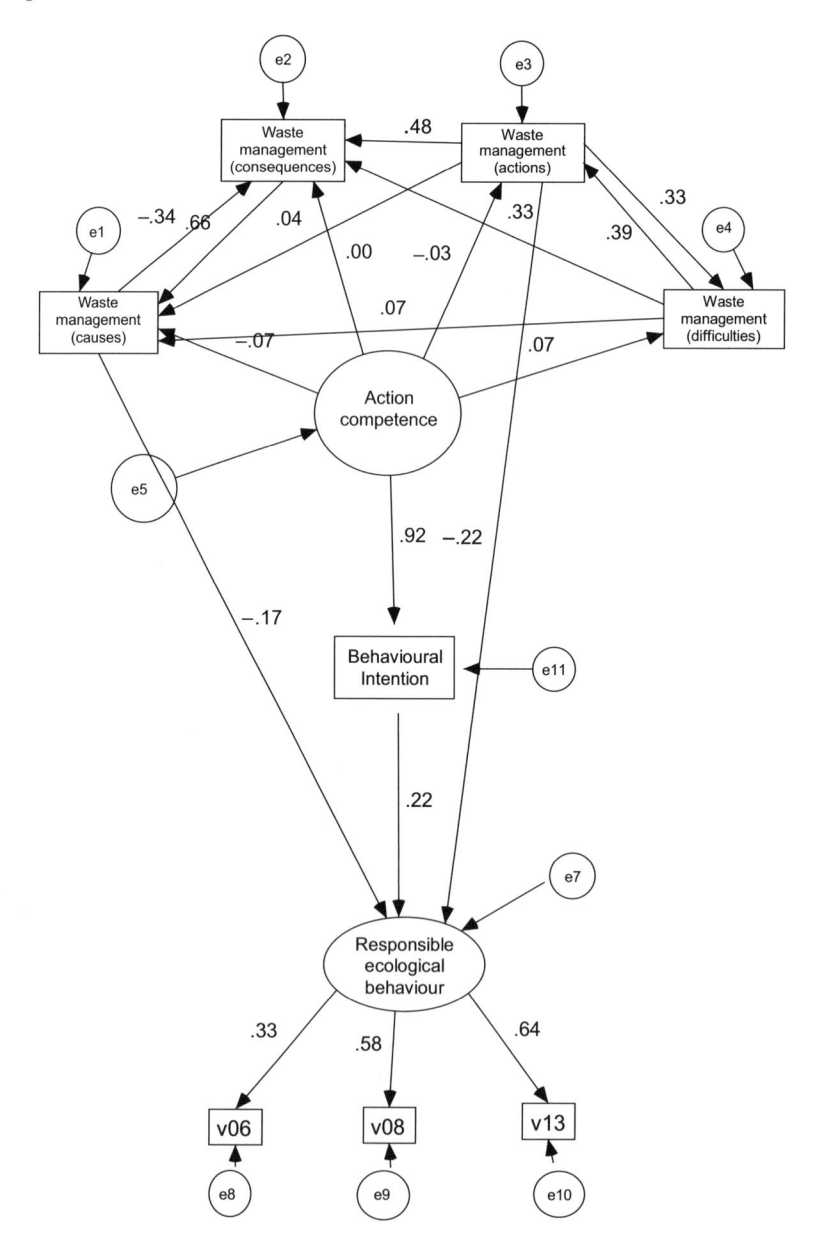

Amos analysis

First, we integrated action competence variables within the Ajzen's planned behaviour model (Ajzen, 1988, 1991), expecting to improve in the results of our previous work (Losada & García-Mira, 2001), and testing the effects on both behavioural intention and responsible ecological behaviours.

The observed variables were considered to be:

(a) **Causes** (identifying causes of non-cooperation with domestic waste management)
(b) **Consequences** (identifying consequences of non-cooperation)
(c) **Actions** (action proposals of individuals in order to promote cooperation)
(d) **Difficulties** (what difficulties these actions may originate)

AMOS derived a fit index, referred to the quotient between minimal discrepance and degree of freedom, which was CMIN/DF = 5.534. This, according to Wheaton et al. (1997) and Marsh and Hocevar (1985), should be an adequate fit (see Figure 1).

A second analysis, considering action competence variables as beeing outside the Ajzen's model, was carried out. We obtained a solution with a chi-square value of 13.53 with 14 *df* and a probability level of $p = .48$. The fit index, referred to the quotient between minimal discrepancy and degree of freedom, was CMIN/DF = .967. This, according to Wheaton et al. (1977) and Marsh and Hocevar (1985), should be an adequate fit (see Figure 2).

Discussion

In this study we introduced *action competence* in relation to domestic waste management within the Ajzen's planned behaviour model and we found that its effect on intention behaviour is very small (.08). Nevertheless, when we considered action competence in relation only to domestic waste management, outside Ajzen's model, we observed a higher effect on behavioural intention. Standardised estimates are higher (.92).

When we used Ajzen's model, the effect of environmental action competence was decreased by the effect of the attitudes, considered in five factors. Our results would appear to indicate that in order to explain both behavioural intention and responsible ecological behaviour, it is easier to predict behavioural intention using a more parsimonious model than a more complex one.

The lower estimate derived in both models for the relation between behavioural intention and responsible ecological behaviour is due to the existence of other variables influencing behaviour that are not explained in this work, but that have been described in other studies. As we have pointed out elsewhere (see García Mira et al., 2002), structural or subjective constraints account for the role of the situational context in its wider sense (social or physical) as predictors of concrete behaviour (see Derksen & Gartrell, 1993; Guagnano et al., 1995; Tanner, 1999). There are also difficulties for explaining and predicting responsible ecological behaviour, because in spite of the number of studies that have been carried out to date, there is insufficient clarification in order to define what variables have what influence – personal and social, political or economical –, or what the relations between them are.

We would like to emphasize the need to continue with the search for explanatory models of responsible ecological behaviour, as well as for instruments to measure this behaviour. We consider as particularly important research into the role of action competence in the prediction of behavioural intention and behaviours towards the environment in the school context, as we have demonstrated in this work, because of the high value that education has both in a formal and non-formal educational context.

We consider the Theory of Planned Behaviour to be a complex model for explaining ecological behaviour in students. The simplicity of the Action Competence model can lead us to a better understanding of an attitude towards a specific target in a given context and help us to predict the behaviour in this respect.

Environmental education is understood as a learning process that increases people's knowledge and awareness of the environment and associated challenges, develops the necessary skills and expertise to address the challenges, and fosters attitudes, motivations and commitments to make informed decisions and take responsible action (UNESCO, Tbilisi Declaration, 1978).

We think the concept of Action Competence should occupy a central position in environmental education and should be included as a subject in its own right in the school curriculum, not just a cross-curricular one, with specific objectives, contents and methodology. Furthermore, environmental education has to be understood as a necessary issue in the community for improving our environmental behaviour in our everyday life.

The Ajzen's model is of particular interest for our study because it explains how psychological processes operate and enables us to know which ones environmental actions depend on. These processes are necessary in order to understand behaviour, but they are not always sufficiently explained. The explanation for this is that environmental behaviour is multidimensional and to study all aspects of human decisions, and in particular social dilemmas, is a very difficult question.

If environmental education can help to develop action competence in students, who are our citizens of the future, it will facilitate the correction of these social dilemmas in the sense of neutralizing negative environmental actions or giving prestige to positive actions.

One of the ways of controlling environmental problems is to promote research clarifying this role, in order to improve environmental education programs both in schools and in the community at large. If we take into account that these problems have been created by mankind, then we have to start from a concept which takes into consideration both the individual and the social group. If students develop the concept of action competence, they will be active and critical citizens, in a democratic society, capable of acting in a more responsible manner towards the environment.

To this end, and from a position of social criticism and transformation, we consider it necessary to introduce environmental action competence in educational programs, as well as in environmental education in general, as one of the main objectives. This will enable students to develop as active persons who are able to analyze problems, and allow them to achieve a more democratic, fair and sustainable society, search for solutions and participate with a greater degree of commitment.

References

Ajzen, I. & Fishbein, M. (1980). *Understanding attitudes and predicting social behaviour.* New Jersey: Prentice-Hall.

Ajzen, I. (1985). From intentions to actions: A theory of planned behaviour. In J. Kuhl &

J. Beckam (Eds.), *Action-control: From cognition to behaviour* (pp. 11-39). Heidelberg, Germany: Springer.
Ajzen, I. (1991). The theory of planned behaviour. *Organizational Behaviour and Human Decision Processes*, 50, 179-211.
Ajzen, I. (2002). Behavioural interventions based on the theory of planned behaviour. *Journal of Applied Social Psychology*, 32, pp. 1-20.
Arbuckle, J. L. (1997). *Amos user's guide: Version 3.6.* Chicago: Smallwaters Corporation.
Benayas del Álamo, J. (1992). *Paisaje y Educación Ambiental: Evaluación de cambios de actitudes hacia el entorno* [Landscape and environmental education: Evaluation of attitudes towards the environment]. Madrid: MOPT.
Breiting, S. (1993). A new generation of environmental education: Focus on democracy as part of an alternative paradigm. In R. Mrazek (Ed.), *Alternative paradigms in environmental education research.* Monographs in Environmental Education and Environmental Studies (Vol. VIII, pp. 199-202). Troy, NY: The North American Association for Environmental Education.
Breiting, S., & Mogensen, F. (1999). Action Competence and Environmental Education. *Cambridge Journal of Education,* 29(3), 349-353.
Byrne, B. M. (1989). *A primer of LISREL: Basic applications and programming for confirmatory factor analytic models.* New York: Springer-Verlag.
Calvo, S. & Corraliza, J. A. (1994). *Educación Ambiental. Conceptos y propuestas* [Environmental education. Concepts and proposals]. Madrid, Spain: CCS.
Catalán, A., & Catany, M. (1996). *Educación Ambiental en la Enseñanza Secundaria* [Environmental education in secondary education]. Madrid, Spain: Miraguano.
Derksen, L. & Gartrell, J. (1993). The social context of recycling. *American Sociological Review*, 58, 434-442.
Eagly, A., & Chaiken, S. (1993). *The Psychology of Attitudes.* New York: Harcourt Brace College Publishers.
García-Mira, R., Arce, C., & Sabucedo, J. M. (Eds.). (1997). *Responsabilidad Ecológica y Gestión de los Recursos Ambientales* [Ecological responsibility and management of environmental resources. A Coruña, Spain: Diputación de A Coruña.
García-Mira, R., Real, J. E., Durán, M. & Romay, J. (2002). Predicting environmental attitudes and behaviour. In G. Moser, E. Pol, Y. Bernard, M. Bonnes, J.A. Corraliza & V. Giuliani (eds): *People, Places and Sustainability.* Göttingen: Hogrefe & Huber.
Gifford, R. (1997). *Environmental Psychology. Principles and Practice* (2nd ed.). Boston: Allyn and Bacon.
Guagnano, G. A., Stern, P. C. & Dietz, T. (1995). Influences on attitude-behaviour relationships: A natural experiment with curbside recycling, *Environment and behaviour*, 27, 699-718.
Jensen, B. B. (1993). *"The concept of Action and Action Competence"*, document prepared for the first international workshop on *Children as Catalysts of Global Environmental Change.*
Jensen, B. B., & Schnack, K. (1994). *Action and action competence as key concepts in critical pedagogy.* Copenhagen: The Royal Danish School of Educational Studies.
Jensen, B. B., & Schnack, K. (1997). The Action Competence Approach in Environmental Education. *Environmental Education Research,* 3 (2), 163-178.
Losada, M. D., & García-Mira, R. (2001). Responsible Ecological Behaviour in the School Context. *Book of Abstracts of the 32nd Annual Conference of the Environmental Design Research Association.* Edinburgh, July, 2001.

Marsh, H. W., & Hocevar, D. (1985). Application of confirmatory factor analysis to the study of self-concept: First-and higher-order factor models and their invariance across groups. *Psychological Bulletin*, 97, *562-582*.
Mogensen, F. (1997). Critical thinking: a central element in developing action competence in health and environmental education. *Health Education Research, 12* (4) (pp. 429-436).
Steele, F. (1980). Defining and developing environmental competence. In C. P. Alderfer & C. L. Cooper (Eds.), *Advances in Experimental Social Processes*, 2, 225-244.
Tanner, C. (1999). Constraints on environmental behaviour. *Journal of Environmental Psychology*, 19, 145-157.
Uzzell, D. (1997). La responsabilidad ecológica en el ciudadano competente en la acción: algunas, cuestiones metodológicas. In R. García-Mira, C. Arce & J.M. Sabucedo (Eds.), *Responsabilidad Ecológica y Gestión de los Recursos Ambientales* [Ecological responsibility and management of environmental resources] (pp. 9-21). A Coruña: Diputación de A Coruña.
Uzzell, D., Joyce, P., Brunn, B., Vognsen, C., Uhrenholdt, G., Gottesdiener, H., Davallon, J., & Kofoed, J. (1998). *As crianças como agentes de mudanza ambiental* [Children as catalysts of global environmental change]. Porto, Portugal: Campo das Letras.
Uzzell, D. (1999). Education for environmental action in the community: New roles and relationships. *Cambridge Journal of Education*, 29 (3), pp. 397-413.
Wheaton, B., Muthén, B., Alwin, D. F., & Summers, G. F. (1977). Assessing reliability and stability in panel models. In D. R. Heise (Ed.). *Sociological methodology* (pp. 84-136). San Francisco: Jossey-Bass
Xunta de Galicia (2000). *Nuevas Propuestas para la Acción. Conclusiones* [New proposals for action]. Santiago de Compostela: Consellería del Medio Ambiente.

Appendix 1.
ACTION COMPETENCE regarding domestic waste management.

Problem: *"Some people do not cooperate with domestic waste management at home".*

1   2   3   4   5

Indicate your degree of agreement or disagreement:

1   agree strongly
2   agree
3   neither agree nor disagree
4   disagree
5   disagree strongly

What do you think are the causes and consequences of the problem?.

Causes:

1..
2..
3..

Consequences:

1..
2..
3..

What actions do you think people should perform in order to cooperate and what are the difficulties they can find?

Actions:

1..
2..
3..

Difficulties:

1..
2..
3..

Participatory action research as a participatory approach to addressing environmental issues

Esther Wiesenfeld, Euclides Sánchez & Karen Cronick

Central University of Venezuela

Abstract. The review of recent publications in journals specializing in the environmental psychology (EP) field, such as *Environment and Behavior* and *Journal of Environmental Psychology*, reflect the prevalence of the positivist paradigm and quantitative methodology in the study of and approach to the topics proper to the discipline. This is noteworthy in view of the formulation of alternative approaches developed from the EP standpoint in an attempt to overcome that paradigm's limitations. Considering the magnitude of the environmental problem and the existence of approaches which, from our point of view, can help overcome those problems, as well as build up theoretical knowledge relevant thereto, we argue in this study in favor of the use participatory action research (PAR).
PAR is a perspective rarely applied in our discipline, but which we consider useful to understand and take action regarding human environmental issues. The argument in favor of this perspective reflects the conception of reality on which it rests, and hence its conception of human-environment transactions, knowledge, social action, the relations between researchers and research subjects, and the purpose of research, among other factors. We then go on to address the definitions, characteristics, and overall guidelines for PAR and its relevance to EP. Finally, we present an example illustrat-

ing the advantages of PAR's application, not only for the solution of environmental problems but also for the construction of knowledge, the strengthening and training of the participants, and the promotion of their well-being.

Keywords: environmental psychology, participatory action research, human environmental issues, communities

Introduction

Slightly over three decades after the birth of environmental psychology (EP), we can assert that the circumstances that drove the discipline's appearance are still in being. We refer to the need for a) vindicating the study of environmental human relations at its different levels (individual and group) and scales (micro and macro), in a holistic and interdisciplinary fashion; b) promoting the understanding and solution of people-environment issues; c) contributing to the development of theories and methods, which will in turn advance our knowledge of the subject and improve practical action in regard to it, being both scientifically relevant and socially useful.

Though there has been a vast theoretical, methodological, and empirical production over the three decades of EP's development, we believe the objectives spelled out at the discipline's outset have been achieved only in part as regards theory, methodology, and application (Saegert & Winkel, 1990).

Important theoretical contributions have indeed been made, such as the transactional approach (Altman & Rogoff, 1987; Saegert & Winkel, 1990; Stokols, 1994; Werner, Brown & Altman, 2002), the ecological approach (Barker, 1963, 1968; Wicker, 1987, 2002), and the sociocultural approach (Saegert & Winkel, 1990). But these theoretical and methodological advances have not generated the expected contributions.

The positivist paradigm still dominates along with its associated methodology and methods, as is evidenced by a review of the articles published in the last six years (1997–2002, Nos. 1 and 2) in the *Environment and Behavior (E&B)* and *the Journal of Environmental Psychology (JEP)* journals: Out of a total of 332 articles (185 in the first of those journals and 147 in the second), only 27 (eight in E&B, representing 4%, and 19 in JEP, making up 13%), yielding a combined total of 8%, reflected a theoretical reliance on any of the alternative approaches mentioned above, and a mere 8 articles in E&B and 15 in JEP used a qualitative research strategy (case studies, ethnographic methods, naturalistic research, qualitative research, participatory planning, and participatory evaluation); four of these combined qualitative and quantitative approaches. Finally, just three of the articles reviewed were oriented toward participatory processes, chiefly with children at the design stage (Chawla & Heft, 2002; Francis & Lorenzo, 2002; Horelli & Kaaja, 2002). The rest of the theoretical and methodological approaches reflected in these articles are explicitly or implicitly grounded in the positivist paradigm.

Based on the three dimensions characterizing paradigms according to Lincoln (1990): ontological, epistemological, and methodological, we can state that the principal features of positivism (among others) are:

    (a) A realist ontology, meaning a conception of reality as an external, independent, observable, measurable, and verifiable entity, governed by causal laws of na-

ture, whose operation is to be discovered in order to predict, explain, and control it.
(b) An objectivist epistemology, since if there is a real world that functions according to the laws of nature, the scientist must approach it in order to know it with the distance and neutrality required to prevent any alteration of its natural course.
(c) A methodology based on the hypothetical-deductive model taken from the natural sciences, commonly designated quantitative methodology. This methodological strategy, long viewed as the only one that can meet the demands of science, seeks to control reality through the application of rigorous quantitative and statistical methods and procedures. They are applicable to every aspect including formulation of the problem and hypotheses congruent with the chosen theoretical bases, fragmentation of reality in terms of variables (likewise based on existing theories that permit their conceptualization and operationalization), design and testing of data collection tools, statistical processing of the data, and analysis and interpretation. This methodology ensures that the knowledge developed will be credible, generalizable, replicable, and objective, and that the theories constructed from it will be true, i.e., in correspondence with reality (Lincoln, 1990; Pérez Serrano, 1994; Torres, 1995).

It should be noted that anything not approachable in this way is ruled out as a subject of scientific inquiry. As stated by Michell (2003):

"The quantitative imperative is the view that studying something scientifically means measuring it. Measurement is thought to be a necessary part of science and non-quantitative methods are thought to be prescientific. This imperative is motivated by the idea that all attributes are fundamentally quantitative." (pp. 6–7).

According to Michell (2003), though it is commonplace to describe Positivism as a quantitative paradigm, the quantitative imperative has been present through the history of western culture and has dominated modern psychology, without the need to describe it as Positivist. Similarly, Reichardt and Cook (1979) and Bryman (1988) stress the technical aspects of methodology. On the other hand, authors like Schwandt (1990) agree with Lincoln (Lincoln, 2001) on the correspondence between epistemology and methodology.

The limitations of positivism for understanding the complexity and interdependence of relations among people as active subjects in interaction with their surroundings – seen as both a dynamic whole and relative to a particular context (Wiesenfeld, 1998) – are among the reasons why the alternative approaches were first proposed. We find the infrequent use of these approaches remarkable.

Moreover, the studies oriented by the positivist perspective have not made the expected contributions to the explanation and/or control of the problems addressed; they have not given rise to actions, nor have they spread to embrace other disciplines, fields, or participants for whom those results and actions might be interesting and relevant, as often as expected. Neither have those studies acknowledged the importance of the features of the general context (economic, political, social) in which the problems under examination play out; those problems are addressed in a fragmentary rather than holistic fashion, and attention is mainly focused on small-scale environments (parks, neighborhoods, buildings, streets), with little or no consideration for other environmental scales. The emphasis in on individual psychological processes (cognition, perception, attitudes, territoriality, personal space) in human-environment transactions, to the detriment of the group processes that make themselves felt in those transactions. These studies' results have not been widely disseminated, especially among decision makers (as in urban planning), and the recommendations emerging from them have not been carried into practice to the expected degree (Churchman, 2002).

We also observe a steady deterioration in the world population's quality of life, and we are responding to that trend with actions – such as economic adjustment programs – that have only worsened the existing social, economic, and environmental problems (Reed, 1996; Ortiz, 1997). Hence, efforts to overcome environmental unsustainability need to take account of that problem's socioeconomic dimension, as a precondition for its solution (Jiménez, 2002). This dimension is a key part of the context – generally ignored in psychoenvironmental works – whose inclusion would help to generate a holistic understanding of the environmental problem.

According to Hamacher (1996), another requirement for the achievement of sustainability is the development of new approaches which encourage the population's participation, along with that of the different actors, in the solution of environmental problems and the conditions that provoke them.

Participation in EP: What, how, and what for?

In recent years the need to actively include the populations in the planning and execution of projects that concern them, has led to the formulation of a variety of participation-oriented proposals.

The general premise that underlies these proposals is the conceptualization of participation as an activity that transforms -in a broad and positive sense- both the people involved and the situations they confront. Group action to help bring about the required change is emphasized (Pol, 1998), even in technical contexts, and hence, there is a focus on the need to foster community processes that promote the community members' organization and participation in the satisfaction of their needs, while helping to strengthen participants' sense of community and solidify their community cohesion and place-related identity (Wiesenfeld, 1994, 2000), and in general contributing to the personal growth of the participants – including the professionals (Sadan & Churchman, 1997) – as the project advances.

In spite of the clear benefits from the inclusion of communities, i.e., the different actors relevant to a project, through their active participation in all its phases, and in spite of the disadvantages of an absence of such participation, few EP works reflect this orientation.

That shortcoming is largely due to the monopoly held by a single conception of science, knowledge, and truth in the training of professionals, which is grounded in the dominant paradigm. Its hegemony has only reinforced the control exercised by the power holders (in the economic, social, political, intellectual, and academic spheres) over the weaker members of society; maintained the separation between theory and action by supporting a preference for technical progress over human well-being; given priority to reason over feeling and action; and lengthened the distance between the researcher and the subject of research, on the ground that the former must be neutral to obtain objective knowledge from the latter, which has left both (the researcher and the subject) atomized and bereft of feelings and values of their own (Wiesenfeld, 1999).

Guidelines for a Participatory Environmental Psychology:

In any attempt to summarize the prevailing situation, we would have to say that many of the problems have not been overcome, i.e., theory has not helped to improve practice to any significant degree, and that there has been no theoretical construction grounded in that practice. Accordingly, we see shortcomings in theory, methodology, and application that are in urgent need of solution.

Some responses stemming from approaches like social constructionism, critical theory, and participatory action research (PAR) have focused on altering the meaning of language, action, knowledge, the researcher-research subject relationship, and methodological strategies to produce knowledge, as well as an affirmation of ethical and political considerations consistent with the taking of positions vis-à-vis the issues outlined above (Fals Borda, 2001; Kemmis, 2001; Lincoln & Guba, 2000; Lincoln, 2001; Treleaven, 2001).

In relation to our field of interest, which is EP, this means that we need to know and understand how citizens interpret and react to events concerning the human-environment interaction, as a necessary precondition for any intervention. Only in that way can an intervention give priority to their interests and not to those of the dominant elites, including the academic elite. An understanding of this kind requires an approach to the authors of these discourses and actions that diverges from the epistemology of distance characterizing the positivist paradigm. The aim is to develop an epistemology that respects the subjects' viewpoints, and treats them as equal to the researcher's in validity. Hence, the results of a study will be the outcome of a negotiation between the two, which acknowledges the subjectivity of both in the construction of their respective versions, as an inseparable dimension of human experience.

At the same time, this type of relationship between researcher and subject requires a methodology and a conception of the subject and his/her knowledge that differs from the traditional ones; i.e., a *hermeneutic/dialectic* (Lincoln & Guba, 2000) and *cooperative methodology*, wherein the researcher interprets the subjects' versions and discusses them with the subjects, giving rise to a reflexive dialog in which those versions are transformed. What is needed is an *objective-subjective* (Reason & Heron, 1995; Heron & Reason, 1997) and *relativistic ontology* (Lincoln & Guba, 2000), which does not ignore the existence of the material world but also acknowledges the role of history, experience, and context in the construction of versions of the world, rather than conceiving the existence of a single true version of reality, and which views the subjects as active participants in the production of these versions. These versions are accepted as a kind of knowledge as valid as scientific knowledge.

These shifts (linguistic, action-oriented, epistemological, methodological, and ontological) require taking an *ethical and political stance* consistent with them. For Rivlin (2002), the *ethical posture* that is required involves giving attention to the value of the research and its short and long-term impact on the participants, analyzing our ability as professionals to exert influence on the objects of research, knowing the subjects' willingness to participate in the research, providing information on agencies and institutions that could help satisfy the needs expressed by the subjects during the course of the research, and putting respect for human dignity above the demands of distance and neutrality common in traditional research methods, etc.

As regards the *political* dimension, Rivlin (2002) calls for a recognition of the values and political postures implicit therein, arguing that the political dimension – no less than the ethical one – is present in all phases of a project (choice of topic, methods, type of analysis, and publication).

The aim, then, is to identify, adapt, and/or develop strategies which, resting on the principles discussed above, make it possible to construct theories and practices consistent with the demands of specific populations in particular contexts, i.e., at the service of human well-being.

From our point of view, critical constructionism, which integrates aspects of social constructionism and critical theory (Wiesenfeld, 1994, 2000), and PAR, are appropriate ways to pursue these goals. Social constructionism is a metatheoretical perspective, which conceives reality as socially constructed through people's interactions and communicative practices in specific cultural and historical contexts. (Gergen, 1999; Guba & Lincoln, 1994). Critical theory or ideologically oriented inquiry includes various perspectives (neo-Marxism, materialism, feminism, Freirerism, the Frankfurt school of critical theory), which aim at transforming the living conditions of oppressed sectors of society, by raising their consciousness on the characteristics (economic, historical, social, political) of their context which fostered them and empowering and facilitating their transformation (Guba, 1990).

Specifically, *participatory action research (PAR)*, the core subject of this article, aims to construct a socially relevant knowledge, oriented toward and grounded in the exercise of practices which transform living conditions viewed as unsatisfactory (or which come to be so viewed in the course of the research process) by the participants – including the academics, professionals, and other people involved in participatory processes. In view of the procedures used by PAR and the principles and values that orient it, the changes achieved tend to go beyond the specific objectives of a given project and express themselves in achievements at varying levels (personal, group, social) and in different spheres (moral, intellectual, technical, environmental, material, spiritual, etc.). All this rests on the foundation of a set of guidelines that incorporate ethical, political, epistemological, and other considerations as well as technical and professional ones.

Participatory action research (PAR): Conceptualizations and characteristics

PAR, under a diversity of names and variants: participatory action research (PAR) (Fals Borda & Rahman, 1991), participatory research, cooperative research (Heron & Reason, 1997, 2001), emancipatory action research (Carr & Kemmis, 1986; Kemmis, 2001), research-intervention (Fryer & Feather, 1994; Serrano García, 1992.), anticipatory planning (Horelli, 2002), appreciative inquiry (Ludena, Cooperrider & Barrett, 2001), and community action research (Senge & Scharmer, 2001) has been understood not only as a methodological strategy but also as a way of life (Fals Borda, 2001) and a world view or paradigm (Heron & Reason, 1997; Fals Borda, 2001).

As a *methodological strategy*, PAR makes it possible to enhance the power of excluded sectors of the population, enabling them to address the tasks required to improve their living conditions (Park, 2001).

As a *philosophy of life*, it includes individual and collection reflection and action by all the participants – including the researchers – on the different problems, needs, and dimensions of the aspects of reality they wish to alter (Fals Borda, 1978, 1986; Fals Borda & Rahman, 1991).

Finally, PAR addresses "... the challenge of constructing *a practical and satisfactory paradigm* in moral terms for the social sciences, to make them congruent with the ideal of service" (Fals Borda, 2001, p.32). The author asserts that this alternative paradigm:

> "... combines practice with ethics, academic knowledge with popular wisdom, the rational with the existential, the regular with the fractal. It breaks down the subject/object dichotomy; it is inspired by participatory values such as altruism, sincerity, confidence, autonomy, social responsibility; and by the democratic and pluralistic concepts of otherhood and service, favoring coexistence with differences and introducing gender, lower-class, and pluriethnic perspectives into projects." (p. 32)

Lincoln (2001) also conceives PAR as a paradigm, and even asserts a number of similarities among it, social constructionism, and critical theory in a series of ontological, epistemological, methodological, ethical, political, and other dimensions.

Though the different conceptualizations of PAR converge in their aim of building a better world, based on democracy, equality, and justice, and of constructing knowledge grounded in, and at the same time strengthening, these emancipatory practices, the conception of PAR as a paradigm coincides with the concepts outlined in the preceding section on a participatory EP, as well as with the basic orientations that have guided our activity as community environmental psychologists (Wiesenfeld, 1998, 2000, 2001).

Principles of PAR: Their place in EP

Let us now examine some principles of PAR and see how they can be understood and applied in EP, in the context of the socioconstructionist and critical theory approaches which, as stated above, are consistent with PAR's principles and values (Lincoln, 2001):

(1) *The concept of reality as a socially constructed totality:* This principle reflects the so-called alternative theoretical approaches, pointing to the need for a holistic conceptualization of the person-environment relationship based on the meanings of the participants and their practices and communicational exchanges, in a particular context and time.

(2) *Every group has resources to carry out transformations, no matter how small:* Emphasis on resources is a strengthening activity which can enhance the participants' power and control over their lives in both individual and collective terms. There are two implicit dimensions: the first refers to the human agency for identification of the intellectual qualities and the social, material, and environmental resources with which to cope with demands of differing kinds. The second focuses on empowerment as the core of PAR, something that gradually develops hand in hand with the process of participation.

(3) *Emphasis on horizontal relationships between researchers and participants:* This principle is of special relevance to EP because horizontal relationships allow the concerns and interests of the different participants (those affected and/or responsible) to make themselves felt, rather than having all the emphasis fall on the

researcher's interests, generally oriented by theoretical concerns and far removed from those of the groups with which he/she works. This requires professionals to familiarize themselves with the members of the groups they work with, in order to share their respective areas of knowledge and spur collective learning processes based on democratic and participatory arrangements.

(4) *Integration of theory and practice:* This principle refers to the emphasis on the production, dissemination, and application of knowledge based on a critical articulation of scientific and people's knowledge. This means that people's experience in connection with the problem in question is critically analyzed so as to be understood as a process influenced by a set of sociohistorical conditions and not as the outcome of individual responsibility. Loss of housing as a result of natural disasters is a good example of this; processes of joint reflection on the reasons for it lead to an understanding informed by a framework that transcends personal responsibilities, i.e., inequality, exclusion, and social injustice. Defining oneself as a social actor from this new epistemological standpoint makes it possible to be a being in relationship, who, together with other participants, can take actions to improve conditions. The proposed actions may be very specific and small-scale at first, but then gradually gain strength from the experience itself and absorb new meanings from the opportunities it provides.

(5) *Recognition of the political implications of social practices, including academic practices:* This means thinking about exclusionary and oppressive processes as the outcome of practices that distort or conceal the mechanisms through which those conditions have emerged and institutionalized themselves. It also alludes to the dominant modes of constructing science and perpetuating a privileged mode of conceiving truth. Reflection on this issue involves the researcher's commitment to the participants and their interests.

PAR in EP: What and How?

From the standpoint of PAR, the environmental issue refers to a holistic understanding of human-environmental transactions, contextualized at two levels: the micro level, at which the project is pursued, and the macro level, that of the social structure in which it is enmeshed. The micro level (community, school, organization) is the one at which the project's purpose is identified, being chosen as the outcome of a diagnostic exercise carried out jointly by the researchers and "subjects" rather than as the researcher's imposition, reflecting the latter's own interests and without regard for the community's.

The joint choice of topic stems from a prior process of *familiarization* by the researcher with the members of the community, who are the co-authors of the project in their context, as well as the community's *diagnosis.*

Familiarization consists of the establishment of the first links, needed to generate confidence and mutual commitment. It involves a review of the available documentation on the community's history, the projects carried out there in the past, its existing organizational structure, etc., as well as making contact with key informants. More important, however, it paves the way for the establishment of horizontal relations between the professionals and the community, thereby fostering an exchange of knowledge and mutual learning.

Familiarization lays the basis for performing the *diagnostic* study, in whose planning and design the community participates; this gives rise to the participatory process and the training of its members in activities of this kind.

The diagnostic activity involves generating a detailed characterization of the community and of its needs and resources, as well as other key points that come up during the familiarization stage, reflecting the particular characteristics of the context. It is necessary here to stress the importance of the identification and appreciation of the community's resources, as a key to strengthening both the individual and the group.

Once the diagnostic phase has been completed, *processing and analysis of the information collected* commences, also done jointly with the members of the community. The analysis of that information makes it possible to identify, rank, and select the need(s) and or issue(s) on which the project will focus, as well as the resources available for its execution.

Though the *reflection–action* cycles need not be viewed in terms of a mechanical sequence, and in practice the process is a fluid one and its stages overlap, the information and discussion of the diagnostic study's results can be viewed as an excellent trigger for reflection, and hence, for *problem definition and awareness raising*. This means that the community members, with the researcher's help, exchange points of view about the nature, origin, and consequences of their needs, which allows them to understand those needs at the two aforementioned levels (micro, i.e., in terms of their implications in daily life, and macro, based on their interpretation in a broader social and economic context). This analysis, which includes what Fals Borda has called the critical recovery of history (1979), makes it possible to imbue needs with new meaning as social products, and hence, alterable through joint effort (problem definition and awareness raising), rather than considering them as natural and encouraging adaptation to them.

Recognition of the possibility of change through group intervention is a key motivator for *participation*, and in general for the *strengthening of community processes* such as organization and community feeling, since people recognize the needs that are detected as their own and recognize themselves in the group and the context with which they share those needs. This process strengthens their commitment to the project and induces continuity in the performance of actions (which may include other research) leading to an improvement of the conditions identified as adverse. Thereafter, the people reflect on the results of their actions, at first in a fragmentary way and limited to the particular context of the group's creation. Regardless of their scope, they are valued as sources of increased resources and individual skills at the individual and group level; hence, in the course of taking action people acquire tools that strengthen them as individuals and as a group, and foster their autonomy.

As the successive reflection–action–reflection cycles develop, understanding (which in PAR means the construction of knowledge based on a critical articulation between scientific and popular knowledge) becomes increasingly profound and action (practice, based on the co-produced knowledge) becomes increasingly complex. The interlinking of reflection and action, which is further enriched in each cycle, satisfies the PAR mandate regarding the union of theory and practice.

This process whereby knowledge becomes more profound and action more complex is what makes it possible to move on from an analysis of the local, community-level situation to an analysis of the social forces that operate in broader contexts, and to specify the relations between the two scales. It also fosters a projection of action from the intracommunity spaces to other spaces and agents (institutions, organizations) directly or indirectly linked to the process as informants, suppliers of resources or services required for the project's success, or potential resources for it. This transit from the private to the public, or from the community to the city and its agents, is an exercise of citizenship and a clearly political action.

Finally, this process needs to be evaluated on the basis of its contribution to the overcoming of human suffering, both in the sphere of the research and in a wider framework (Kemmis & McTaggart, 2000).

PAR in EP: An illustrative case

Having described the principal characteristics of PAR and discussed its relevance to environmental intervention, we can now examine a case that illustrates its application in EP.

The example we present, developed by Karen Cronick, has the town of Maitana in Venezuela as its environmental scenario. The aim is to show how the environment cannot be studied without consideration for the unbreakable relationship between the physical and social environments, and the characteristics and needs of the inhabitants. In fact, the well-being of the people and the integrity of their space cannot be addressed in isolation.

This is a case of participatory intervention, grounded in an active and committed conception of facilitation from an ecological point of view. According to Cronick, efforts to solve environmental problems often give rise to other problems, which in the case at hand include dumping of refuse and sewage into streams (to solve individual hygiene problems) or construction of urban shacks (to meet families' housing need). The new problems appear when the cumulative outcomes of isolated behaviors provoke new social difficulties. This unintended generation of environmental disorganization leads to phenomena such as landslides when constructions are built on unstable terrain. Other collective phenomena have complex origins stemming from a combination of local and global factors.

In this respect, *habitat* is a global phenomenon and its study cannot be approached as a politically neutral academic exercise. In addition, EP can play an important role in the re-evaluation of ideological parameters like individual solution of environmental problems and the construction of the social subject as an "agent" who plays an active role in the diagnosis, change, and evaluation of his/her environmental problems.

The following example illustrates the fact that what appeared to be a solution to the problem of housing construction and occupation of a territory has resulted in other problems for the community's inhabitants. These problems arose in two senses: a) as collective phenomena related to the presence of vectors for endemic diseases such as dengue fever, Chagas' disease, and leishmaniasis; and b) as a malaise stemming from contamination of water sources. Contagious disease is an environmental problem because it relates to the conditions of the *habitat* and the people who live in it. And pollution is the consequence of dumping sewage into the sources of drinking water.

The intervention to which we refer, facilitated by Cronick, is currently under way, and the case study should be read in that light. The community in which the intervention is taking place consists of approximately 250 families and is located in the southern part of the city of Caracas. Some of its members work in the city center, and must commute to it every day by public transport, which can take as long as two hours. The community is comprised of several sectors occupying mountainous terrain at an altitude of approximately 1,000 meters above sea level. The houses are made of varied construction materials, but most are of cement block and *bahareque* (similar to adobe), wood panels, and sheet metal.

It was initially agreed with the community to work on the problem of leishmaniasis contagion. This is a tropical disease that expresses itself in several forms, but in Maitana it tends to take the form of severe skin lesions. Since it is spread by the bite of an insect called "flebotoma," the project is now looking into the relationship between the insect, conceived as an environmental factor, and the local inhabitants' lifestyle. The residents are helping to capture insects for an entomological study, and are participating in discussions on how their living habits (their way of dressing and building their homes) contribute to their vulnerability to this disease.

At the present time, and with the support of the PAR guidelines, new fields for collective reflection and action are emerging, suggested by the community members' concerns about environmental deterioration. This damage is related to a predatory style of utilization of land and water, which is similar to those observable in many residential areas. Hence, the facilitation is following an ecological orientation. The ecological point of view stresses: a) recycling of resources within socially and biologically interdependent systems (because a change of resources at one ecological level will affect the other levels); b) individuals' adaptation to the needs of group survival; c) respect for social and biological diversity; and d) change over time which will affect all the participants in the system.

In addition to a set of practical activities (such as capturing flebotomas), the aim is to achieve a collective conversion in which the new ideas introduced by the facilitation team are discussed. Disagreements are expressed and addressed, and an attempt is made to achieve a dialectic resolution of the different points of view in collective fashion. This dialectic resolution occurs when different perspectives are combined to produce new options, which are in turn defined as problems by the participants. In this way, knowledge contributed by science is combined with common sense and local experience to bring about practical changes that are appropriate for the community. For Maitana, the ecological orientation of the facilitation team is subjected to the test of the local residents' requirements. The following areas of problem definition (from the facilitation team's standpoint) can be mentioned:

(1) A concept of exploitation of the environment which gives no concern for its fragility. The community members dump their sewage into the sources of drinking water, cut down trees indiscriminately, and prepare earth for agriculture without regard for the long-term consequences of their actions.
(2) A tradition of housing construction which takes no account of the need for protection against tropical disease vectors.
(3) The practice of using insecticides to eliminate disease vectors. Not only are they applied incorrectly (through indiscriminate spraying along the edge of the highway), but this practice fails to overcome the problem of contagion and destroys the ecosystems where it is used.

The intervention is structured in three participatory phases (not necessarily consecutive):

(1) Identification of the sociocultural factors that foster the spread of leishmaniasis and their relations to the vector's life cycle and behavior, with particular emphasis on the participants' reflection on housing construction-related factors.
(2) Identification of housing construction techniques and materials that are effective, economical, and culturally adequate for the population.

(3) An assessment of the sources of stream pollution is also under way at the present time, and among the proposed solutions is the construction of "restorative ponds" which create micro-ecosystems, along the entire course of the streams. A community member who is now finishing his undergraduate thesis involving a characterization of the plants in the area is drawing up recommendations for choosing the plant species to be planted in these ponds.

This community intervention reflects a recognition of the cyclical stages in which the intervention plan's progress is evaluated and the necessary corrective measures and modifications are decided upon. This cyclical aspect of collection action is, indeed, one of PAR's characteristics. Also contemplated is the exchange of information between the researchers and the participants, regarding how the community can continue the project (observations, considerations, technical contributions). This information provides a basis for participatory discussions, in which the community members reflect on their needs, as they construct them, alter their meaning, and derive implications for practice and daily life in collective fashion. In this respect, when the researcher interprets and suggests actions to the other participants, he/she is merely providing one more input in a community process, and not a professional expertise that defines and imposes the most appropriate activities. The process is expected to continue as long as the interested parties are willing to keep it in motion.

As a result of the process, the facilitation is expected to foster an empowerment of the participants and of their ability to continue resolving the psychosocial and environmental problems they confront in an autonomous and well-informed way, with respect for the community's social needs and for the environment in which it is located.

In a nutshell, this is a case of problem definition by the participants in relation to the situations they confront, in the presence of researchers whose aim is not only to help the community to overcome specific problems but also to facilitate the strengthening and training of the community members, thereby equipping them to confront other problems, through a deepening of knowledge pertinent to their needs and a collective and participatory planning and implementation of actions to improve living conditions. These processes require a self-managing and dialectic confrontation and resolution of three types of knowledge: a) that stemming from scientific knowledge; b) that stemming from common sense and the community's own traditions; and c) that stemming from reflection and experience in the sense that awareness springs from community practice. The combination of these types of knowledge produces something new for the participants, which cannot be transferred to other social groups although certain aspects of the solutions achieved can serve them as examples.

References

Altman, I. & Rogoff, B. (1987). World views in psychology: trait, interactional, organismic and transactional perspectives. In D. Stokols & I. Altman (Eds), *Handbook of Environmental Psychology* (pp.7-40). New York: John Wiley and Sons.

Barker, R. (1963). On the nature of the environment. *Journal of Social Issues,* 19(4), 17-38.

Barker, R. (1968). *Ecological psychology.* Standford: Standford University Press.

Carr, W. & Kemmis, S. (1986). *Becoming Critical: Education, Knowledge and Action Research* (3rd. Edition). London, Falmer.

Chawla, L. & Heft, H. (2002). Children's competence and the ecology of communities: A functional approach to the evaluation of participation. *Journal of Environmental Psychology,* 22(1/2), 201-216.

Churchman, A. (2002). Environmental psychology and urban planing: Where can the twain meet? In R. Bechtel & A. Churchman (Eds), *Handbook of Environmental Psychology,* pp. 191-202. New York: John Wiley and Sons.

Fals Borda, O. (1978). *Por la praxis: el problema de cómo investigar la realidad para transformarla [For the praxis: the problem on how to investigate reality for transforming it].* In O. Fals Borda (Ed.), *Crítica y Política en las Ciencias Sociales: del Debate Teoría y Práctica [Critique and Politics in Social Sciences: On the Theory-Practice Debate].* Simposio Mundial de Cartagena [World Symposium of Cartagena], pp. 209-249. Bogotá: Editorial Guadalupe.

Fals Borda, O. (1986). *Conocimiento y Poder Popular [Knowledge and Popular Power].* Bogotá: Siglo XXI.

Fals Borda, O. (2001). Participatory (action) research in social theory: origins and challenges. In P. Reason & H. Bradbury (Eds.), *Handbook of Action Research*, pp. 27-37. London: Sage Publications

Fals Borda, O. & Rahman, M. (Eds.). (1991). *Action and knowledge: Breaking the monopoly with PAR.* New York & London: Apex Press and Intermediate Technology Publications.

Francis, M. & Lorenzo, R. (2002). Seven realms of children's participation. *Journal of Environmental Psychology,* 22(1/2), 171-189.

Fryer, D. & Feather, M.T. (1994). Interventions techniques. In C. Cassell & G. Symon (Eds.), *Qualitative Methods in Organizational Research,* pp. 230-247. London: Sage Publications.

Gergen, K. (1999). *An Invitation to Social Construction.* London: Sage ublications.

Guba, E. (1990). The alternative paradigm dialog. In E. Guba (Ed.), The Paradigm Dialog, pp.17-31. London: Sage Publications.

Guba, E. & Y. Lincoln. (1994). Competing paradigms in qualitative research. In N.F. Denzin & Y. Lincoln (Eds.), *Handbook of Qualitative Research*, pp. 105-117. Thousand Oaks, CA: Sage publications.

Hamacher, W. (1996). *Manejo de Conflictos en el Area de Medio Ambiente. Instrumento de Política Ambiental en los Paises en Desarrollo [Conflict Management in the Environmental Area. Instrument of Environmental Policy in Developed Countries].* Germany: Deutsche Gesellschaft für technische Zusammenarbeit.

Heron, J. & Reason, P. (1997). A participatory inquiry paradigm. *Qualitative Inquiry,* 3(3), 274-294.

Heron, J. & Reason, P. (2001). The practice of cooperative inquiry: research 'with' rather than 'on' people. In P. Reason & H. Bradbury (Eds.), *Handbook of Action Research,* pp. 179-188. London: Sage Publications.

Horelli, L. (1997). A methodological approach to children's participation in urban planing. *Scandinavian Housing and Planning Research,* 14, 105-115.

Horelli, L. (2002). A methodology of participatory planing. In R. Bechtel & A. Churchman (Eds), *Handbook of Environmental Psychology,* pp. 607-628. New York: John Wiley and Sons.

Horelli, L. & Kaaja, M. (2002). Opportunities and constraints of 'internet-assisted urban planning' with young people. *Journal of Environmental Psychology,* 22(1/2), 191-200.

Jiménez-Domínguez, B. (2002). Which kind of sustainability for a social environmental psychology. In P. Schmuck & W. Schulz (Eds.), *Psychology of Sustainable Development*, pp. 257-276. Boston: Kluwer Academic Publishers.

Kemmis, S. (2001). Exploring the relevance of critical theory for action research: emancipatory action research in the footsteps of Jürgen Habermas. In P. Reason & H. Bradbury (Eds.), *Handbook of Action Research*, pp. 91-102. London: Sage Publications.

Kemmis, S. & McTaggart, R. (2000). Participatory action research. In.N. Denzin & Y. Lincoln (Eds.), *Handbook of Qualitative Research*, pp.567-606 (2nd. Ed). Thousand Oaks, CA: Sage Publications.

Lincoln, Y. (2001). Engaging sympathies: relationships between action research and social constructivism. In P. Reason & H. Bradbury (Eds.), *Handbook of Action Research*, pp. 124-132. London: Sage Publications.

Lincoln, Y. & Guba, E. (2000). Paradigmatic controversies, contradictions and emerging confluences. In N. Denzin & Y. Lincoln (Eds.), *Handbook of Qualitative Research*, pp.163-188 (2nd ed.). Thousand Oaks, CA: Sage Publications.

Ludema, J., Cooperrider, D. & Barrett, F. (2001). Appreciative inquiry: the power of the unconditional positive question. In P. Reason & H. Bradbury (Eds.), *Handbook of Action Research*, pp. 189-199. London: Sage Publications.

Michell, J. (2003). The quantitative imperative: Positivism, naive realism and the place of qualitative methods in psychology. *Theory and Psychology*, 13(1), 5-32.

Ortiz, R. (1997). *Globalización y Conflictos Socio Ambientales [Globalization and Socio Environmental Conflicts]*. México, D.F.: Editorial Manaral, Ayayala.

Park, P. (2001). Knowledge and participatory research. In P. Reason & H. Bradbury (Eds.), *Handbook of Action Research*, pp. 81-90. London: Sage Publications.

Pol, E. (1998). Evoluciones de la psicología ambiental hacia la sostenibilidad: tres propuestas teóricas y orientaciones para la gestión [Evolutions of environmental psychology towards sustainability: three theoretical proposals and orientations for management]. In D. Páez & S. Ayestarán (Eds.), *Los Desarrollos de la Psicología Social en España [Developments of Social Psychology in Spain]*, pp. 105-120. Madrid: Infancia y Aprendizaje.

Reason, P. & Heron, J. (1995). Co-operative inquiry. In J. A. Smith, R. Harré & L. Van Langenhove (Eds.), *Rethinking Methods in Psychology*, pp.122-142. London: Sage Publications.

Reed, D. (1996). *Ajuste Estructural, Ambiente y Desarrollo Sostenible [Structural Adjustment, Environment and Sustainable Development]*. Caracas: Fondo Mundial para la Naturaleza, CENDES, Nueva Sociedad.

Reichardt, C. S. & Cook, T. D. (1979). Beyond qualitative versus quantitative methods. In T. D. Cook & C. Reichardt (Eds.), *Qualitative and Quantitative Methods in Evaluation Research*, pp. 7-32. Beverly Hills, California: Sage Publications.

Rivlin, L. (2002). The ethical imperative. In R. Bechtel & A. Churchman, (Eds.), *Handbook of Environmental Psychology*, pp.15-27. New York: John Wiley and Sons.

Saegert, S. (1987). Environmental psychology and social change. In D. Stokols & I. Altman (Eds), *Handbook of Environmental Psychology*, Vol. 1, pp. 99-128. New York: John Wiley and Sons.

Saegert, S. & Winkel, G. (1990). Environmental Psychology. *Annual Review of Psychology*, 41, 441-477.

Sadan, E. & Churchman, A. (1997). Process focused and product focused community planing: two variations of empowering professional practice. *Community Development Journal,* 32(1), 3-16.

Scwandt, T. (1990). Paths to inquiry in the social disciplines: scientific, constructivist and critical theory methodologies. In E. Guba (Ed.), *The Paradigm Dialog,* pp. 258-276. London: Sage Publications.

Senge, P. & Scharmer, O. (2001). Community action research: learning as a community of practitioners, consultants and researchers. In P. Reason & H. Bradbury (Eds.), *Handbook of Action Research,* pp. 238-249. London: Sage Publications.

Serrano García, I. (1992). Intervención en la investigación: su desarrollo [Intervention in research: its development]. In I. Serrano García & W. Rosario (Coords.), *Contribuciones Puertorriqueñas a la Psicología Social Comunitaria [Puerto Rican Contributions to Community Social Psychology],* pp. 211-282. San Juan de Puerto Rico: EDUPR.

Stokols, D. (1994). *Environmental psychology: past accomplishments and future challenges.* Paper presented at the 23rd International Congress of Applied Psychology, Madrid, Spain.

Torres Carrillo, A. (1995). *Aprender a Investigar en Comunidad II. Enfoques Cualitativos y Participativos en Investigación Social [Learning to Research in the Community II. Qualitative and Participatory Approaches in Social Research].* Santa Fé de Bogotá: Facultad de Ciencias Sociales y Humanas de Unisur.

Treleaven, L. (2001). The turn to action and the linguistic turn: towards an integrated methodology. In P. Reason & H. Bradbury (Eds.), *Handbook of Action Research,* pp. 261-272.Lodon: Sage Publications.

Werner, C., Brown, B. & Altman, I. (2002). Transactionally oriented research: examples and strategies. In R. Bechtel & A. Churchman, (Eds.), *Handbook of Environmental Psychology,* pp. 203-221. New York: John Wiley and Sons.

Wicker, A. (1987). Behavior settings reconsidered: temporal stages, resources, internal dinamics, context. In D. Stokols & I. Altman (Eds.), *Handbook of Environmental Psychology,* Vol. 2, pp.6 13-653. New York: John Wiley and Sons.

Wicker, A. (2002). Ecological psychology: historical context, current conception, prospective directions. In R. Bechtel & A. Churchman, (Eds.), *Handbook of Environmental Psychology,* pp. 114-128. New York: John Wiley and Sons.

Wiesenfeld, E. (1994). La teoría crítica y el construccionismo: hacia una integración de paradigmas [Critical theory and constructionism: towards an integration of paradigms]. *Revista Interamericana de Psicología,* 28(2) 251-264.

Wiesenfeld, E. (1998). Desarrollo teórico en psicología ambiental: el enfoque construccionista crítico [Theoretical development in environmental psychology: the critical constructionist approach]. *Revista AVEPSO,* 21(2), 33-62.

Wiesenfeld, E. (2000). Researcher's place in qualitative inquiry: (un)fulfilled promises? *FQS (Forum Qualitative Studies) 1,* (2). (Online Journal)

Wiesenfeld, E. (2000). *La Autoconstrucción: un Estudio Psicosocial del Significado de la Vivienda [Self Help Building: a Psychosocial Study on the Meaning of Housing].* Caracas: CONAVI.

Wiesenfeld, E. (2001). Environmental problems from a psychosocial community perspective: toward an environmental psychology of change. *Medio Ambiente y Comportamiento Humano,* 2(1), 2-20.

Wiesenfeld, E. & Sánchez, E. (2002). Sustained participation: a community based approach to addressing environmental problems. In R. Bechtel & A. Churchman, (Eds), *Handbook of Environmental Psychology,* pp. 629-647. New York: John Wiley and Sons.

Neighbouring, sense of community and participation: Research in the city of Genoa

Laura Migliorini, Antonella Piermari & Lucia Venini

University of Genoa, Italy

Abstract. This paper presents a study of the effects of sense of community, neighbouring ties and inhabitants' participation on local activities. Sense of community is considered a central concept in psychological theory concerning the impact that the individual must bear when living in an urban reality. Sense of community, at the empirical level, has been studied in relationships to territory, the neighbouring and the local community. The research of Chavis and Wandersman (1990), falling within this conceptual framework, explores an empirical model in which sense of community can act as a catalyst for local action such as, for example, participation in neighbouring association. Within this reference frame, the ecological model of Perkins, Brown and Taylor (1996) considers the relationship between physical environment, social environment, community cognition and community behaviour: these elements are considered to be the main predictor of participation in local activities.

A preliminary analysis on secondary data about all the neighbourhoods of the city of Genoa was conducted. The results of this analysis led to the choice of two suburbs as subjects of the second phase of this research, on the basis that educational level was almost the same within both areas and thus would minimise the influence of this index on the results. Social and spatial features of these two suburbs (such as public green areas, number of local associations, number of children) were used as indexes within the present work. The second phase of research was conducted on two suburbs ($N = 100$ subjects) in order to investigate the differences in neighbouring, sense of community and participation.

Keywords: neighbouring, sense of community, participation, environmental variables.

Theory

Neighbourhood social ties and environment variables

In the last few decades, a theoretical perspective has been developed by several researchers concerning the study of the "individual in context"; that is, in the social systems and physical environments he/she lives within, which offer clear opportunities and constraints (Bronfenbrenner, 1979; Trickett, 1996). Talking of context in our culture necessarily means talking about a series of contexts in which the individual finds himself. Within this theoretical background, Cohen and Siegel (1991) specify various aspects which help us investigate this question in more depth; namely, contexts as social systems, contexts as physical places and contexts as they evolve in time. Social systems influence the developmental outcomes and behaviours of subjects both directly – through face-to-face relationships in concrete everyday environments – and indirectly – through exchanges which still influence the context in which the subject lives even when they do not take place in it. The physical and structural characteristics of places stand in close relationship to any possible behaviours that may occur in these places. The structural and architectural configuration of a school, a neighbourhood, or a city can influence the ways in which a person acts and relates to that context.

Scholars' interest in neighbourhood ties started in the 1970s with Sarason (1974) and has continued in the 1980s and 1990s with the consolidation of the conviction that "the needs and vicissitudes of individuals are rooted in a specific context of relationships, opportunities and bounds, defined and delimited spatially" (Chaskin, 1997, p. 521). Initially scholars' interest focused on the way the formation of neighbourhood social ties was influenced by individual characteristics (for example, how neighbourhood social ties are facilitated by a similar socio-economic status, or by other characteristics such as having children, belonging to a minority group, the fact of having lived in the same place for a certain number of years). The neighbourhood can be seen as a set of individuals but also as a well-defined space, for example, the district. Is it then possible to argue that neighbourhood social ties are only an expression of relationships with people? To what extent does physical-spatial configuration play a role in the transformation of a group of unrelated individuals into a real community? In this connection, Perkins, Florin, Rich and Chavis (1990) studied the effects of the physical-spatial environment on the participation of residents in neighbourhood organisations and in turn how these related to neighbourhood social ties. Environmental psychologists and environmental designers have long been interested in the way residential architecture, the arrangement of buildings and the characteristics of public and semi-public spaces can facilitate the formation of stronger communities. In particular, scholars have tried to identify the characteristics of an urban landscape which may promote the development of a community.

Within this line of research, we find the studies by Skyaeveland, Garling, Maeland (1996) and Skyaeveland and Garling (1997) to be significant contributions, since here neighbourhood social ties are analysed and measured by combining traditional studies with a new, more qualitative approach, and through a multidimensional analysis. The authors (Skyaeveland & Garling, 1997) set out to examine the physical characteristics of a district in relationship to social interactions among residents. The basic idea behind this investigation is that the set of spatial-environmental characteristics constitutes a factor that can impact favourably on social interaction between people. The authors suggest the term *interaction space,* which includes not only the visual and functional factors of the physical environment but also the social activities and representations of those who occupy it. In particular, the authors maintain that social contacts between neighbours appear to be favoured by three elements: the opportunity for passive social contact, proximity to others and an appropriate space for interacting. With regard to the first factor, Porta says that:

"The influence of the physical environment on social activities is indirect; these have different degrees of intensity, from the simplest form of passive contact (seeing and hearing others) to friendship, and the transition from the lower to the higher levels takes place spontaneously: most human contacts observable in public space are low-intensity, simply seeing or listening to other people, or exchanging comments, giving or receiving information, but this level is fundamental in setting up a varied network of interpersonal relations." (Porta, 2000, p. 14)

For Skyaeveland et al. (1996) neighbourhood ties are characterised by "positive and negative aspects of social interaction, expectations and feelings of attachment both towards the individuals living beside them and in relation to the place where they live" (p. 418).

The study by Skyaeveland and Garling (1997) points to specific, significant "models of social-spatial relations" in the district. Environmental variables have been found to be more effective predictors than demographic variables on each of the four dimensions of neighbourhood ties considered by the authors. Physical environment factors can thus be considered as attributes of interactional space, which have the power to influence relations between neighbours.

Kuo, Bacaicoa and Sullivan (1998) have found that, thus far, research into the characteristics of the urban environment (such as, for example, semi-public space) which encourage or inhibit social interactions between inhabitants has focused excessively on the structural characteristics of buildings. The authors ask themselves whether there might not be other characteristics of the urban environment capable of promoting casual contacts between neighbours and stronger ties with the inhabitants and the local area, given that it is not always possible to change already present architectural features owing to lack of resources. This question represents the starting point of a series of studies which seek to explore the possibility that the presence of natural elements (trees, lawns, flowers) in open spaces next to dwelling-houses can make up for the lack of semi-public spaces and promote well-being and social ties.

Similarly, a study conducted by Kuo, Sullivan, Coley and Brunson (1998) confirms that one of the most important features of such common spaces is the presence of trees and lawns. These studies have shown that residents do not appreciate and are afraid of areas devoid of any vegetation; and the simple fact of planting trees and cultivating lawns is enough to transform the reaction of the inhabitants towards previously arid and bare spaces. It has been observed that people who live in an urban area with green spaces lead a much more socially active life and know their neighbours, even though this acquaintance may only be superficial; these individuals talk about how they help each other and are conscious of a feeling of belonging. The amount of time that inhabitants spend exploiting the common spaces outside their houses, which depends on the presence and the arrangement of trees, influences the behaviour of adults and young people. The closer to the buildings the trees have been planted, the more time people spend near them. One can therefore conclude that, since people stay close to each other longer in these spaces, these experiences provide opportunities for social encounters between neighbours. So, through a series of mechanisms, such as an increase in face-to-face encounters and an increase in residents' satisfaction with their neighbours, the presence of trees can contribute to the social cohesion of the community, developing not only strong social ties but also a sense of safety and adjustment among residents (Kuo, Bacaicoa & Sullivan, 1998).

Sense of community

Recently, the concept of *sense of community* has come to occupy a central position in the literature about the relationship between individual and urban contexts. The term refers to a personal quality which characterises a strong attachment that people feel towards their "communities", whether these are geographical entities, neighbours or functional entities such as clinical treatment groups or jobs. Sarason (1974) defines sense of community as the feeling that enables an individual to perceive himself as part of a structure that is always there, supportive and loyal. The sense of community transcends individualism because, to maintain such a relationship of interdependence, a person treats others in the way he would expect to be treated by them. Sarason's view, comments Amerio, "develops from the idea that contemporary society appears radically marked by social and cultural practices that have not only produced a weakening of those ties that are based on a sense of belonging and active participation in collective life, but also – as the final outcome of the progressive destruction of the psychological sense of community – isolation, anomie and segregation" (Amerio, 2000, p. 420).

Since the introduction of the concept, the meaning of sense of community has been looked at in various ways; the theory developed by McMillan and Chavis (1986) is the one which is thus far considered to be the most complete. They define sense of community as having four basic aspects: belonging, which creates feelings of emotional security together with a sense of belonging and identification; influence, which implies both exercising influence on the community and accepting the influence of the community on oneself; integration and satisfaction of needs, which reinforce one's own appropriate behaviour in relation to the community; shared emotional connection, which creates the positive affective aspect of membership of the community. In particular, the consciousness of belonging to a community – belonging – has precise boundaries, which make it possible to define, who belongs to a community and who does not. These boundaries provide the emotional security necessary for expressing feelings and needs and developing intimate relationships.

For McMillan and Chavis influence means the perception by the members of the community of their importance as well as their personal power and that of the community. Shared emotional connection is based on a common history. Although it is not necessary for the members of the group to have taken part in that history, they must identify with it. Interactions between members in common events and the specific attributes of such events can foster or inhibit the strength of the community. The factors which combine to determine an emotional connection are: the contact hypothesis (the more people interact the more likely it is that their ties will strengthen); the quality of the interaction (the more positive the experiences and the relationships experienced together, the more significant the tie, in that success facilitates the relationship); the definition of relationship (if the interaction between people is ambiguous and if the tasks of the community are left incomplete, group cohesion is weakened); investment (the level of investment of a community correlates positively with the level of connection). Investment, indeed, determines the importance of the history of a community, defines its values and heightens affective involvement.

Felton and Shinn (1992) put forward the idea of sense of community as an "extra-individual" construct which needs to be understood at a system level. In this way, a community, defined in terms of setting (neighbours, physical-social environment) can be experienced as supportive even if a person is not able to identify any particular individuals in the community or their role in creating this psychological sense. Moreover, a community can still be experienced as supportive even when the individuals in that setting have changed. In conclusion, one can say that people feel a sense of community not necessarily because they have experienced supportive transactions on an individual level, but, as Sarason (1974) had argued previously, because they have a feeling that the community is there for them.

Sense of community has been investigated on the empirical level mainly in relation to territory, to ties to the neighbourhood and to the local community. In particular, sense of community has often been studied in the literature in relation to residents' satisfaction with the community and to sense of belonging. Satisfaction with the community has been defined and studied both in relation to environmental characteristics (for example, climate and green space) and to the quality of local facilities and to other indicators; these elements function as real mediators in the perception of the environment.

Perception quality of environment and participation

How can a sense of community be fostered? Some of the special measures suggested by the literature include: the strengthening of social ties, the creation of supportive social networks and the promotion of forms of participation. Participation is an active feeling which can be expected to produce a sense of community and a feeling of belonging. Basically, taking part in local activities helps to make one feel part of a "community". However, it needs to be pointed out that reference is also made by forms of participation which are the indirect result of living everyday social relations inside the community.

The investigation carried out by Chavis and Wandersman (1990) fits into this framework: here we find an empirical exploration of a model which sees sense of community as having a true "catalytic effect" on local action such as, for example, membership of district associations. This model is made up of three main components, namely: the perception of the environment, social relations and the perception of community through the active participation of its members.

This is a very complex investigation which not only considers the three main components but also the perception of the problems existing in an apartment block, the evaluation of the characteristics of the apartments, the satisfaction of the residents, as elements which come into play in a model which talks about the determinants of local action. Here our attention is focused only on the main components (the interested reader can consult the original book for an analysis of the overall model).

Perception of the environment involves judgements about the environment (for example, the perceived quality of the environment, satisfaction or problems with the environment). These judgements are formulated in relation to the degree to which the environment or a specific aspect of it is considered to be positive or negative by the individual. An analysis of the literature suggests that, in general, there are significant relations between the quality of the physical environment, the social environment (for example, social interactions and the sense of belonging) and residents' satisfaction. The term *social relations* refers to any type of interaction between neighbours, such as asking to borrow something, paying informal visits or asking for help in emergencies. Through these interactions, neighbours provide each other with social support. When people perceive a sense of community, they are more inclined to interact with their neighbours. At the same time, ongoing face-to-face contacts foster a shared emotional connection which helps to maintain the sense of community. In the development of neighbourhood ties and supportive ties, the social network favours the efforts of neighbourhood associations to share information about the association, and also promotes co-operation in the creation of services (for example, security facilities) by means of informal social control (in this context a negative correlation has been established between informal social control and crime). In general, one can say that residents who interact more with their neighbours are also more inclined to be involved in and be members of local associations. This fact has been found to be particularly significant in the case of people who have several friends in the district and close ties with neighbours whom they turn to for socio-economic or practical support.

The results of the research carried out by Chavis and Wandersman confirm the proposed model: sense of community modifies perceptions and behaviours encouraging participation and other forms of local action. If an individual has a strong sense of community, environmental conditions are judged more positively and in general the feeling of satisfaction about the environment grows. Sense of community therefore plays an important role in fostering neighbourhood ties and increasing personal and group perception of empowerment – in other words, the power to influence in some way what happens around us. One can thus conclude that neighbourhood social ties contribute to a significant extent towards the prediction of high levels of participation.

Physical and social environment: An ecological model

When looking for significant links between neighbouring, sense of community and participation, it's really useful to refer to ecological model of Perkins, Brown and Taylor (1996). The model attempts to investigate which aspect of community life could influence participation behaviours. In particular, the authors consider physical environment (green areas, lighting, dirtiness etc), economic environment (resources, property etc) and social environment (race, length of residence, neighbouring, satisfaction, participation etc). Our work aims to study in-depth the social relationships between people living in two different districts of Genoa, considering also the incidence of some of the most relevant structural elements.

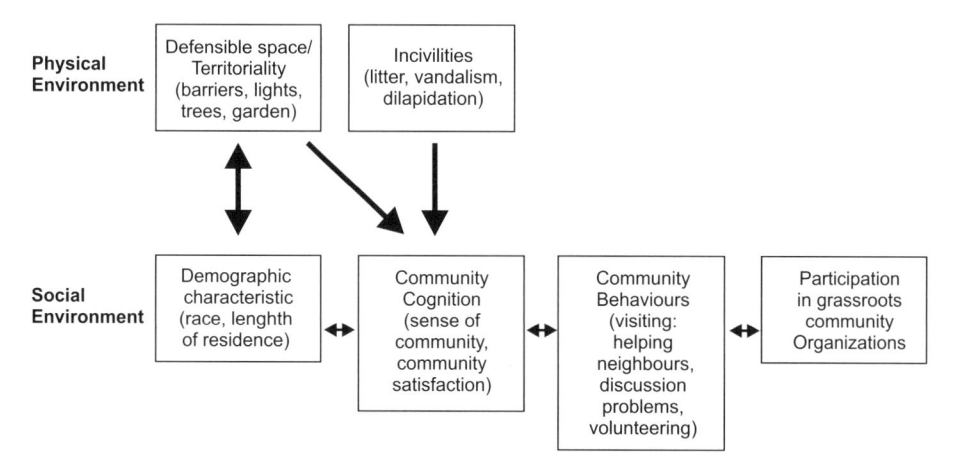

Figure 1. Ecological model of Perkins, Brown and Taylor, 1996 (modified)

```
Physical          Defensible space/         Incivilities
Environment       Territoriality            (litter, vandalism,
                  (barriers, lights,        dilapidation)
                  trees, garden)

                  Demographic       Community         Community         Participation
Social            characteristic    Cognition         Behaviours        in grassroots
Environment       (race, length     (sense of         (visiting:        community
                  of residence)     community,        helping           Organizations
                                    community         neighbours,
                                    satisfaction)     discussion
                                                      problems,
                                                      volunteering)
```

Assumptions

The general aim of the present study is to investigate possible relationships between physical and social features of neighbourhood and social interaction among the residents. In order to extend the literature about this research field, we were interested in the possibility of deepening knowledge about correlations between places of residence and a person's perception of social ties. Our hypothesis is that people of the same city may enjoy different relationships with neighbours and feel different level of sense of community according to their residence zone.

Taking Perkin's model as a frame of reference, several relationships among considered variables may be assumed. First, we assume that neighbourhood ties, sense of community and participation are associated. We also assume that green areas, presence of children, number of local associations and retail shops act on the considered variables, so we decided to carry out a preliminary analysis on every district of Genoa city and to compare two ones chosen as they were characterised by differences in some of the most relevant social and structural variables. We proposed also to examine if the length of residence lifetime can influence neighbourhood ties, sense of community and participation. We expected, furthermore, that the perception of urban incivility signs may decrease the sense of community of residents.

Method

Characteristics of neighbourhoods of Genoa: A preliminary analysis

According to theoretical elements considered before, we have defined some particular indices for the research such as: districts population and area, demographic information about population, social elements, family structure, existence and extent of green areas, the distribution of retail stores and associations in the different part of the city. In order to analyse the urban context, we used secondary data found in three publications by the city municipal corporation (Annuario statistico – Città di Genova, 1999; Atlante demografico della città, 2000; Notiziario statistico, April, 2001).

We studied 25 districts of the city of Genoa, which are all the districts in the urban area. After analysing all the information, we decided to focus attention on the most meaningful demographic and social indications: educational level of residents, quantity of children/adolescents (0–13 year olds), retail stores, presence of local associations in relation to the number of inhabitants and extent of public green areas (municipal gardens, park, flower border, trees, etc).

In particular, according to literature, the variable educational level has basic importance, because it generally negatively correlates with both neighbouring ties and sense of community (Prezza & Pacilli, 2002). So we decided to take this variable as the main discriminating element.

On the contrary, other indices mentioned above, positively correlate with social relationships in the place where people live as reported by literature. So, significant differences about these indices have been very important in order to choose the districts to be analysed.

According to these elements, we have decided to choose districts characterised by intermediate educational level. Four districts have this characteristic, Pegli, Prè Molo Maddalena, San Martino and San Fruttuoso. Among them we excluded Prè Molo Maddalena because it's too peculiar to compare with other districts; this zone, in fact, is the historical centre of the city and has structural and urban features not comparable with any other.

S. Martino and S. Fruttuoso are bordering districts and so we decided to choose one of them to compare with Pegli; we therefore selected S. Martino because it has more similar values of educational level.

Table 1. Districts analysis of considered variables

Districts	% having a university degree	Age	Population 0-13 years%	Public green (m² per person)	Shops per 10,000 persons	Associations per 10,000 persons
S. Martino	8.1%	47.2	8.8%	2.33	151	8.92
Pegli	7.5%	46.5	9.5%	9.07	132	24.46

The characteristics of the urban environment of the two districts are different because San Martino is bordering on the centre of the city while Pegli is a peripheral area; in particular, Pegli is bounded by the sea instead of San Martino internalised within the city.

The main differences between these two zones are the extension of green public area and the number of local associations (Table 1). As it is possible to notice in the table, in Pegli there are much more m²/ person of green public area than in San Martino. Also the number of local associations is higher in Pegli than in San Martino; on the contrary the two districts chosen are homogeneous with regards retail shops. Furthermore the number of children (up to 13 years old) is slightly higher in Pegli. We expected that these elements would influence the neighbourhood ties, sense of community and participation.

In order to expand knowledge about these two zones, we have applied in reasoned reading of Municipal Urbanistic Plane, as suggested by Amerio (2000), from which we have obtained information about the prevalent uses of different parts of districts, their global structure, the viability net and the presence of various services (rail station, sports centre, health service etc).

Participants

Participants were 100 subjects, 50 living in S. Martino and 50 living in Pegli.
- *S. Martino:* 14 men and 36 women were interviewed; frequencies of the age classes are: 15–18 years old = 2%, 19–30 = 16%, 31–40 = 8%, 41–50 = 8%, 51–60 = 18%, 61–70 = 22%, over 70 years old = 26%. We also used classes to codify the length of residence in the same district and the frequencies are: 0–5 years = 8%, 6–10 = 6%, 11–15 =8%, over 15 years = 78%. The average family size is 2.48, and only 6 persons have at least one child or adolescent under 18 years old living at home. In the sample, 62% of subjects are housewives or pensioners, 46% are workers and 12% are students.
- *Pegli:* 15 men and 35 women were interviewed; frequencies of the age classes are: 15–18 years old = 2%, 19–30 = 12%, 31–40 = 16%, 41–50 = 20%, 51–60 = 18%, 61–70 = 14%, over 70 years old = 18%. Classes about length of residence in the same district are distributed in the following way: 0–5 years = 18%, 6–10 = 4%, 11–15 =10%, over 15 years = 68%. The average family size is 2.94 and 16 persons have at least one child or adolescent under 18 years old living at home. In the sample, 48% of subjects are housewives or pensioners, 40% are workers, 8% are students and 4% are looking for a job.

Measures

The questionnaire is composed by 5 sections:
- *Demographic information:* a brief part of questionnaire was used to obtain information concerning subject's age, sex, educational level, kind of work, social information about his/her family composition, presence/absence of children and duration of residence in that district.
- *Questions about participation:* according to the theory of place (Canter, 1988; Bonnes & Bonaiuto, 1996), which assumes activities to be one of the main constituent components of the place itself, a series of activities performed by subjects (districts activities, local associations, sports groups etc and time spent) within the residence suburb was investigated.
- *Sense of Community Italian Scale (Prezza, Costantini, Chiarolanza & di Marco, 1999):* taking the theory of McMillan and Chavis as a reference frame, this questionnaire investigates sense of community both at the subjective level and at the macro-level. The total number of items is 21, but we used only 18 questions because authors have observed low coherence for three items. The answer modality is a 4-point Likert scale (highly degree, degree, disagree, highly disagree).
- *Neighbourhood Ties Scale (Prezza & Pacilli, 2002):* this is a measure of the intensity and quality of neighbourhood ties. It's composed of 7 items about different

kinds of relationships with neighbours: how long do they spend time together (talking, doing activities and staying at each other homes), how often do they help each other and how many people do they consider to be available and friends. The first 5-item answer is on a 5-point Likert scale, whereas in the last 2, subjects indicate the number of persons they consider to be a "good neighbour."
- Questions about the subjective perception of urban incivility in one's own living zone and describing it (e.g., litter in the streets, writing and graffiti on building walls, parking in forbidden areas etc).

Overall, the final questionnaire included 39 questions; the answer procedures are different in different parts of the research and the answering would produce both close and open answers.

Procedure

The subjects were contacted by telephone according to the contact protocol; researchers used a street list and telephone guide in order to find people living in S .Martino and Pegli. After briefly presenting the aim of the study, subjects were interviewed. The questionnaire is anonymous, in fact the researchers knew only the answered telephone number and address, but no name. While people were answering the questionnaire, researchers filled in a paper copy and, later data were recorded on a PC for analysis.

Results and discussion

Table 2 shows the mean and standard deviation values of the variables included in the study concerning the comparison of the two suburbs in these variables.

Table 2. Comparisons (t test, phi, cramer's V) between San Martino and Pegli

	San Martino	Pegli	Comparison
Neighbouring ties	$M = 16.94$ $SD = 5.95$	$M = 16.68$ $SD = 4.73$	$t = 0.242$ ns
Sense of community	$M = 49.32$ $SD = 7.21$	$M = 52.54$ $SD = 6.22$	$t = -2.392$ *
Quantity of activities	$M = 0.5$ $SD = 0,789$	$M = 1.14$ $SD = 0.904$	$t = -3.772$ **
Valuation of incivilities	YES = 41 NO = 9	YES = 30 NO = 20	phi = -0.267 ** cramer's V = 0.267 **

** correlation is significant at the 0.01 level (2-tailed)
* correlation is significant at the 0.05 level (2-tailed)

First, the two samples reported no differences concerning the neighbouring ties. This is probably due to the strong relationship between neighbouring tie and educational level. It's worth remembering that the latter has been assumed a key criterion in order to minimise its influence on choosing the suburbs to be investigated. Thus the first result of this research confirms the literature since the two suburbs show the same average educational level.

Conversely, taking into account the sense of community, the t test indicates significant differences between the two districts; Pegli's residents perceive a higher sense of community than those in San Martino, thus confirming the assumption that social-environmental variables considered could affect the feeling toward the community one belongs. According to the model of McMillan and Chavis, the feeling of belonging to a community is influenced also by the well-identifiable borders of the districts. In this regard, the more peripheral location of Pegli makes it possible to recognise in a clear way who belongs to this area and who does not. This same distinction is more difficult in San Martino because it's more internalised within the city and it's contiguous with the centre of the city.

Also, with regards the quantity of activity performed by the subjects within the residence suburb, the t test indicates significant differences between the two districts showing higher people participation in Pegli than San Martino. This data is consistent with number of local associations, which is higher in Pegli. Participation and the number of associations influence each other; we can suppose that where there are a lot of participation opportunities people are more active and, vice versa, people who desire to be committed in the districts life try to create plurality of occasions to meet each other. The linkages among persons sharing common interests or activities build the community, which is not only considered as a territorial unit but also as a relational one.

Moreover, looking at results related to the evaluation of incivilities (as it defined by Perkin's model), the same discrepancy is revealed. It's worth reporting that the incivilities evaluation is based on the presence or absence in suburb life. The relationship among incivilities perception and the other variables is not still clear, here and in the general research and literature about this topic. Certainly we can observe, however, that also this element seems to be coherent with the other ones.

In sum, these data show that Pegli's residents are more satisfied than those in San Martino. In Pegli people express better relations with both other people and place, they are more active in local participation and they perceive less frequently, than those who live in San Martino, incivility signs. With regards the latter result, we don't know if people perceive incivility signs in different ways between the two districts or if really the two areas are characterised by significantly different number of signs. It's probably the both elements together that cause the result (Moser, 1995).

Table 3 reports the correlation matrix among the main variables taking into account the whole sample. This is useful in order to verify the assumption of this research about the relationships between neighbouring ties, sense of community, participation, urban incivilities perception and residence lifetime in the same suburb.

As expected, the residence lifetime duration is strongly related with the neighbouring ties, confirming the literature (Bonaiuto, Aiello, Perugini, Bonnes & Ercolani, 1999). Also, the same variable is not related to the other ones investigated. The other demographic information doesn't correlate significantly; however, further data analysis could be needed. A significant correlation between neighbouring ties and sense of community can be found, thus showing that the stronger the ties involved, the higher the sense of community and belonging. Furthermore these two variables are related to the quantity of performed activities in the suburb, therefore confirming that taking part in suburb social life could increase relationships.

Table 3. Correlations (Pearson's *r*) and levels of significance between investigated variables in the Genoa sample

	Neighbou-community	Sense of	Participation	Incivilities residence	Length of
Neighbou-ring ties	1	0.444 **	0.202 *	0.034 ns	0.259 **
Sense of community	0.444 **	1	0.316 **	–0.262 *	0.172 ns
Participation	0.202 *	0.316 **	1	–0.253 *	–0.120 ns
Incivilities	0.034 ns	–0.262 *	–0.253 *	1	0.168 ns
Length of residence	0.259 **	0.172 ns	–0.120 ns	0.168 ns	1

** correlation is significant at the 0.01 level (2-tailed)
* correlation is significant at the 0.05 level (2-tailed)

Participation, defined as the activities performed within the suburb, is significantly correlated to neighbouring ties and sense of community. Also, the same variable is negatively and significantly correlated to the evaluation of incivilities, which is inversely related to sense of community. Thus, this result could provide experimental evidence for the study of the relationships between social and physical environment described in Perkin's model.

Concerning the incivilities evaluations reported in Table 3, these are demarcated not only by the evaluation of their presence/absence, as described in Table 2, but also by their level (low/high).

According to the literature, the data confirm that all these variables are linked, but there are still several ways to read the possible causal relations among them. The open questions that arise are:
– Is it possible to define a causality order among sense of community, neighbouring ties, participation and valuation of incivility signs?
– Which element is the source of the other ones?

The Table 4 is reported in order to compare the whole Genoa sample to the Italian sample, which is drawn from the measures validation work performed by Prezza, Costantini, Chiarolanza and di Marco (1999).

The results carried out are not so different to the sample investigated in this work. However, it's worth mentioning the differences concerning the *Neighbouring Ties Scale* that can be investigated into many cultural aspects typical of Genoa country, describing it's people as reserved and bashful ones.[1] Also the quantity of neighbours considered helpful by the Genoa sample is very small.

The sense of community is slightly higher in Genoa than in Italy on a whole, but this result may depend on the procedure used (telephone interview), which could have influenced subjects towards social desirable answers.

---

[1] The poetry itself sung by famous regional poets (Montale, Campana, Sbarbaro and Barrile) uses the description of the typical harsh and inaccessible landscape to symbolise the temperament of the inhabitants.

Table 4. Mean and standard deviation of investigated variables in Genoa sample and Italian sample

	Genoa sample	**Italian sample**
Neighbouring ties	$M = 16.81$ $SD = 5.35$	$M = 17.73$ $SD = 6.02$
Sense of community	$M = 50.93$ $SD = 6.68$	$M = 48.24$ $SD = 6.69$

Moreover, it's useful to remember that this variable can show striking variations depending on which context has been applied (town, village, country, suburb). Also Chaskin (1997) takes into account the influence of territorial dimension on the construct of sense of community.

In conclusion, the picture described by the results underlines the need to consider sense of community, both as an individual variable and as an environmental variable.

Conclusions

Finally, we can state that the main aim of the present work has been achieved, since the possible relationships among social and physical suburb features on one side and neighbouring ties and sense of community on the other side have been demonstrated. Nevertheless, a further and more in-depth investigation is required in order to strengthen the basic assumption that the physical and natural environments are a source of influence on neighbouring ties and sense of community.

McMillan and Chavis, in their theoretical framework defining sense of community, underline the distinction between two main uses of the term community. The first is the "geographical" aspect of community, the relation to a place such as a neighbourhood or a town; the second is "relational", in connection with the quality of human relations. McMillan and Chavis suggest applying their definition of sense of community equally to places and people, given that the nature of human interactions within the boundaries that create sense of community is very important in local communities. The authors argue that people "make" places. Although there are some justifications for this view, we suggest that the opposite effect should also be studied and that it can be argued that places "make" people. According to this point of view, the environmental context can foster contacts, emotional sharing and feelings of belonging.

Participation in local activities seems to play a double function: on one side it can be helpful in promoting neighbouring social ties, on the other it can be a way of increasing the sense of belonging and attachment to one's place of residence. According to the theory of place, activities are one of the main components of place itself; they also represent the opportunity for social contact and for the creation of new "weak ties".

Perception of the environment involves judgements about it, as mentioned above. These judgements are formulated in relation to the degree to which the environment, or a specific aspect of it, is considered to be positive or negative by the individual. In the present work concerning place, the lower the incivility signs perception, the higher the sense of community. Future research could be conducted about the perception of urban environment quality; it would be interesting to analyse which elements are more influential in this field of research, both structural variables of urban planning and public space, and specific variables such as traffic, streets cleanliness, parking etc.

In conclusion, our data point to the central importance of relationships among environmental variables and sense of community, participation and perception of signs of incivility. We hope, in the future, to improve the understanding about these relationships.

References

Amerio, P. (2000). *Psicologia di comunità [Community Pschology]*. Bologna: Il Mulino.
Annuario statistico. Città di Genova *[Statistical yearbook – City of Genoa]*. (1999). Publicatzione del Sistema Statistico Nazionale – Comune di Genova – Servicion Statistica *[Edited by Genoa Municipality]*.
Atlante demografico della città. *[Demographic Atlas of the city]*. Publicazione des Sistema Statistico Nazionale – Comune di Genova – Unità Organizzativa Statistica *[Edited by Genoa Municipality]*. (2000).
Bonaiuto, M., Aiello, A., Perugini, M., Bonnes, M., & Ercolani, A.P. (1999). Multidimensional perception of residential environment. Quality and neighbourhood attachment in the urban environment. *Journal of Environmental Psychology*, 19(4), 331-352.
Bonnes, M., & Bonaiuto, M. (1996). Multiplace analysis of the urban environment A comparison between a large and small Italian city. *Environment and Behaviour*, 28(6), 699-747.
Bronfenbrenner, U. (1979). *The ecology of Human development: Experiments by Nature and Design*. Cambridge: Harvard University Press.
Canter, D. (1988). Action and Place: an existential dialectic. In Canter, D., Krampen, M., Stea, D. (eds). *Ethnoscapes: environmental perspectives*, pp. 1-18. Aldershot: Gover.
Chavis, D.M., Wandersman, A. (1990). Sense of Community in The Urban Environment: A Catalyst for Participation and Community Development. *American Journal of Community Psychology*, 18(1), 55-81.
Chaskin, R. (1997). Perspectives on neighbourhood and community: a review of the literature. *Social Service Review*, 4, 521-547.
Cohen, R., & Siegel, A.W. (1991). *Context and development*, Hillsdale, NJ: Erlbaum.
Felton, B.J., & Shinn, M. (1992). Social Integration and Social Support: Moving "Social Support" Beyond and Individual Level. *Journal of Community Psychology*, 20, 103-115.
Kuo, F.E., Bacaiocoa, M., & Sullivan, W. (1998). Transforming inner-city landscapes. *Environment and Behaviour,* 3(1), 28-59.
Kuo, F.E., Sullivan, W., Coley, R., & Brunson, L. (1998). Fertile Ground for Community: Inner-City Neighbourhood Common Spaces. *American Journal of Community Psychology*, 26(6), 823-851.
McMillan, D.W., & Chavis, D.M. (1986). Sense of Community: A Definition and Theory. *Journal of Community Psychology*, 14, 6-23.
Moser, G. (1995). *Gli stress urbani [Urban stress]*. Milano: Edizioni Universitarie di Lettere Economia e Diritto.

Notiziario statistico *[statistical bulletin]*. Publicatzione del Sistema Statistico Nazionale – Comune di Genova – Servicion Statistica *[Edited by Genoa Municipality]*. (2001).

Perkins, D., Brown, B., & Taylor, R.B. (1996). The ecology of empowerment: Predicting Participation in Community Organization. *Journal of Social Issues*, 52(1), 85-110.

Perkins, D., Florin, P., Rich, R., & Chavis, D. M. (1990). Participation and the social and physical environment of residential blocks: Crime and community context. *American Journal of Community Psychology,* 18, 83-115.

Porta, S. (2000). La riconquista dello spazio pubblico urbano [Reclaiming urban public space]. *Quaderni di architettura*, 22.

Prezza, M., Costantini, S., Chiarolanza, V., & di Marco, S. (1999). La scala italiana del senso di comunità [The Italian sense of community scale]. *Psicologia della Salute*, 3(4), 135-159.

Prezza, M., & Pacilli, M.G., (2002). Il vicinato e la presentazione e validazione di uno strumento [Neighborhood and the presentation and validation of a measure]. In Prezza, M., Santinello, M., *Conoscere la comunità. L'analisi degli ambienti di vita quotidiana [Knowing the community. Everydas life context analysis]*. Bologna: Il Mulino.

Sarason, S.B. (1974). *The psychology sense of community.* San Francisco: Jossey-Bass.

Skyaeveland, O., & Garling, T. (1997). Effects of Interactional Space on Neighbouring. *Journal of Environmental Psychology*, 17, 181-198.

Skyaeveland, O., Garling, T., & Maeland, J.G., (1996). A Mutidimensional Measure of Neighbouring. *American Journal of Community Psychology*, 24(3), 413-435.

Trickett, E. (1996). A future for community psychology: the contexts of diversity and the diversity of contexts. *American Journal of Community Psychology*, 24(2), 209-234.

Environmental values and behavior in traditional Chinese culture

Chin-Chin (Gina) Kuo[1] & A. T. Purcell[2]

[1]Feng Chia University, Taichung, Taiwan and [2]University of Sydney, Australia

Abstract. Environmental values have been found to be good predictors of environmental behavior in the West by Western scholars. However, it is not clear whether values predict behaviors in other cultures. It is also not clear whether there are values and behaviors that are important to the environment in other cultures but are not part of Western cultures. The initial stage of this research was to study Chinese values and behaviors. In the research, we first identified twenty-seven environmental beliefs, values, and norms about nature and thirty-three environmental behaviors from ancient Chinese texts, scholarly works, and anecdotal materials such as newspapers in Taiwan. Experts in Chinese/Taiwanese culture, ecology, environmental design, and educators were then surveyed to establish whether they considered these beliefs, values, and behaviors do reflect traditional values and behavior in Chinese and Taiwanese culture. The survey data showed a high to medium level of agreement between the experts about the existence of environmental beliefs, values, and behaviors in ancient Chinese culture, from which five factors of values and six factors of behaviors were derived. The research to be reported represents the first step in a research program aimed at assessing the value and behavior relationship in a different cultural and geographic context: Chinese culture and different social groups located in Taiwan.

Keywords: Chinese environmental values, Chinese environmental behavior, environmental value and behavior relationship, value and behavior measuring scales, cultural factors

Introduction

Human behavior has been widely considered as playing a central role in the deterioration of our living environment (Maloney & Ward, 1973). If destructive behavior is to change it is necessary that what influences such behavior be understood. A considerable amount of research has attempted to address this question by looking for relationships between environmental behavior and variables such as attitudes (Ajzen & Fishbein, 1980) towards and knowledge (Hines et al., 1986/87) of environmental issues. These variables have been shown not to be good predictors of environmental behavior. The best predictor of environmental behavior has been found to be environmental values (Stern et al., 1995; Karp, 1996; Schultz & Zelezny, 1998).

One characteristic of this research on the values–environmental behavior relationship is that it has been carried out predominantly in western cultures. This raises the general issue of environmental values and behaviors in other cultures. This issue would appear to be particularly interesting in relation to specific cultures that have a traditional and rich relationship to nature. One such culture where this may apply is traditional Chinese culture. The aim of the research to be reported was to take a first step in addressing the question of the structure of traditional Chinese values and behavior related to their natural beliefs. This was part of a larger program that will examine the relationship between traditional values, if they are found, the typical western environmental values and environmental behavior in a number of different social groups in Taiwan. Taiwan represents a particularly interesting setting for such research as the Taiwanese culture is deeply embedded in Chinese culture, yet it suffers from environmental problems resulting from development. Parts of the traditional Chinese values and behavior might still exist in Taiwanese people, but may have altered slightly. A very significant aspect of the research therefore is whether different groups in this society hold different environmental values and engage in different environmental behaviors.

A two-stage procedure was employed to identify potential Chinese values and Taiwanese behavior items to be used in the research. In the first stage, traditional ancient Chinese texts /Taiwanese anecdotal materials such as newspapers were examined (for example, Confucian Analects) to identify potential values and behaviors about nature and the environment. In addition, Western and Chinese scholarly work relating to beliefs about nature in ancient Chinese culture were also examined (for example, Fang, 1973; Ip, 1983; Murphey, 1967). Considerable evidence was found relating to environmental beliefs, values, and behaviors in this literature. Careful analysis of the content of these beliefs, values, and behaviors indicated that they could be condensed into twenty-seven value and thirty-three behavioral statements. In the second stage, forty-nine experts in areas relating to traditional Chinese beliefs and culture in Taiwan were surveyed to establish whether they considered the statements did reflect traditional Chinese values. Seventy-seven experts in areas relating to Taiwanese culture, environmental design, and education, as well as ecology in Taiwan were surveyed to establish whether they considered the statements reflected traditional Taiwanese behaviors related to their nature beliefs. The finding of this research corresponds to the various constructs and theories found in the review of ancient Chinese texts as well as literature of Western and Chinese scholarly research.

Part A:
Existence of environmental values in Chinese culture

Method

The purpose of the first part of the research was to find the structure of environmental values in Chinese culture. In the first stage, abundant evidence was found in the search of Chinese environmental values. Table 1 shows the list of values that were identified (The list also includes the source for the items). A number of items were included that relate to values that are familiar to Taiwanese people but where no direct supporting evidence was found (numbers 4 and 10). Similarly, a number of items were included (numbers 20-24) that relate to salient beliefs and values in Chinese culture, such as altruistic value orientation which have no overt environmental content but directly or indirectly affect the environment and environmental behavior.

Survey of experts' opinions

In the second stage, the opinions of the experts were surveyed. The list of 27 belief and value items identified in Table 1 were converted into a statement questionnaire format. It was supplemented by two western value statements (Nos.28 and 29) as check items to be judged by Chinese experts against Chinese values without increasing the item list too much. Item No.29 *scientific study of nature* was included because it had been regularly shown to be important in Western literature, while item No. 28 *environmental law* was included because there is a growing concern in the West of the citizens' right to use law to control environmental issues. The questionnaire used the 5-point Likert agree/disagree scale with responses *strongly agree, agree, neutral, disagree, strongly disagree* (and uncertain). The experts were asked to indicate on the scale how much they agree with the existence of each of the beliefs and values in traditional Chinese culture. A score of 5 means *strongly agree* and a score of 1 means *strongly disagree*. In addition, the experts were also asked to write down any other traditional values that they thought were missing from the list.

Producing a Chinese version of the questionnaire

The survey instrument was written in English. It was then translated into Chinese as accurately as possible. In order to minimize any translation biases, native English speakers as well as an English-speaking linguist evaluated the English version of the questionnaire. The translated versions were evaluated by a number of independent bilingual experts and an environmentalist who all speak and write fluent English and Chinese. The Chinese version was reviewed by a Chinese scholar to avoid ambiguity. A preliminary test of the questionnaire was made among several Taiwanese and Chinese postgraduate students at the University of Sydney. A small number of changes in wording resulted from this process.

Table 1. List of Chinese beliefs, values and norms[1] related to nature

1	Belief of nature as a self-sustaining organism (Baynes, 1968; Ip, 1983)
2	Belief of nature's ability to keep balance (Baynes, 1968)
3	Belief of man-nature interdependence (Chung,1999)
4	Belief of unlimited natural resources
5	Value of harmony and balance with nature (Murphey, 1967; Lin Jun-Yih, 2000)
6	Value of Feng Shui[2] for own children and descendants (Han, 1983; 1987)
7	Value of beauty of nature (Liu, 1970)
8	Belief of sacredness of nature and value of reverence of nature (Liu, 1970)
9	Value of spiritual and affective interaction with nature (Fang, 1973)
10	Value of approach to nature based more on emotion than rationality
11	Value of contact with nature for the body and soul (A common theme in arts)
12	Value of love, respect and appreciation of nature (Murphey, 1967)
13	Value of sense of belonging, place-attachment, and land identity (A theme commonly reflected in literature)
14	Norm of observing natural laws (Chung, 1999)
15	Value of limited exploitation of nature (Murphey, 1967; Chung, 1999)
16	Value of adaptation to the environment-an attitude of wu-wei or avoiding interference with the way nature operates (Marshall, 1994; Tuan, 1970)
17	Value of the general practice of Feng Shui (Han, 1983; 1987)
18	Value of equality of all living forms (Gardner & Stern, 1996)
19	Value of respect for all life forms (Gardner & Stern, 1996)
20	Values of compassion, benevolence, and mercy (Marshall, 1994)
21	Values of self-restraint of desires and needs (Chinese cultural connection, 1987)
22	Values of contentedness and appreciation for things (Fang, 1973)
23	Values of simple lifestyle, minimalism of material life (Confucian Analects)
24	Values of a denial of selfishness, little self-interest, individual subservience to public goods, altruism (Murphey, 1967; Marshall, 1994)
25	Value of order and integrity in the social environment (Murphey, 1967)
26	Norm of acting according to natural laws rather than human law (Marshall, 1994)
27	Norm of regulation of behavior based on morality and ethics (Fang, 1973)

---

[1] Conceptual distinction among beliefs, values, and norms. An environmental belief is what one accepts as true in his/her relationship with environment. It can exist separately from behavior, i.e. it may or may not cause behavior to change (Milbrath, 1984). An environmental value is an environmental belief and a guiding principle for behavior in the environment (Schwartz, 1994). A norm is a customary behavior. There are two kinds of norms: 1) A descriptive norm is what most people do 2) An injunctive norm is what most others approve or disapprove. It refers to rules or beliefs as to what constitutes morally approved and disapproved conduct (Cialdini et al.1990). Moral, in the discipline of ethics, refers to questions concerning the well-being of other human beings or other living creatures, not just oneself (Kempton et al., 1995). Both values and norms have behavior components, but a belief does not have a behavioral component, although a forced behavior change might cause a change in belief (Milbrath, 1984). It is hard to distinguish values from norms here as Chinese are highly concerned with moral conduct and emphasize the consistency of knowledge and behavior (Fang, 1973).

[2] Originally Feng Shui (geomancy) is the Chinese view of the cosmological system but later it is developed into an art of organizing the surroundings to enhance one's health, wealth, success, and happiness. Literally, Feng means wind, Shui means water. They are the cosmic energy. The Chinese believe that the cosmic energy flows everywhere like wind and water. The art teaches you how to use the energy to maximize your health, fortune, and happiness.

Procedure

The drop and collect method was adopted in this survey. After the preliminary test, the questionnaires were given to a professor or an administrative assistant in the departments of Chinese literature or humanities in 5 different universities in Taiwan to help distribute and collect the questionnaires. A total of 74 questionnaires were distributed at the end of February, 2001 and collected within a month. Because of the relationship of the professors or assistants in charge of distribution with the participants, the return rate was quite high. A total of 60 questionnaires were returned (a return rate of 81%). The respondents who were not specialists in Chinese literature and philosophy (for instance Chinese history, or other majors) or who did not fill in their educational background were excluded from the data. This left a total of 49 valid questionnaires by respondents that specialize in either Chinese literature or philosophy. Out of the 49 respondents, 49% of the participants had a PhD degree in either Chinese literature or philosophy, 36.7% a master degree, and 7% a bachelor degree; and 77.6% of the respondents specialized in Chinese and/or Taiwanese literature; 20.4% in Chinese philosophy, and 2% in other field.

Results

Descriptive statistics

The data for this study was the responses of 49 respondents of highly educated experts in Chinese literature and philosophy. Each participant is represented by a score on each of the 29 belief and value items. To process the data, means and standard deviations for each of the sample items were calculated and are arranged from high to low according to the rank of their mean scores in Table 2. On the 5-point Likert scale, means above 4 are taken as indicating high level of consensus between the experts on the existence of the beliefs and values in Chinese culture; means between 4 to 3 indicate a medium value and a fair amount of experts' agreement of the existence of the items in Chinese culture; means between 3.00 to 2.50 indicate a neutral response or low value and that the items might exist only in a certain number of people. Items with mean scores below 2.50 indicate that the items are perceived by the experts as barely existent in the Chinese culture. Among the 29 items, 19 items (Nos.1, 7, 2, 13, 11, 14, 4, 9, 15, 20, 22, 12, 6, 23, 21, 5, 10, 8, 17) have high mean scores above 4.00, and 8 items (Nos. 16, 27, 24, 26, 25, 3, 19, and 18) have medium mean scores between 4 and 3.00. The two supplementary items, no. 28 environmental laws and 29 scientific study of nature, adopted from Western values as check items had the lowest mean scores, with scores at 2.55 and 2.33, respectively. In general, we have found a significant degree of consensus among the experts regarding the existence of the values in Chinese culture with the exception of the two check-items, which clearly shows that they are not salient traditional Chinese values.

The experts were also asked to write down any other environmental values relating to their culture that were missing from the list. The purpose of this open-ended question was to find out environmental beliefs and values which may have been overlooked. The majority of the experts did not respond to the question. However, one important norm and behavior relating to their beliefs and values of nature- the religious ritual of worshipping nature was suggested. The ritual of worshipping nature, which was actually mentioned quite often in old Chinese texts such as Li Chi will therefore be incorporated into the next survey of the existence of Chinese environmental behavior. A list of Chinese environmental beliefs, values and norms was thus created based on experts' opinions.

Table 2. Means and *SD*s of experts' opinions of the existence of traditional Chinese values

Rank	Item No.	Value and belief items	Mean	SD
1	1	Belief of nature as a creative living organism	4.74	.44
2	7	Value of beauty of nature	4.58	.58
3	2	Belief of nature's ability to keep balance	4.57	.61
4	13	Value of sense of belonging, land identity and place attachment	4.57	.50
5	11	Value of contact with nature for the body and soul	4.52	.58
6	14	Norm of observing natural laws	4.51	.58
7	4	Belief of man-nature interdependence	4.46	.77
8	9	Value of spiritual and affective interaction with nature	4.44	.68
9	15	Value of limited exploitation of nature	4.38	.64
10	20	Values of compassion and benevolence	4.31	.71
11	22	Values of contentedness and appreciation of things	4.27	.71
12	12	Value of love, respect and appreciation of nature	4.25	.86
13	6	Value of Feng Shui for own children and descendants	4.21	.85
14	23	Value of simple lifestyle	4.18	.60
15	21	Value of self-restraint of desires and needs	4.17	.71
16	5	Value of harmony and balance with nature	4.15	.97
17	10	Value of approach to nature based more on emotion	4.15	.69
18	8	Belief of sacredness of nature	4.07	1.08
19	17	Value of the general practice of Feng Shui	4.06	.82
20	16	Value of adaptation to nature	3.98	.79
21	27	Norm of morality and ethics	3.98	.87
22	24	Value of self-sacrifice and altruism	3.98	.94
23	26	Norm of regulation of behavior on natural laws	3.96	.89
24	25	Value of the order in the social environment	3.92	.90
25	3	Belief of unlimited natural resources	3.72	1.25
26	19	Value of respect of all living forms	3.67	1.02
27	18	Value of equality of all living forms	3.14	1.10
28	29	Value of scientific study of nature	2.55	1.19
29	28	Values of regulation of behavior on man-made laws	2.33	1.12

Factor analysis

The 29 items were then factor analyzed using principal axis factoring with a varimax rotation. The results of the factor analysis revealed that nine factors had eigenvalues larger than 1 after varimax rotation. However, after careful analysis of the interrelation of the constructs in the nine factors, only 5 factors or sets of values have Alpha values greater than 0.7, each accounting for 15.05, 10.66, 9.62, 8.93, and 6.80 percent of the matrix variance respectively and 51 percent of the total variance. The scree plot was also used to check their presence. The five factors are labeled as: Factor 1: Harmony and balance with nature; Factor 2: Simple lifestyle; Factor 3: Spiritual communion with nature; Factor 4: Altruistic and benevolent norms; and Factor 5: Feng Shui. Table 3 shows the five factors and items associated with each factor together with the factor loadings.

Table 3. Factor analysis of experts' opinions of the existence of Chinese Environmental Values

	Environmental belief, value factors and items	Factor loading	Cronbachs' Alpha
Factor 1:	**Harmony and balance with nature**		.89
1	Respect of all living forms	.79	
2	Harmony and balance with nature	.77	
3	Man-nature interdependence	.71	
4	Equality of all living forms	.71	
5	Natural laws as behavior regulations	.62	
6	Love and respect of nature	.55	
Factor 2:	**Simple lifestyle**		.80
1	Simple lifestyle and minimalism of material life	.81	
2	Contentedness and appreciation of things one possesses	.67	
3	Self-restraint of desires and material needs	.63	
Factor 3:	**Spiritual communion with nature**		.86
1	Spiritual and affective interaction with nature	.80	
2	Limited exploitation of natural resources	.70	
Factor 4:	**Altruistic and benevolent norms**		.86
1	Self-sacrifice and altruism	.71	
2	Compassion, benevolence and mercy	.66	
3	Nature's ability to keep balance	.66	
4	Nature as self-sustaining organism	.55	
Factor 5:	**Feng Shui**		.78
1	The general practice of Feng Shui	.77	
2	Practice of Feng Shui for own children	.65	

Developing traditional Chinese value scale

The first two values from each factor that had mean scores above 3.00 and factor loadings greater than 60 comprise the final set of Chinese value scale for future study. These items are:
(1) Respect of all living forms
(2) Harmony and balance with nature
(3) Simple lifestyle and minimalism of material life
(4) Contentedness and appreciation of things

(5) Spiritual communion with nature
(6) Limited exploitation of natural resources
(7) Self-sacrifice and altruism
(8) Compassion, benevolence, and mercy
(9) The general practice of Feng Shui
(10) Practice of Feng Shui for children

PART B:
Existence of environmental behavior in Chinese culture

Method

Procedure

A two-stage procedure similar to the one developed for the value study was employed in the study of the existence of Taiwanese behaviors. The following differences are variations to the basic procedure that were introduced for the behavior study.

(1) As environmental behavior is a field seldom looked at in the academic scholarly works in Taiwan, and there are fewer materials in English, more Taiwanese texts and anecdotal materials such as newspapers were searched. Observations were made of peoples' behaviors which may still carry traditional ways of behavior. Table 4 shows the list of the behavior items identified.

(2) The survey was written in Chinese first rather than English, using the Western questionnaire format. It was then reviewed by a number of scholars in Taiwan, specializing in Taiwanese culture, ecology or other disciplines, who also had been concerned for a long time with Taiwanese environment.

Taiwanese culture has undergone a drastic change during the past 50 years as a result of political pressure and the rapid development of the economy. As a result, the majority of subjects chosen for this survey were in an age greater than 40 years or were people who grew up with the culture, and knew the traditional culture well. Also scholars with specializations in Chinese and Taiwanese culture, environmental design, and education as well as scholars that were often engaged in environmental activities were chosen for this study. This survey was conducted in the departments of Chinese/Taiwanese literature, humanities, and architecture in 7 different universities. A total of 112 copies of the questionnaire were distributed in early December and were returned around the end of December, of which 90 were returned (a return rate of 80%). Out of the valid 77 respondents, 61% of the participants had a PhD, 28.6% a master degree, and 10.4% a bachelor degree, and 50.6% of the respondents specialized in Chinese and/or Taiwanese culture; 22.1% in environmental design or education, 9.1% in natural science or ecology, and 18.2% in other fields. Among these respondents, 75.4% of the respondents are over 40-years old meaning that quite a high percentage of respondents that had grown up with the traditional culture.

(3) The responses of some items (such as domestic worshipping behavior) were scanned to check if the subjects know the culture well. The responses which indicated a lack of knowledge of the culture were treated as invalid data.

Table 4. Evidence of traditional Chinese and Taiwanese behaviors related to their beliefs about nature

1. Worshipping rituals to heaven (nature), ancestors and deities to protect the family and descendants (a common practice still carried out in Taiwan now)
2. Worshipping rituals to heaven (nature), ancestors and deities to protect residence and work place (small-scale environment) (a common practice still carried out in Taiwan now)
3. Worshipping rituals to heaven (nature), ancestors and deities to protect the community (There used to be a temple in each community or a family temple in the extended family compound in ancient time, and worshipping rituals were carried out by the elderly in the community)
4. Worshipping rituals to heaven (nature), ancestors and deities to protect their country (A practice used to be performed by the emperor in China, Murphey, 1967)
5. Worshipping rituals to heaven (nature), ancestor and deities to protect the environment of humans-A practice used to be performed by the emperor in China, and now sometimes still practiced by government officials in Taiwan (The Commons Daily, June 7, 1999)
6. Donating money to temples (a common practice still carried out in Taiwan now)
7. Burning paper money to heaven, deities and the dead (a practice still carried out in Taiwan)
8. Ecological engineering and house construction methods
9. Practice of Feng-shui (the geomancy) in building houses (Han, Pao-The, 1983)
10. Practice of Feng-shui in choosing graveyards for the deceased (Han, Pao-The, 1987)
11. Decorating with symbols of good fortune and symbols to expel evil spirits (Guo, 1995)
12. Following traditional farmers' calendar for important family activities and events such as moving, getting married etc. (Chung, 1999)
13. A thorough cleaning of the house at the end of each year (a customary practice)
14. Tomb sweeping (a common practice still carried out in Taiwan now)
15. Practicing vegetarianism (a practice by certain religious groups)
16. Engaging in activities to maintain their good health and prolong their lives (a practice by many people in the general community)
17. Performing acts of releasing animals caught by humans back to nature as a practice of good deeds (a common Buddhists' practice in ancient time)
18. Wu-Wei-doing nothing to interfere with nature (Gardner & Stern, 1996, Tuan, 1970)
19. Adapting to the environment (a traditional Chinese norm, Chinese Culture Connection, 1987)
20. Leading a simple lifestyle (Confucian Analects)
21. Acts of love and appreciation of the things one possesses (Confucian Analects)
22. Close contact with nature in order to appreciate its beauty (a common theme in arts)
23. Close contact with nature in order to find emotional balance (a common theme in arts)

Table 4. continued

24 Close contact with nature to revitalize their health (a common practice in general community)
25 Communion with nature to find spiritual inspiration (a common theme in arts)
26 Practicing meditation (a practice by certain religious groups)
27 Cleaning the community (voluntary acts performed by certain people)
28 Performing acts of community service (voluntary acts performed by certain people)
29 Identifying with ancestors' origins
30 Identifying with one's birthplaces
31 Identifying with the places where one had once lived
32 Maintaining good neighborhood relationship and community watch
33 Returning to homeland

Results

Descriptive statistics

The data was the responses of 77subjects of highly educated experts in Taiwanese culture, environment and ecology. Each participant is represented by a score on each of the 33 behavior items. To process the data, means and standard deviations of the 33 items for each of the statements were calculated and are arranged from high to low according to the rank of their mean scores in Table 5. On the 5-point Likert scale, means above 4 are taken as indicating high level of consensus between the experts on the existence of the behaviors in Taiwanese culture; means between 4 to 3 indicate medium values and a fair amount of experts' agreement of the existence of the items in the Taiwanese culture; means between 3.00 to 2.50 indicate neutral responses or low values and that the items might exist in only a small number of people. Items with mean scores below 2.50 indicate that the items were perceived by the experts as barely existent in the Taiwanese culture.

Among the 33 items, 10 items (Nos. 13, 1, 14, 2, 6, 10, 12, 7, 9, and 11) had high mean scores above 4.00; 20 items (Nos. 29, 32, 30, 21, 33, 16, 3, 31, 17, 24, 20, 8, 22, 28, 23, 5, 4, 15, 19, 25) had medium mean scores between 3.99 to 3.00; and 3 items (No. 27, 18, and 26) had low mean scores from 2.79 to 2.55. In general, we have found a reasonable degree of consensus among the experts regarding the existence of the environmental behaviors in Taiwanese culture with the exception of three low-scored items which might exist only in small amount of people in the culture. Thus, a picture of the existence of Taiwanese behaviors related to their belief about nature was revealed.

Factor analysis

The 33 behavioral items in the questionnaire were factor analyzed to identify underlying structures of behaviors, using principal axis factoring with a varimax rotation. The results of the factor analysis, shown in Table 6, revealed that nine factors had eigenvalues larger than 1 after varimax rotation. However, after careful analysis of the interrelation of

Table 5. Means and *SD*s of experts' agreement of the existence of Taiwanese environmental behaviors

Rank	Item No.	Value and belief items	Mean	SD
1	13	A thorough cleaning of the house at the end of the year	4.61	.57
2	1	Worshipping for one's family and descendants	4.56	.55
3	14	Tomb sweeping	4.51	.77
4	2	Worshipping for one's residence and work place	4.45	.62
5	6	Donating money to temples	4.35	.72
6	10	Practice of Feng-Shui in choosing graveyards	4.35	.71
7	12	Following traditional farmers' calendar	4.32	.77
8	7	Burning paper money	4.29	.92
9	9	Practice of Feng-Shui in the house	4.08	.74
10	11	Decorating with symbols	4.00	.86
11	29	Identifying with ancestors' origins	3.91	.91
12	32	Maintaining neighborhood relationships.	3.81	.87
13	30	Identifying with one's birthplace	3.79	.81
14	21	Love and appreciation of one's possessions	3.77	1.18
15	33	Returning to homeland	3.77	1.07
16	16	Maintaining good health to prolong life	3.68	.95
17	3	Worshipping for one's extended family and community	3.65	1.20
18	31	Identifying with the places where one had lived	3.61	.89
19	17	Liberating creatures back to nature	3.43	.97
20	24	Contact with nature for health	3.42	1.15
21	20	Leading a simplistic lifestyle	3.42	1.12
22	8	Ecological building design and construction	3.40	1.29
23	22	Contact with nature for its beauty	3.37	1.13
24	28	Performing acts of community service	3.26	.88
25	23	Contact with nature for emotional balance	3.27	1.22
26	5	Worshipping for human living environment	3.21	1.18
27	4	Worshipping for the country	3.17	1.18
28	15	Practicing vegetarianism	3.14	1.13
29	19	Adapting to the environment	3.13	1.24
30	25	Communion with nature for inspiration	3.09	1.09
31	27	Help cleaning the community	2.79	1.08
32	18	Doing nothing to interfere with nature	2.70	1.39
33	26	Practicing meditation	2.55	1.03

the constructs in the nine factors, only 6 factors or sets of values have Alpha values greater than 0.7. The scree plot was also used to check the presence of the factors. The six factors are labeled as: 1. Religious Rituals for Family Environment; 2. Contact with Nature; 3. Place Identity; 4. Religious Rituals for the Large-Scale Environment (e.g., country, universe); 5. Simple Lifestyle Practices; 6. Religious Altruistic Practices, which accounts for 13.35, 12.82, 8.61, 7.20, 7.20, and 6.74 percent of the matrix variance respectively and 55.90 percent of the total variance. Under these six factors, a number of behavior items related to Taiwanese belief about nature and the environment were identified and the factors associated behavior items together with the factor loadings are shown in Table 6.

Table 6. Factor Analysis of experts' opinions of the existence of Chinese and Taiwanese environmental behaviors

Environmental belief, value factors and items	Factor loading	Cronbachs' Alpha
Factor 1: Religious rituals for family environment		.83
1 Worshipping for one's family and descendants	.86	
2 Worshipping for one's residence and work place	.83	
3 Following traditional farmers' calendar	.72	
4 Practice of Feng-Shui in choosing graveyards	.72	
5 Donating money to temples	.56	
Factor 2: Contact with nature		.92
1 Contact with nature for emotional balance	.92	
2 Contact with nature for health	.88	
3 Contact with nature for its beauty	.83	
4 Communion with nature for inspiration	.75	
5 Adapting to the environment	.54	
Factor 3: Place Identity		.78
1 Identifying with the places where one had lived	.82	
2 Identifying with one's birthplace	.66	
3 Returning to homeland	.60	
4 Maintaining neighborhood relationships and community watch	.55	
Factor 4: Religious rituals for larger-scale environment		.85
1 Worshipping for the country	.88	
2 Worshipping for one's extended family and community	.77	
3 Worshipping for human living environment	.65	
Factor 5: Simple lifestyle practices		.78
1 Leading a simple lifestyle	.80	
2 Acts of love and appreciation of things one possesses	.60	
3 Ecological building design and construction	.57	
Factor 6: Religious and altruistic practices		.76
1 Practicing meditation	.70	
2 Help cleaning the community	.60	
3 Performing acts of community service	.51	
4 Practicing vegetarianism	.50	

Developing traditional Taiwanese behavior scale

From each factor we adopted the first two items that have a high factor loading greater than 6 and a mean greater than 3.00 to form the Taiwanese behavioral scale with some exceptions. Three items were taken from Factor 2 (Contact with Nature) because the concepts of contact with nature for health, for emotional balance and for beauty usually are a concept in Chinese culture. One item was taken from Factor 4 (Ritual Practices for Large-Scale Environment) as the second behavior-worshipping ritual for the extended family and family community are no longer existent because of the destruction of the old family structure and communities in modern era. Two items practicing meditation and help cleaning the community were taken from Factor 6 (Religious and Altruistic Practices). The only reason to adopt these two items from factor 6 which had quite low means but high factor loadings is that these behaviors might not exist in the majority of the people in the culture, but they might exist in certain minority groups of people (e.g., Zen Buddhists). It would be interesting to find out which minority groups of people would be involved in these kinds of behaviors and how their behaviors correlate with environmental values and other pro-environmental behaviors. On the basis of these results, a set of behaviors relating to Chinese beliefs and values about nature was constructed. This set of items, shown below, will now be used in combination with the behavioral scales from Western research to measure the environmental behavior of different social groups in Taiwan.

(1) Worshipping for one's family and descendants
(2) Worshipping for one's residence and work place
(3) Contact with nature for emotional balance
(4) Contact with nature for health
(5) Contact with nature for its beauty
(6) Identifying with the places where one had lived
(7) Identifying with one's birthplace
(8) Worshipping for the country
(9) Simplistic lifestyle practices
(10) Acts of love and appreciation of one's possessions
(11) Practicing meditation
(12) Help cleaning the community

Discussion

A difficult part in doing this cross-cultural study is the use of Chinese language in Western questionnaires. Although abundant evidences of values could be found in Chinese texts, several experts expressed the difficulty in responding to the questions on the existence of the values in Chinese culture. This was due to the following facts: China had such a large area of land, diversity of ethnic groups and extensive history that some values might have existed only in certain areas of the country, or among certain groups of people or for a certain time period. The respondents seemed to be puzzled by the question concerning the existence of the ideology in Chinese culture and the generality of the values in the real life of Chinese people. Nevertheless, the majority of the experts seemed to have less difficulty and reached an agreement in their minds about the existence of the values in Chinese culture throughout its extensive history after the question was rephrased as a statement of the existence of the concept. Another problem encountered in the survey is that several experts did not answer appropriately to the literal translation of the "existence" of certain popular behaviors. For example, the answer to the question "Do you agree with the existence of the worshipping behavior/Feng Shui in Taiwanese culture?" should be quite clear since the behavior was very popular in traditional culture. However, quite a few experts disagreed or strongly disagreed with the behavior. This resulted in several invalid questionnaires in the preliminary test and formal survey. The inappropriate answers might be due to the following:

(1) The subjects did not read the instructions carefully. They responded as if they didn't agree with the moral aspect of the behavior, instead of the existence of the behavior.
(2) Problems with the usage of the Chinese language in Western questionnaire format. The Taiwanese do not seem to be used to the sentence structure of the agreement/disagreement of the existence of the value or behavior items. The words *agree or disagree* in Chinese seemed to be used directly with opinions, conduct or statements of facts. In this case, some people directly responded to whether they agree with the moral aspects of the behavior instead of the existence of the behavior. For example, the original question *"Do you agree with the existence of the following environmental behavior (Feng Shui practice for instance) in traditional Taiwanese culture?"* was revised several times in Chinese until it was reworded to sound like a statement.
Do you agree with the following statement about the existence of environmental behavior in traditional Taiwanese culture?
"In traditional Taiwanese culture, people practiced Feng Shui."

This research has identified the existence of environmental values and behaviors in Chinese culture. However, some dimension of values (such as value orientation for self, for others, or for biosphere) may require further study, for some people may exhibit different environmental value and behavior depending on the objects of their values. For example, the mean score of item no.17, the general practice of Feng Shui, ranked 19 (with a mean of 4.06), but item no. 6, Feng Shui for own children, with value orientation added it ranked 13 (with a mean 4.21). Similarly, the value of *compassion and benevolence* received a rather high mean score of 4.31. However, value item no. 19 *respect of all living forms* and item no. 18 *equality of all living forms* received a somewhat lower score of 3.67, and 3.14 respectively. One would question if the compassion and benevolence were extended to human beings rather than to other living forms, i.e. the value of anthropocentrism would rank higher than that of eco-centrism. In other words, a hierarchy of valuing self than others and the biosphere seems to exist in Chinese culture.

Although we have identified a set of behaviors related to Taiwanese beliefs about nature, some of the items such as worshipping the deities do not seem to be pro-environmental in the modern sense. In some cases, some of them are even environmentally unfriendly. For example, behavior such as contact with nature may be healthy to human beings, however, excessive mass activities might be destructive to the natural environment. Whether they are all pro-environmental and in what way can they be of help to the environment needs to be studied. A further study to compare these traditional Chinese and Taiwanese environmental behaviors with the Western pro-environmental behaviors is suggested.

Summary and implication of the study

Culture, both Eastern and Western, are important dimensions of the study. In each culture, there are unique characteristics which may complement each other. Using the evaluation rating scales consisting of both Chinese and Western value and behavior items (those developed in this study and those adopted from the Western scholars), we are now ready to proceed with the testing of different social groups in Taiwan.

References

Ajzen, I. & Fishbein, M. (1980). *Understanding attitudes and predicting social behavior.* Englewood Cliffs, NJ: Prentice-Hall.
Baynes, Cary F. (1968). *The I Ching, or, Book of changes: the Richard Wilhelm translation.* Melbourne : Routledge & Kegan Paul.
Chinese Culture Connection (1987). Chinese values and the search for culture-free dimensions of culture. *Journal of Cross-Cultural Psychology,* 18, 143-164.
Chung, Ding-Mao (1999). *The analysis of Chinese environmental ethics.* Taipei: Royal Library Co. Ltd. [in Chinese].
Claldini, R. B., Reno, R. R., & Kallgren, C. A. (1990). A focus theory of normative conduct: recycling the concept of norms to reduce littering in public places. *Journal of Personality and Social Psychology,* 58, 1015-1026.
The Commons Daily (1999), June 7.
Confucian Analects. (1956). (Pound, Ezra. Trans.). London: Peter Owen Limited.
Fang, Thome H. (1973). A philosophical glimpse of man and nature in Chinese culture. *Journal of Chinese Philosophy.* (1):3-26. Honolulu, HI.
Gardner, G. T., and Stern, P. C. (1996). *Environmental problems and human behavior.* Needham Heights, MA: Allyn & Bacon.
Guo, Daiheng (1995). Chinese traditional architecture and culture, *Proceedings: the International Conference on Chinese Architectural History*, Hong Kong: The Chinese University of Hong Kong.
Han, Pao-The (1983). A study of Feng Shui as a Chinese concept of the environment. *Bulletin of Environmental Studies,* National Taiwan University, (2)1, 123-150. (In Chinese).
Han, Pao-The (1987). A study of taboos in Feng-Shuei practice of home building *Bulletin of Architecture and City Planning (*National Taiwan University, (3), 5-55. (in Chinese)
Hines, J. M. Hungerford, H.R. & Tomera, A.N. (1986/87). Analysis and synthesis of research on responsible environmental behavior: a meta-analysis. *Journal of Environmental Education,* 18, 1-8.
Ip. P. (1983). Taoism and the foundations of environmental ethics. *Environmental Ethics*, 5, 335-343.
Karp, D. G. (1996). Values and their effects on pro-environmental behavior. *Environment and Behavior,* 28, 111-133.
Kempton, W., Boster, J.S. & Hartley, J. A. (1995). *Environmental values in American culture.* Cambridge, MA: MIT Press.

Lin, Jun-Yih (2000). In search of the concept of the harmony between nature and man in traditional China: A critique. *Journal of the Studies in Dialectics of Nature*, (9) [in Chinese]

Liu Wu-Chi (1970). Moral and aesthetic values in Chinese literature: An historical survey. *Tamkang Review,* (1), 1.

Maloney, M. P. & Ward, M. P. (1973). Ecology: Let's hear from the people. An objective scale for the measurement of ecological attitudes and knowledge. *American Psychologist*, 28, 583-586.

Marshall, Peter. (1994). *Nature's Web: Rethinking our place on earth.* N.Y.: Paragon House.

Milbrath, Lester W. (1984). Environmentalists: vanguard for a new society. Albany: State University of New York Press.

Murphey, Rhoads (1967). Man and nature in China. *Modern Asian Studies*, I, 4, 313-333.

Schultz, P. Wesley & Zelezny, Lynnette (1998). Values and pro-environmental behavior. A five-country survey. *Journal of Cross-Cultural Psychology*, 29(4), 540-558.

Schwartz, S. H. (1994). Are there universal aspects in the structure and contents of human values? Journal of Social Issues, 50, 19-45.

Stern, P.C., Dietz,T., Kalof, L., & Guagnano,G. (1995). Values, beliefs and pro-environmental action: Attitude formation toward emergent attitude objects. *Journal of Applied Social Psychology, 25,* 1611-1636.

Tuan, Y.-F. (1970). Our treatment of the environment in ideal and actuality. *American Scientist*, 58, 244-249.

Some sociodemographic and sociopsychological predictors of environmentalism

Maaris Raudsepp
Tallinn Pedagogical University, Estonia

> **Abstract.** The study aims to reveal the relative impact of various sociodemographic and sociopsychological variables on empirical indicators of environmentalism (ecological behavior, environmental attitudes, and beliefs). Standard multiple regression models were used for analyzing the data of a questionnaire study of a representative sample of an Estonian rural subpopulation ($N = 440$). Among socio-demographic variables, *age, sex, education*, and *subjective religiosity* were significant predictors of environmentalism. Among socio-psychological variables, *general values*, perceived *control* over the environment, *local identity*, as well as *nature experiences in childhood* had a significant impact on various indicators of environmentalism. The level of family *income* and perceived local pro-environmental *norms* had insignificant effects. Different measures of environmentalism were predicted by different patterns of independent variables.
>
> **Keywords:** values, environmental attitudes, ecological behavior, local identity, nature experiences

Theoretical framework

I proceed from a broad definition proposed by Milton (1996, p. 33): "environmentalism is a concern to protect the environment through human effort and responsibility (…) wherever and in whatever form it exists." Environmentalism is usually operationalized as a certain way of thinking about the environment (beliefs and attitudes concerning nature and human-environment relations) and/or a practical way of relating to it (ecological behavior, e.g., self-restriction in consumption, participation in ecological movements, willingness to sacrifice for environmental quality).

In psychology models for studying environmentalism are most often based on various attitude theories. One group of models concentrates on the mechanisms that connect attitudes and behavior, assuming a causal chain starting from most general beliefs, proceeding to more specific attitudes, which cause behavior intentions and/or actual behavior, e.g., theory of reasoned action, theory of planned behavior (Ajzen & Fishbein, 1980), and norm activation model (Schwartz, 1977). Another group of theories concentrates on the content and structure of general regulators of environmentalism, e.g., theories of values (Schwartz, 1992) and models of environmental beliefs (Dunlap & Van Liere, 1978). The main organizing principles of contemporary environmental worldview can be described as the opposition of the so-called new environmental paradigm (NEP) to the dominant paradigm. The latter is a system of beliefs in limitless economic growth, human exceptionalism, and justified domination over the natural world, as well as beliefs in abundance of ecological resources and resilience of nature. An alternative environmental paradigm is based on an image of a fragile and threatened environment that needs human protection and requires the restriction of human expansionist activities; environment is regarded as an ecological whole with humans as only a small part of it. These opposing orientations may also be described as anthropocentric and ecocentric worldviews (Eckersley, 1992). Models of attitude-behavior relationships link these general systems of beliefs to ecologically significant activities.

Some authors (e.g., Corraliza & Berenguer, 2000) claim that attitude theories are insufficient for explaining environmentalism. In addition to attitudinal variables, situational and macrosocial factors should also be included into explanatory models.

Recently an integrative model of environmental behavior has been presented by Stern (2000). It includes attitudinal factors, personal capabilities, contextual (interpersonal), and social structural forces as determinants of environmentalism. The model postulates several distinct hierarchically related classes of variables: 1) social structural factors (variables that reflect position in the social structure, as well as institutional constraints and early socialization experiences); 2) general worldview, values and general beliefs about human-environment relations; 3) specific beliefs and attitudes about environmental issues; 4) behavior commitments, intentions, and environmentally relevant behavior. The causal chain inside the attitudinal domain moves from relatively stable and general elements of personality and belief structure to more specific beliefs about the environment, to beliefs about the consequences of an action, personal responsibility, and personal norm to take pro-environmental action. Different types of causal factors may interact (e.g., contextual or personality factors may promote or constrain the attitude-behavior associations). It is supposed that different types of ecological behavior are associated with a specific set of determinants. This approach regards environmentalism as a joint product of social structural, socialization, and social psychological processes, considering both attitudinal and contextual factors as relevant. Dietz et al. (1998) have made the first effort to assess inductively the relative impact of various social structural and sociopsychological factors in shaping environmentalism, using regression analysis of the data from a representative national sample. In general, social psychological variables appeared to have greater explanatory power than social structural variables, but their effect varied for different indicators of environmentalism (self-reported behavior and environmental beliefs).

Although some authors (e.g., Dietz et al., 1998) argue that there is little theoretical argumentation about why various **sociodemographic variables** would influence environmentalism, there are plenty of empirical evidence about the links between sociodemographic variables and the environmental concern or ecological behavior. Such variables like age, education, gender, place of residence, and political preferences have often shown strong and consistent relations with environmentalism (younger, more educated, women, members of minority groups, urban residents and politically liberal expressing usually more environmental concern), whereas income, class membership, occupation, and religious beliefs have shown weak and inconsistent relationships with environmentalism (Dietz et al, 1998; Zelezny et al, 2000). Other authors (e.g., Greenbaum, 1995; Olli et al., 2001; Weaver, 2002) report controversial results concerning the relationships of age, gender, education, socio-economic status, and occupation with environmentalism. On the other hand, Brand (1997) claims that nowadays there are no more consistent relations between traditional sociodemographic categories and environmental attitudes or behavior.

There are many **sociopsychological factors** that influence a person's pro-environmental orientation: environmental locus of control (Allen & Ferrand, 1999), authoritarianism (Schultz & Stone, 1994), personal norms (Blamey, 1998; Widegren, 1998), emotional affinity towards nature (Kellert, 1996; Kals et al, 1999), self-esteem (Geller, 1995), type of motivation (e.g., De Young, 2000; Green-Demers, 1997), ecological attitudes and knowledge (Olli et al., 2001), general value orientation (Boehnke et al., 1998; Schulz & Zelezny, 1998; Stern & Dietz, 1994), group identity (Bonaiuto et al, 1996), group norms of environmental friendliness (Widegren, 1998), to mention a few.

Among individual level sociopsychological factors perceived control over the state of the environmental situation is considered as a significant determinant of ecological activity. A person has to see the effectiveness of his activity. If a person believes that his activities can make any difference (to the state of the environment), then he is more likely to engage in some kind of pro-environmental activity. Among interpersonal factors, group identity and perception of shared norms of pro-environmental behavior in one's in-group are considered as conditions that promote environmentalism. Perception of shared norms may be conceptualized as an aspect of external support for the environmentally friendly behavior. Environmentally relevant individual actions are influenced by other people through perceived social norms and personal contact with people who already act pro-environmentally. If a person believes that pro-environmentalism is a norm in his or her in-group, (s)he will be more likely to engage in pro-environmental activities.

Several studies have shown that different forms of ecological behavior are caused by different determinants (Kaiser et al., 1999; McKenzie-Mohr et al., 1995) and that in different sociocultural circumstances various determinants of environmentalism may dominate (Brand, 1997; Levy-Leboyer et al., 1996; Nas & Dekker, 1996; Weaver, 2002).

Research questions and hypotheses

The present study starts from the conceptual approach developed by Stern (2000).
We use a simplified model that includes attitudinal factors (values, perceived control), socialization experiences (including nature experiences in childhood), contextual

factors (perceived norm of pro-environmentalism, local attachment), and sociodemographic factors (age, sex, education, income, subjective religiosity) as independent variables, and different varieties of environmentalism (self-reported habitual pro-environmental behavior, environmental concern, general environmental beliefs, and specific attitudes to forest) as dependent variables in multiple regression analysis. The data are based on a questionnaire study of a representative sample of adult population of a rural island community in Estonia.

Our study aims to clarify the relative impact of social psychological and sociodemographic variables in determining various indicators of environmentalism. Based on the model by Stern (2000) and earlier empirical findings we suppose that the pattern of relevant predictors will be different for different indicators of environmentalism.

Sample

A representative sample of the adult population of Hiiumaa consisting of 440 persons (46% men and 54 % women) was taken. Of the sample, 27 % are 15-29 years old, 45 % are 30-54 years old and 28 % are 55-74 years old. The sample represents age, sex, and geographical distribution of the population of the island with the precision of 1%. Compared to the general profile of Estonians, there are slightly more older persons and slightly fewer younger persons in this sample.

The site

Hiiumaa provides an interesting site for studying how global ecological concern has been contextualized in a relatively isolated rural community with strong local identity. Hiiumaa is the second largest island in Estonia, situated 22 km west of the mainland (1023 square km, 12 000 inhabitants). It is relatively isolated and has well-preserved nature, being rich in wildlife and forests. Differently from the rest of Estonia, Hiiumaa is ethnically homogeneous (99% of inhabitants are ethnic Estonians). Hiiumaa is characterized by the combination of isolated rural way of life (with traditionally strong relations to nature), and openness to the world. Hiiumaa is linked to the international environmental protection program Man and Biosphere (MAB) and is part of the West-Estonian Archipelago Biosphere Reserve, whose main objective is to create pre-conditions for community-based nature protection practice.

Method

A structured questionnaire "Hiiumaa and its inhabitants" was constructed by the TPU environmental psychology research group (M.Heidmets, G.Tamm, J.Uljas, and the author) in collaboration with the Biosphere Reserve Center (R. Post). The survey was carried out in June 1999 on Hiiumaa Island by trained interviewers in respondents' homes.

Measures

A. Independent variables:
1. Sociodemographic variables: age (years) ($M = 43.6$, $SD = 18.07$); sex (0 = male, 1 = female), education (years of schooling) ($M = 11.7$, $SD = 3.5$); income (mean monthly income per family member in Estonian crowns) ($M = 1620.2$, $SD = 922.3$).

Subjective religiosity (self-assessment on a 5-point rating scale of the importance of religion in one's life) ($M = 2.9$, $SD = 1.4$).

2. Sociopsychological variables
Local attachment was measured with an abridged collective self-esteem scale (Luhtanen & Crocker, 1992), which was modified to measure the sense of worth related to the local Hiiumaa community (assessment of 8 items on a 5- point scale). A composite *index of local self-esteem* (LSE) was computed (á = 0.809, $M = 2.48$, $SD = 0.74$).

The index of *perceived norm* of pro-environmental behavior was constructed as a mean of 2 items: "How many of your neighbors/inhabitants of your village try to act pro-environmentally?" and "How many of your friends/relatives try to act pro-environmentally?" (assessment on a 4-point scale) ($\alpha = 0.854$, $M = 3.06$, $SD = 0.68$).

The index of *perceived control* over environment was constructed as a mean of 5 items ("How much can you do for the environment at home, in home community, in Hiiumaa, in Estonia, in the world?"), assessment on a 5-point scale ($\alpha = 0.8509$, $M = 2.88$, $SD = 0.99$).

Values were measured, using an abridged version of the Schwartz value survey (44 items). Principal component analysis with Varimax rotation extracted 7 factors that explained 49.7% of the total variance. The first rotated factor (10%) with the highest loadings on 3 environmental values (*protecting environment, beauty of the nature, unity with nature*), as well as value items *tradition, peace, social justice, honoring parents, politeness, national security (*loadings from 0.44 to 0.7) was used in the regression models.

Nature experiences in childhood: a summation index based on the mean scores of 7 items (assessed on a 5-point scale) which concern associations of childhood with the natural environment ($M = 2.28$, $SD = 0.96$).

B. Dependent variables
Index of ecologically oriented activity: A summation index of 15 items (assessed on a 4-point scale), encompassing pro-environmental behavior at home (using compost, recycling paper, saving water and electricity, separating hazardous waste etc.) ($\alpha = 0.776$, $M = 3.02$, $SD = 0.45$). From similar measures used previously (e.g., Schultz & Zelezny, 1998; Diekmann & Preisenderfer, 1998; Kaiser et al., 1999) we chose the activities that are relevant in Estonia and have intuitively clear pro-environmental meaning in the local context.

Index of environmental concern: A summation scale, based on the mean response across 5 items (e.g., to what extent you are worried about the changes in the Earth's ozone layer, about the health of Hiiumaa forests, etc.) measuring self-assessed interest in specific (locally and globally relevant) ecological problems ($\alpha = 0.7946$, $M = 2.44$, $SD = 0.38$).

General environmental beliefs were measured with several question batteries (using items from the scales proposed by Grendstad & Wollebaek (1998) and Dunlap & Van Liere (1978). Summation indexes based on the first 2 rotated factors were used in the regression models. The first factor (20.6% of variance) has the highest loadings on items that characterize belief systems labeled as "soft ecocentrism" or "preservationism", or "balance of nature/limits to growth" beliefs. Various environmental values (cognitive, spiritual and aesthetic value of nature, rights of the animals), as well as justifications related to human welfare (health, interests of the next generations) are united in this factor. We labeled this factor "general pro-nature beliefs". The second factor (12.2 % of variance) unites items characteristic of anthropocentric environmental beliefs – natural resources exist for the benefit of humans, "cornucopian" beliefs of resilience of nature, belief in human rightful mastery over nature, reliance on man-made environment and technological optimism, practical exploitation of natural resources. This factor is labeled "utilitarian beliefs".

Specific *attitudes* were measured as attitudes *towards forest*. 10 attitudinal items were factor analyzed and after the rotation 3 factors explained 59,4% of the total variance. Mean scores of the first factor (labeled as general positive attitude to forest) were entered into the regression models.

Data analysis

The aim of the empirical analysis was to find out the extent to which a set of sociodemographic and sociopsychological variables account for the variance in different measures of environmentalism. The same group of predictors (10 variables) was used for 5 dependent variables.

We performed several multiple regression analyses with various environmental indexes as dependent variables. The following indexes described in the measures section were included as independent variables in the regression models: *age, sex, education, subjective religiosity, income, index of collective (local) self-esteem, value factor score, index of childhood nature experiences, index of environmental control and pro-environmental norm*. Five different environmental indexes (*index of ecological activity, index of environmental concern, pro-nature and utilitarian belief indexes, and attitudes to forest factor score*) were successively used as dependent variables.

Results

Variability *of ecological activity* and *environmental concern* are to the greatest extent (23% and 25.6% respectively) predicted by this set of variables. Five of the 10 predictors affected significantly the degree of everyday *ecological activity*. Age, collective (local) self-esteem, values, and religiosity appear to be the strongest predictors compared to other variables included in the model. The older, more religious and more locally attached a person is, the more often (s)he is engaged in environmentally friendly activities at home. Nature experiences in childhood have also a positive contribution to ecological activity.

Environmental concern is significantly predicted by 5 independent variables, values being the most important contributor. Women, more religious and more educated persons who perceive that they can control some aspects of the environment are more likely to be concerned about the environmental situation.

Table 1. Standard multiple regression of sociodemographic and sociopsychological variables on measures of environmentalism (standardized beta coefficients)

	Ecological activity	Environment concern	Pro-nature beliefs	Utilitarian beliefs	Attitude to forest
Age	0.162**	0.090	0.037	0.078	0.043
Sex	0.002	0.185***	0.086	–0.073	0.007
Education	0.055	0.185***	–0.055	–0.222***	–0.017
Religiosity	0.138**	0.165***	–0.099	–0.056	0.078
Income	–0.056	–0.028	–0.079	0.005	–0.092
Values	0.139**	0.283***	0.290***	–0.046	0.327***
Local self-esteem	0.155**	0.023	0.091	0.218***	0.086
Control	0.061	0.187***	0.098*	–0.135**	0.025
Norm	0.067	0.022	–0.018	0.079	0.053
Nature experiences	0.129*	0.066	0.048	–0.027	0.134*
R	0.480	0.506	0.428	0.423	0.475
R square	0.230	0.256	0.183	0.179	0.226

*$p < 0.05$, ** $p < 0.01$, *** $p < 0.001$

General pro-nature beliefs are most strongly predicted by values and subjective control.

General utilitarian beliefs are most strongly predicted by education, the level of local attachment, and perceived control. Persons less educated, locally attached, who perceive less subjective control over the environment tend to hold more utilitarian beliefs concerning the environment.

The *specific attitude to forest* was significantly predicted by two variables: values and nature experiences in childhood.

Almost all the variables included in the model have significant explanatory power for some of the environmental indicators when other variables are controlled for- (the exception is the level of income and perceived norm of environmental friendliness that did not reach significance in relation to any environmental indexes).

Comparing the *relative impact of various independent variables* we can see that the effects of explanatory variables vary across different environmental indicators.

Sociodemographic variables on the whole contribute less to the prediction of environmentalism than sociopsychological variables (attitudes and beliefs). *Age* predicts significantly 1, *education* 2, *religiosity* 2, and *sex* only 1 of the five environmental indicators. With the increasing *age* everyday ecological activity and environmental concern increase. Sex is an important predictor for environmental concern. *Women* are significantly more likely than men to be concerned with environmental problems. More *educated* persons tend to be more concerned about the environment and think less about the environment predominantly in utilitarian terms. Similarly to Grendstad & Wollebaek (1998) who report inverse relations of *education* with specific varieties of environmentalism, we observed significant negative relation of education with utilitarian beliefs concerning nature. On the other hand, education was positively related to environmental concern.

In contrast to earlier studies (Boyd, 1999) where various r*eligion* variables appeared to be weak predictors of environmental attitudes and behaviors, our study showed that subjective religiosity was a significant predictor of both the environmental concern and ecological behavior. Dietz et al (1998) refer to the possibility that religiosity and environmentalism are mediated by the assignment of sacredness (of whatever origin - intrinsic or God-related) to nature. Our question on subjective religiosity (without specifying the denomination or kind of practicing) may capture this general tendency. Besides, a significant correlation ($r = 0.366**$) between subjective religiosity and the belief in supernatural forces in nature indicates the same relationship in our study. Similarly Weaver (2002) reports that belief in the sacredness of nature consistently promotes pro-environmental attitudes, regardless of region.

Among **sociopsychological variables** environmental *values* and *local self-esteem* appeared to be the most significant predictors of environmentalism.

A noteworthy result of our study is that in exploratory factor analysis the *values* related to environment and its protection did not form a separate factor of biospheric values (as it has been hypothesized by Stern & Dietz (1994), but were grouped together with several items from universalism, tradition, and security value types. The resulting group of values was the most significant predictor of various forms of environmentalism, compared to sociodemographic and contextual factors used in our study. *Value* complex containing environmental values contribute most to the prediction of ecological activity, environmental concern, pro-nature beliefs and positive attitude to the forest. The only negative association of this group of values was with utilitarian beliefs concerning the environment.

Among contextual factors *subjective affinity* to the local community significantly promotes ecological activity and utilitarian beliefs. The relation of the index of local self-esteem is positive with all indicators of environmentalism.

Perceived control significantly promotes environmental concern and pro-nature beliefs, and is negatively related to utilitarian beliefs. Persons who feel that they can somehow affect the environmental situation, are more environmentally concerned and tend to hold more pro-nature beliefs. On the other hand, people who feel relatively more helpless in relation to the environment, tend to hold more utilitarian views concerning the environment.

Childhood nature experiences are a significant predictor of ecological behavior and positive attitude to the forest, indicating to the importance of emotional dimension in the environmentalism.

The level of income and *perceived norm of environmental friendliness* did not reach significance in these regression models. The fact that the perception of social context as holding *pro-environmental norms* did not affect environmentalism may be explained by orientation to widespread cultural norms and disregarding behavioral models in the local communities.

Discussion

The results demonstrate that, on the whole, sociodemographic characteristics have relatively minor importance for explaining the variability in environmental attitudes/beliefs and ecological behavior compared to a set of attitudinal variables.

Among sociodemographic variables *age* has strong impact on ecological activity. Environmental friendliness increases with advancing age: the peak of environmental concern is in the age range of 40–54, the peak of everyday pro-environmental habits is in the age range of over 65. Younger age groups tend to be more passive in this respect. This result is in variance with the tendencies observed in several previous studies of environmentalism where reverse relation with age has been observed (see overviews by Brand, 1997; Dietz et al, 1998; Greenbaum, 1995; McKenzie-Mohr, 1995). On the other hand, when ecological behaviors are disaggregated into distinct types, similar results with ours have been obtained by several authors (Dietz et al., 1998; Olli et al., 2001) where habits of frugality and pro-environmental consumer behavior characterized the oldest cohort. In our study the ecological activity was operationalized as the frequency of everyday activities at home. Although we can assume deliberate choices to some extent, these activities reflect at the same time a certain kind of lifestyle that is, in all probability, largely shaped by contextual factors – objective constraints and opportunities (rural way of life, relative poverty). However, such attitudinal factors as *values* and *collective self-esteem* had a strong impact on the frequency of such behavior. Similarly, Grendstad & Wollebaek (1998) report that increasing age leads to increased ecocentrism. In our study, although not reaching significance, age is positively related with all indicators of environmentalism. Possible explanations of this result may include both cohort effects and life-cycle effects. Most environmentally friendly age groups in our sample are those who have personal experience of scarcity and rural traditional way of life, which corresponds to many criteria of behavioral environmental friendliness. At the same time old and middle-aged persons have witnessed Estonian mass environmental movement in the 1980s, which has probably had an impact on their environmental consciousness. Younger cohorts, on the contrary, have grown up in the context of lessened public concern for the environment. Our results indicate also that socialization experiences that include *direct contacts with nature* (which are more common among the older persons) have certain impact on the formation of environmental mentality.

In our study *sex* is significant predictor for environmental concern, which is in accordance with several previous studies that found a consistent link between sex and environmentalism (Dietz et al., 1998; Olli et al., 2001; Weaver, 2002). Positive effect of *the level of education* on environmental concern and lack of its effect on ecological behavior was also observed in Norway (Olli et al., 2001) but not in the USA (Dietz et al., 1998). Probably not only the level of education but also its type and content have impact on environmentalism.

Value complex that unites environmental values with tradition, conformity, and security values was the most important predictor of environmentalism as it was operationalized in this study, confirming the assumption that environmentalism belongs to a large extent to moral domain (see Kaiser et al., 1999; Thogersen, 1996).

Environmentalism was related to *subjective affinity to the local community*, but not to perceived *norm* of environmental friendliness in the locality. This result indicates that the norm of environmentalism has a diffused character and is not linked to certain group identities or local social control.

Although we used different indicators of environmentalism, our results support the general conclusion by Dietz et al. (1998) about the primacy of sociopsychological factors over sociodemographic variables in predicting environmental attitudes, beliefs, and behavior. Inconsistencies in the significance and direction of some sociodemographic correlates of environmental concern and behavior indicate that the impact of demographic and attitudinal factors on environmentalism varies in different sociocultural contexts.

The relatively small amount of explained variance in all our analyses (from 17.9 to 25.6%), which, however, is comparable to previous results (e.g., Dietz et al., 1998; Weaver, 2002) indicates that there are also other factors responsible for the unaccounted variance in environmental attitudes, beliefs, and behavior.

On the whole, our study demonstrates variability of the determinants of different forms of environmentalism. Specific indicators of environmentalism are related to different combinations of independent variables. This result refers to the inherent heterogeneity of environmentalism.

References

Ajzen, I. & Fishbein, M. (1980). *Understanding attitudes and predicting social behavior.* Englewood Cliffs, NJ: Prentice Hall.
Allen, J.B. & Ferrand, J. (1999). Environmental locus of control, sympathy and proenvironmental behavior. *Environment & Behavior*, 31(3), 338-354.
Blamley, R (1998). The activating of environmental norms. Extending Schwartz's model. *Environment and Behavior*, 30(5), 676-709.
Boehnke, K., Stromberg, C., Regmi, M. P., Richmond, B.O., & Chandra, S. (1998). Reflecting the world "out there": A cross-cultural perspective on worries, values, and well-being. *Journal of Social and Clinical Psychology,* 17(2), 222-247.
Bonauito, M., Breakwell, G. M., & Cano, I. (1996). Identity processes and environmental threat: The effects of nationalism and local identity upon perception of beach pollution. *Journal of Community and Applied Social Psychology*, 6, 157-175.
Boyd, H. H. (1999) Christianity and the environment in the American public. *Journal for the Scientific Study of Religion,* 38(1), 36-45.
Brand, K.-W. (1997). Environmental consciousness and behavior: The greening of lifestyles. In M. Redclift & G. Woodgate (Eds.), *The handbook of environmental sociology* (pp. 204-217). London: Edward Elgar.
Corraliza, J.A. & Berenguer, J. (2000). Environmental values, beliefs and actions. A situational approach. *Environment and Behavior*, 32(6), 832-848.
De Young, R. (2000). Expanding and evaluating motives for environmentally responsible behavior. *Journal of Social Issues,* 56(3), 509-527.
Diekmann, A. & Preisendörfer, P. (1998). Environmental behavior: Discrepancies between aspirations and reality. *Rationality and Society,* 10(1), 79-103.
Dietz, T., Stern, P. C., & Guagnano, G. A. (1998). Social structural and social psychological bases of environmental concern. *Environment and Behavior,* 30(4), 450-472.
Dunlap, R. & Van Liere, K. (1978). The "new environmental paradigm": A proposed instrument and preliminary results. *Journal of Environmental Education*, 9(4), 10-19.
Eckersley, R. (1992). *Environmentalism and political theory*. London: UCL Press.
Geller, E.S. (1995). Integrating behaviorism and humanism for environmental protection. *Journal of Social Issues,* 51(4), 179-195.
Green-Demers, I., Pelletin, L.G. & Menard, S. (1997) The impact of behavioral difficulty

on the saliency of the association between self-determined motivation and environmental behaviors. *Canadian Journal of Behavioral Science,* 29(3), 157-166.

Greenbaum, A. (1995). Taking stock of two decades of research on the social bases of environmental concern. In M. D. Mehta & E. Ouellet (Eds.), *Environmental sociology: Theory and practice,* pp. 125-151. North York: Captus Press.

Grendstad, G. & Wollebaek, D. (1998). Greener still? An examination of Eckersley's ecocentric approach. *Environment and Behavior,* 30(5), 653-676.

Kaiser, F., Ranney, M., Hartig, T. & Bowler, P. A. (1999). Ecological behavior, environmental attitudes, and feelings of responsibility for the environment. *European Psychologist,* 4(2), 59-74.

Kals, E., Schumacher, D. & Montada, L. (1999). Emotional affinity towards nature as a motivational basis to protect nature. *Environment and Behavior*, 31(2), 178-202.

Kellert, S. R. (1996). *The value of life: Biological diversity and human society.* Washington, D. C.: Island Press.

Levy-Leboyer, C., Bonnes, M., Chase, J., Ferreira-Marques, J., & Pawlik, K. (1996). Determinants of pro-environmental behaviors: A five-countries comparison. *European Psychologist,* 1(2), 123-129.

Luhtanen, R. & Crocker, J. (1992). A collective self-esteem scale. Self-evaluation of one's social identity. *Personality and Social Psychology Bulletin,* 18, 302-318.

McKenzie-Mohr, D., Nemiroff, L. S., Beers, L. & Desmarais, S. (1995). Determinants of responsible environmental behaviors. *Journal of Social Issues,* 51(4), 139-156.

Milton, K. (1996). *Environmentalism and Cultural Theory.* Routledge.

Nas, M. & Dekker, P. (1996). Environmental involvement in four West European countries: A comparative analysis of attitudes and actions. Innovation. *The European Journal of Social Sciences*, 9(4), 509-536.

Olli, E., Grendstad, G. & Wollebaek, D. (2001). Correlates of environmental behaviors. Bringing back social context. *Environment and Behavior, 33*(2), 181-208.

Schultz, P. W., & Stone, W. F. (1994). Authoritarianism and attitudes towards the environment. *Environment and Behavior,* 26(1), 25-48.

Schultz, P. W. & Zelezny, L. C. (1998). Values and proenvironmental behavior: A five-country survey. *Journal of Cross-Cultural Psychology,* 29(4), 540-558.

Schwartz, S. H. (1977). Normative influences on altruism. In L. Berkowitz (Ed.), *Advances in experimental social psychology.* Vol. 10, pp. 221-279. New York: Academic Press.

Schwartz, S. H. (1992). Universals in the content and structure of values: Theoretical advances and empirical tests in 20 countries. In M. Zanna (Ed.), *Advances in experimental social psychology*, Vol. 25, pp. 1-65. Orlando: Academic Press.

Stern, P. C. (1992). Psychological dimensions of global environmental change. *Annual Review of Psychology,* 43, 269-302.

Stern, P. C. (2000). Towards a coherent theory of environmentally significant behavior. *Journal of Social Issues,* 56(3), 407-425.

Stern, P. C. & Dietz, T. (1994). The value basis of environmental concern. *Journal of Social Issues,* 50(3), 65-84.

Thogersen, J. (1996) Recycling and morality. A critical review of the literature. *Environment and Behavior,* 28(4), 536-559.

Weaver, A. A. (2002) Determinants of environmental attitudes. A five country comparison. *International Journal of Sociology,* 32(1), 77-108.

Widegren, O. (1998) The new environmental paradigm and personal norms. *Environment and Behavior*, 30(1), 75-101.

Zelezny, L., Pheng-Chua, P. & Aldrich, C. (2000) Elaborating on gender differences in environmentalism. *Journal of Social Issues,* 56(3), 443-458.

Sustainable technology and user participation:
Assessing ecological housing concepts by focus group discussions

Michael Ornetzeder
Centre for Social Innovation, Vienna, Austria

> **Abstract.** For the social acceptance of new technological solutions it is of decisive importance to deal with users' needs and experiences at the earliest possible stage. In the following article the possibilities and limitations of focus group discussions as a method of participatory assessment of technology is discussed in the light of two examples from the area of ecological housing construction. A comprehensive model for participation, comprising different methods of user participation for different phases of development, serves as the theoretical framework of the discussion. The application of this model in exemplary fashion in the form of focus group discussions demonstrates that, even during a very early phase of development, it is possible and worthwhile to let experienced users assess innovative housing concepts. Three kinds of results are differentiated: concept assessments, desired qualities and reports involving users' individual experiences.
>
> **Keywords:** technology assessment, focus groups, participation, ecological housing, sustainable development

Introduction

Ecologically optimised housing must not only fulfil criteria derived from environmental technology, it should in addition meet widespread acceptance within the population. Particularly within the area of large volume housing projects, however, drawing on the knowledge gained from users' experience already at an early stage in the project has been attempted in only a few exceptional cases up to now. In the cases of single family dwellings and ecological group housing projects, technological innovations are implemented only after consciously involving the later users of these dwellings. Yet in the case of large volume housing construction, dwellings are produced for a certain standard user or just generally for the "housing market". Even in the case of concepts involving sophisticated environmental technology, later users are normally confronted with the buildings once they have been completed and are in the process of being marketed. Nor is experience with dwelling in similar projects referred to as a rule either.

This situation is peculiar to the extent that social scientists in the context of innovation research have repeatedly pointed out that users' experience represents a significant pool of knowledge for technological innovation processes. For, on the one hand, paying attention early on to users' specific requirements improves the social acceptance of new technology. On the other hand, certain solutions are initiated only due to the active role of certain groups of users. This new aspect has influenced views as to when and with which strategies the development of technological innovations may be carried out. Classical concepts of technology assessment based primarily on the knowledge of experts have increasingly become complemented by participatory elements.

The possibilities and limitations of focus group discussions as a method of participatory technological assessment are discussed in the present article in the light of two concrete examples. The two discussions were carried out as part of a research project on the topic of social acceptance of innovative, environmentally friendly housing units within the framework of the programme *"Haus der Zukunft"* ("Building of Tomorrow") initiated by the Austrian Federal Ministry for Transport, Innovation and Technology.[1] Experienced residents of ecological buildings discussed two innovative building concepts for constructing ecologically oriented multi-story dwellings.

In the first part of this paper, the programme "Building of Tomorrow" is presented briefly. For this primarily technological research programme a model for participation was developed, aiming to involve various groups of users in the different phases of development during the innovation process. The experiences gained in dealing with focus groups which are presented in the second part of this article represent a first application by way of example of this method of participation.

Ecological construction within the programme "Building of Tomorrow"

In 1999 the Austrian Ministry of Science launched an impulse programme on the subject of sustainable economic activity, originally stated for five years. The first emphasis dedicated to the topic of the "Building of Tomorrow" aims to develop and implement innovative, sustainable residential and office buildings. As defined within the impulse programme, sustainability refers to: an increasingly efficient consumption of energy with respect to

---

1 The project entitled "users' experiences and attitudes as a basis for developing sustainable dwelling concepts with a high degree of social acceptance" was commissioned by the Federal Ministry for Transport, Innovation and Technology and carried out by the Centre for Social Innovation (Vienna) in co-operation with the Inter-university Research Centre for Technology, Labour and Culture (Graz). The project was completed in April 2001. The following individuals were involved in the task: Uli Kozeluh, Bernd Kumpfmüller, Irene Schwarz, Michael Ornetzeder, Harald Rohracher.

the entire life cycle of the building; a greater use of renewable sources of energy (especially the use of solar energy); the greatest possible use of organically renewable raw materials as well the efficient use of materials; and increased attention to service and use from users' point of view. Conceptually speaking, this broader perspective finds its foremost expression in social science research projects which accompany the project. The research emphasis "Building of Tomorrow" aims primarily at developing and promoting the market diffusion of components, construction elements and methods for residential and office buildings which conform to the highest possible degree to the guidelines for sustainable development (BMWV, 1999). Drawing on the basis of developmental principles encompassing a broad field of subjects, the authors designing the programme were, however, aware that they would have to seek solutions oriented not only on technological criteria. Research and development with the goal of sustainable development requires balancing a variety of interests. The resulting search for new priorities deemed necessary is described as follows:

> "Co-ordinating these requirements is highly demanding and entails conflicts in goals for which solutions need to be found which will meet with a consensus of approval. On the other hand, integrating social, economic and ecological goals represents a significant opportunity, whereby the key to success lies in innovation, yet not alone in a technological sense, but rather here once again in combination with social, economic and institutional innovation. Particularly by combining the afore mentioned criteria, the door is opened for achieving technological advances with great potential for marketability." (BMWV, 1999, p. 3)

One possible strategy for achieving the objectives stated here is to open up the process of technological development to new players. Thus, attention could be paid in particular to users' perspective at the earliest possible stage in order to avoid acceptance problems when developing and implementing innovative building concepts. Within the framework of the above-mentioned research project a model for participation for the entire research programme was therefore developed on the basis of international experience. This model included discussing different options for involving the various user groups in the process of technological development. Two innovative building concepts were selected and assessed by experienced users within the framework of focus group discussions toward the end of putting the model into practice in exemplary fashion. This procedure made it possible on the one hand to gather practical experience with user participation by means of focus group discussions, and on the other hand to produce the first tangibly useful results of project-related feedback.

User participation in the case of innovation processes

Recent studies in the area of innovation research assume that technological development processes can derive substantial profit from direct user participation (e.g., Akrich, 1995; Schot, 1998; Weyer et al., 1997). In particular, users are expected to present arguments serving to improve the social acceptance and user friendliness of new technologies. Beyond this, one may also reckon with encountering unconventional and for the most part novel suggestions for improvement. This assumption is based on the premise that technological innovations are essentially to be interpreted as the result of social processes deriving from economic interests, political power constellations and cultural value concepts. Yet the interests and experience of users are taken into account to only a small degree as a rule, even though each user of a technology possesses valuable expertise in his or her specific area. Users are experts on the user context of technologies (see Bijker 1996). Building upon this insight, new approaches in the field of technology assessment (TA) have been developed in recent years, no longer concentrating on the evaluation of new technologies, but rather focussing on the processes by which technologies arise and, beyond this, attributing to the potential users of such technologies a significant role (e.g., "Constructive Technology Assessment" in the Netherlands or *"innovationsorientierte TA"*[2] in Germany). By thus broadening the design process, possible problems arising only within the corresponding contexts of use are to be recognised at an early stage and are to be minimised by making appropriate changes. Technological assessment thus becomes increasingly a participatory process, systematically nurtured not only by scientific know-how but also by the everyday experiences of technology users (see Bröchler & Simonis, 1998).

As a rule, experts in planning (architects, building technology planners, energy experts, structural engineers) are exclusively involved in the early phases of building development. For this reason, users along with their experiences and needs are taken into account only indirectly or in biased fashion. Madeleine Akrich (1995, p. 173ff) speaks in such cases of implicit user representation, discerning three typical approaches:

(1) Designers see themselves as "normal users," attempting in this way to anticipate the possible needs and experiences of future users.
(2) Certain experts for user questions are consulted (e.g., marketing experts).
(3) Trust is placed in the (market) success of similar technologies and concepts.

These three widespread strategies for representing users and possible forms of use during technological development processes are subject to considerable uncertainties. Engineers are not able to negate their professionally trained approaches at will; marketing experts' knowledge of users is usually confined to the latter's role as consumers; and even the success of comparable solutions is nothing more than a vague guideline when developing innovative solutions. As long as technological development is embedded primarily in a culture of scientific engineering, users exist for the most part in the shape of "user images" (Hofmann, 1997). The imminent danger of planning for a fictitious standard user that in reality does not at all exist thus remains.

To what extent one succeeds, parallel to technological development, in constructing socially acceptable contexts of use is, therefore, a question of paramount importance for the success of ecological architecture and technology. "Only if one succeeds," Weyer writes, "in accommodating a new technology to the behaviours of potential users in advance and, in turn, in taking into account users' interests and potential behavioural patterns in the design of the new technology does a socio-technological innovation stand a chance of succeeding" (Weyer, 1997, p. 50, see also Schot et al., 1994). The goal must be to "find out during the construction process of a new technology how later users will deal with that technology and which strategies and interests they will pursue in dealing with it. Only this type of early feedback between the user and the producer makes it possible to develop approximately realistic models of socio-technological systems" (ibid.). The manner

---

[2] Innovation-oriented TA.

in which technological innovations are perceived and the question of whether or not they are accepted as solutions depends for the most part on their "social embedding"; it thus hinges on the degree to which a new technology is adapted to society and, conversely, on the extent to which behavioural changes are required on the part of users. The manner in which this interactive accommodation process is organised socially and during which phase in development the process is initiated are also essential factors in this context.

Elements of a participation model

In this context, user participation is viewed not only as a specific form of participation in the sense of direct democracy, it is also recommended specifically as a strategy for technological policy. In general it may be assumed that only when those traditionally responsible for the development, planning and construction of dwellings actively support the innovation process does stronger user participation lead to sound results. This entails not just the willingness to make planning results and professional expertise available, it also means the will to integrate users' feedback and needs in further stages of the job. Interactive learning processes can taken place only when on the part of both professionals and users the willingness to accept the other's perspective is given. Beyond this, planners must accept users in their role as experts in certain matters.

The participation model[3] suggested here is tailored specifically to involving users in the development and implementation of innovative dwellings. With respect to content the model seeks to:
– confront technical development even at an *early stage* with aspects of use, thus providing incentives for socially acceptable solutions;
– discuss and implement in a broad manner the principles of *sustainable development* in the field of housing construction;
– give *future users* a voice in development, leading to the expectation that positive effects on the way individuals deal with innovative technology during the use phase will be achieved.

In the course of the impulse programme "Building of Tomorrow," building concepts are not only to be investigated but also realised and used in the form of dwellings. For this reason our considerations cover the four phases of 1) research and development, 2) planning, 3) construction and 4) use of the buildings. For each of these phases, concrete tasks, methods of implementation and potential groups of participants were discussed. The results of these considerations are presented in overview in table 1. With respect to those involved in the forms of participation recommended, it may be generally assumed that it is appropriate to have experienced users participate in early phases of development. The farther the project progresses toward realisation, the more sense it makes to involve future users in the planning process. Both forms of participation serve to encourage the social embedding of innovative building concepts.

Depending on the development phase of the project, various groups of users may deal with different topics using suitable methods in each case. The following details on participation models are limited to the phase of research and development, since focus group discussions were carried out in this area by way of example[4].

---

[3] We would like to thank Johan Schot for helpful discussions.
[4] An extensive presentation of the participatory model may be found in Ornetzeder, M. et al. 2001.

Table 1. User participation toward developing and implementing building concepts

Phase	Topics	Methods	Participants
Research and development	• future needs • achieving sustainability • evaluating architectural concepts	• Open Space • Future Workshop • target group participation • Planning Cells • focus groups • series of focus group sessions	• experienced users (lead users) • representatives of interest groups • intermediate users
Planning	• development of construction and usage concepts • evaluating construction details • evaluating energy concepts	• planning for real • moderated planning workshops • focus groups • moderated construction groups • residents' committee	• experienced users (lead users) • intermediate users • future users
Construction	• choice of materials • floor plans of dwellings • concepts for free space • community facilities	• residents' committee • moderated construction groups	• future users
Usage	• information for residents • feedback from users	• information workshop • feedback on experience • post-occupancy evaluation	• users of the building

While the project is in the development stage, the main goal is to clarify fundamental questions. According to Sclove (1995), R&D processes profit from participation in a threefold manner: first, by the participation of a large number of individuals, representing a broader spectrum of viewpoints; this increases the odds of discovering creative solutions with a high degree of innovative potential. Second, it ensures that existing social needs and experiences are reflected and respected. And third, the participation of quite different players makes it more likely that ideas are transferred from one area of society to another (see Sclove, 1995, p. 181). Possible topics in this stage are related to the future needs of users, to paying greater attention to sustainability criteria in the conception of dwellings and to evaluating building concepts which have already been developed. Other topics could be developed while working with the individual project groups. Sometimes concrete decisions irreversibly influencing the direction of further research and development activities have to be taken very early in the process. In the case of a reasonably priced passive solar house, for example, it must be determined whether users would be prepared to do without an additional source of heat in the bathroom. This question can only be clarified through immediate interaction with potential residents. Experiences gained from dwellings which have already been built and the results of research on heating needs in private homes – inasmuch as such results even exist – can only be classified as additional information. Gathering current opinion at an early stage in such cases increases the efficiency of research and development work and gives way to the expectation that the concepts realised will meet with greater acceptance.

The choice of methodological approaches will be determined substantially by the topic at hand. Approaches like the Future Workshop (Jungk & Müllert, 1989) or Open Space (Owen, 1997) are suited to gathering concept ideas and future needs. If the aim is to pay more attention to sustainability, methods such as the Planning Cell (Dienel, 1993) and target group participation (Sperling, 1999, p. 49) are appropriate. Focus groups or series of focus group sessions (Dürrenberger & Behringer, 1999) are suitable for discussing and evaluating concepts which have already been developed.

Three user groups in particular may be considered for participation during this phase: experienced users (lead users, final users) who have been able to gather tangible experience with similarly designed buildings and are interested in technological questions; representatives of interest groups, particularly of ones involved in-depth in the subject of "sustainable development"; and intermediate users such as construction financiers, housing bureaux and construction companies which take an intermediate role between planning and usage during the planning process, offering specific expertise.

An example of user participation by means of focus group discussions

The focus group discussion is an investigative method which has rarely been used up to now in empirical social research. Even though this form of qualitative group interview was already used in the 1950s by Paul Lazarsfeld and Robert Merton to investigate audience reactions, its main field of usage for a long time was commercial marketing and opinion research. Yet in English-speaking countries in particular a great deal of experience has been gathered in applying this method. Due to the general trend toward revaluating qualitative research methods in recent years, the focus group discussion has, among others, increasingly come to be regarded as a serious instrument for social scientific investigation and to be used in scientific studies (cf. Catterall & Maclaran, 1997; Dürrenberger & Behringer, 1999; Gibbs, 1997; Hörning, Keck & Lattewitz, 1999; Littig & Wallace, 1998).

Focus groups consist of six to 12 individuals who discuss a specific topic under controlled conditions. The participants all have a common experience which serves as the starting point for the discussion. In our case this was the oral presentation of a building concept by each of the project managers. The moderator plays an essential role in the focus group. This individual is responsible for maintaining an agreeable atmosphere during the conversation, for channelling incipient conflicts and for ensuring that as many participants as possible become involved, while maintaining a neutral attitude in topical matters. The one- or two-hour discussions are audio- or video-recorded as a rule, transcribed and then evaluated using methods of content analysis (see Dürrenberger & Behringer, 1999).

One of the main advantages of focus group discussions, compared with individual interviews for example, is according to the special literature on the method the opportunity to interact (see Gibbs, 1997, p. 2). Participants can ask each other questions, reconsider existing views in the light of new arguments and if necessary revise these views. A group discussion allows a learning process to take place among the participants and usually offers a more complex and differentiated picture of the points of view inherent among them.

Two focus group discussions with experienced residents of ecological dwellings were carried out in the course of the project. Two new building concepts were selected whose scientific and technical development is being subsidised within the framework of the impulse programme "Building of Tomorrow". Both projects belong to the area of large volume residential housing and each evidences a high degree of potential for innovation. Specifically, the projects are as follows:

(1) *HY3GEN – a renewable house:* a hybrid building (HY) of the so-called third generation (3GEN) is planned. The concept foresees the use of renewable materials to the greatest degree possible as well as the implementation of various solar technologies. The hybrid character of the building is found in two aspects: on the one hand, a combination of different forms of usage (commercial/residential) is planned, while on the other hand a building concept is to be implemented in which the infrastructure (i.e. utility lines, building technology, dividing walls) is independent of the building shell. This is in order to heighten the flexibility of usage as well as increasing the period of use of the building shell. Due to the combination of shopping centre, office wing and living quarters, the project requires a central location in the inner city. For marketing reasons a share of 75% commercial to 25% residential usage is to be achieved. This constellation entails 60 to 70 dwellings.

(2) *Application of the passivhouse technology in social housing in Vienna:* The objective of this research project is to develop an especially cost-efficient housing project complying with the passivhouse standards with around 100 dwelling units for a location in Vienna. The term passivhouse refers to a building concept entailing extremely low heat energy requirements, rendering a conventional heating system entirely superfluous (hence the term "passive"; a passivhouse does not require a conventional "active" heating system, heating itself passively for the most part). Remaining heat energy requirements are covered by means of a ventilation system, controlling the flow of warm air into the rooms of the dwelling. To date, no passivhouses have been erected in the framework of social housing projects in Austria.

Both of the projects began as building concepts, meaning that at the point in time when discussion rounds took place no building property had been acquired and thus no building plans or any other drawings existed either. Essentially, therefore, project ideas were evaluated which were presented orally at the beginning of the discussions by representatives of the two research projects.

The participants in the discussions were without exception users, each with their specific backgrounds of experience. All those involved had amassed several years of experience in the usage of ecological dwellings, displaying interest in the topic of discussion even in the preparatory phase. The participants may be classified as experts on the usage of ecologically optimised dwellings. While selecting the participants in the discussion rounds, care was taken to ensure the greatest possible degree of heterogeneity within a generally homogeneous target group. With respect to socio-structural characteristics the participants represented for the most part the general population living in ecological dwellings. Typical characteristics of this group are above all a high level of education, often coupled with an above-average income level. The participants were aged 35 to 50 years and most of them are employed in educational, social or technical jobs. All of the participants looked back on several years (between one and 12 years) experience with ecologically optimised dwellings. With respect to individual participants' specific background of experience, varying but in each case typical living situations were intentionally selected. Residents of large volume residential buildings, group dwelling projects (eco-villages) and in each group one of single family houses participated in the two discussion rounds. Homeowners and tenants were represented in each of the groups. Three women and three men were involved in each of the discussion groups, that is, six people in each group. Both of the discussions lasted about two and a half hours including the presentation which was focussed more sharply by means of detailed questions on the part of participants. The ensuing discussions were based on a rough question catalogue.

Results of the focus group discussions

Principally speaking, similar topics were discussed in both focus groups. The statements made concerned the concepts themselves as well as individual aspects of them and even took in possible construction details, which the participants discussed and evaluated on the basis of their own experiences.

The discussion of *HY3GEN* centred on the topic of the mixed usage of the building. The high degree of commercial usage which the operators strove for (a share of as much as 75% was planned) was repeatedly criticised as being too high. Yet not only the share of commercial usage affects the attractiveness of the project, the type and selection of commercial tenants is also of intrinsic importance. In the participants' view, the quality of the site could be greatly increased with the aid of mixed usage as planned. The participants described their vision of an exciting centre in the city quarter, including short distances to facilities and attractive offerings for residents. An urban ambient conducive to strolling, stimulating while at the same time offering opportunities to rest, should be created in the public area of the property. Specifically, the participants expressed a desire for shopping facilities, a market hall, a coffee house, a fitness club, a day-care centre and a school. The share of commercial tenants in the entire area of the complex should not exceed 50%. Production firms not able to rule out noise and other emissions are to be largely banned. The hybrid building concept was viewed positively for the most part. The very low energy standard targeted for the building shell was judged by several participants as being an additional positive aspect. On the other hand, the share of renewable building materials was criticised as being too small. More than the planned 10% total tonnage should be derived from renewable materials in order to satisfy the project's expectations.

With respect to the *passivhouse*, the lack of a social concept and the overly high degree of concentration on technical aspects and low building costs were criticised foremost. All discussion participants were of the opinion that the concept, of itself positive, would be handicapped by the cost restrictions for social housing. Problems with the building's technical systems and/or cost reduction measures while erecting and furnishing the dwellings would result, greatly reducing the attainable living quality in the finished dwellings. The discussion group concluded that the project would only succeed if residents can be found early enough who are largely able to identify with the basic idea and goals of a passivhouse. In the view of some of the discussion participants, the attractiveness of the project could be significantly increased with the aid of a suitable social concept (marked by a careful selection of tenants, participation in planning decisions, technical training, counselling and advice during the residential phase). None of the participants could imagine that the building technology to be implemented would function in a fashion neutral to users' behaviour, that is, regardless of the way users' reacted the desired effects could be achieved. The participants backed up their scepticism with a number of examples from their own living experience.

Current practice with respect to participation in large volume housing projects and its significance was broadly treated in both discussion groups. All participants were of the opinion that a voice in planning is very important particularly in innovative model projects. The success of innovative projects does not, therefore, depend exclusively on planning know-how derived from engineering science and architecture and on the exact execution of building plans. With their behaviour, later residents, too, determine to what extent the original objectives (e.g., goals for saving energy) are reached. Particularly in the case of pilot and model projects, therefore, which are followed with keen professional interest, it is important to involve users in the concept. If in the framework of such processes one succeeds in stimulating learning processes on the part of those involved, this will also affect users' behaviour later on.

If the attempt is made to classify the results on a more general level, three types of results may be differentiated. First, the concepts presented were evaluated both positively and negatively. Second, participants named qualities desirable for significantly increasing the attractiveness of the projects according to a user's point of view. Third, in the course of the discussions users related their own concrete experiences, serving to substantiate critical evaluations as well as to illustrate desired positive characteristics.

Evaluation of the building concepts

The critical stances in the discussion on the passivhouse are a good example for the evaluation of building concepts. In this case, criticism was levelled at the, in participants' opinion, overly strong concentration on the topics of saving energy and how to reach that goal technically. Participants were of the impression that, in comparison with the technical parameters mentioned above, potential users played only a minor role. In users' view a liveable dwelling was of foremost importance, realising a building with extremely low heat energy requirements was not accepted as an isolated objective – even though all participants placed great importance on ecological living and themselves live in such buildings. One of the participants described this dilemma using a question:

> A: "The question is whether this is about quality of living or technical matters. As far as we [the participants] are concerned, we all want a high living quality, and if they built long corridors like that, then there would simply be no living quality; just because they want us to use little energy, they will accept anything, that is what you mean, isn't it?"

Another participant criticised the basic approach of the project group, focussed too strongly on optimising costs and thus on technical questions. And that, even though he was aware that a research project is concerned foremost with plotting possibilities for realising the architecture and construction of the project.

B: "This is about social housing, but for me the technical part has got out of hand. But that's your part, of course, I know that. ... The problem is construction costs are limited. You've got a fixed amount, and if you spend a lot of money on technology, then that technology will take up a fair part of the budget. The result is that the general contractor looks for very cheap companies; that also happened with our building, for instance. Of the different subcontractors who built part of our building, four out of seven of them went bankrupt, so that is about standard. ... You just can't find a firm that can build it for a lower price, which is the aim in social housing projects, so that the whole thing works."

Stating desirable qualities

In stating desirable qualities, mostly topics are concerned to which attention is hardly given in the early stages of building development, generally speaking. Yet from the users' point of view they are very important because they decisively influence the usability of the building and its immediate environment on a day-to-day basis. The following quotations are derived from the discussion about the *HY3GEN* project. The participants developed some very definite notions with respect to the mixed usage of the building, the decoration of internal and external areas as well as to possible service offerings.

A1: "I would very much like to live in an area that is not just a residential area. Car sharing in the garage would not be bad either. I like many-sidedness too, various kinds of free space and so forth. ... If it's in the city, then with as much greenery as possible inside."

B1: "I would prefer nice, bright rooms, good acoustical separation between the different functions [commercial/residential], a view with some green, although I don't care whether it grows from the side, below or above. Cocktail tomatoes on the window still would be nice."

C1: "40% firms and 60% flats would be all right. I'd like a market hall. Maybe 10% of all companies in there could be technical. There should be a Billa [supermarket], a coffee house, things, that is, that I use myself everyday."

Reports of users' concrete experiences

In the course of the discussion about the passivhouse concept the question was raised as to what influence in general users' behaviour has on the functioning of buildings' technical systems. Despite the view of technicians in the project team that in multi-story dwellings technology should be implemented which reacts as neutrally as possible to users' behaviour, users were convinced that residents' behaviour is of great significance in any case. In the users' view the technology implemented should allow for various uses without fundamentally endangering the objectives of the concept (e.g., extremely low energy requirements). For this reason it would be important for users to have a certain affinity toward the objectives of the project from the very start, for them to be selected, informed adequately and advised. For wilful forms of use can never be ruled out, as the following quotation about the winter garden illustrates:

> A3: "The winter garden is not generally used in line with its original purpose. Due to the great need for space, since there are no basements, the winter gardens are used for all kinds of things, as an office, as a heating room, an additional bedroom etc. A winter garden should not be used for anything which requires heating. People are already starting to put in some kind of heating, electric heating or something else. Those then are my experiences, what really happens with a good concept [which the participant himself lives in], I would say with a good project is simply that the residents just make of it what they think is the best for them. ... That is my worry too, that this social housing, the passivhouse that they want to build or that is supposed to be built will naturally depend very much on the tenants whether it is a success or not, on each of them."

Wilful forms of use can, however, also arise when those responsible for planning deliberately attempt to block users' influence while at the same inadequately communicating their reasons for doing this. One participant, for example, reported how residents treated a ventilation system forced on them in a new building they had recently moved into.

> B3: "I have my own personal experience with that. We have a fan in the bathroom and the toilette. ... We have to have it, it's a regulation, we can't turn it off. ... I moved in and I thought, something is broken. It goes all the time. Some people have shut it down by themselves, tinkered with it. Turning it off and on, you've got to be able to influence it yourself."

Technology which cannot be influenced by its users, in this case a ventilation system working around the clock and installed by the developer in order to fulfil a guarantee, provokes some tenants to engage in "wilful self-help" (destruction), while causing many people to feel themselves hampered by it.

Conclusion

Only rarely are users' needs and experiences directly respected, even during the planning of innovative housing projects. Usually, architects, planners for house technology and representatives of the builder take the necessary decisions during such a building project. Users come into contact with the building and its dwellings only once construction work is already underway. The participatory model outlined here presents ways of involving experienced user groups at least intermittently in the innovation process even at an early stage.

The two focus group discussions showed that it is feasible and worthwhile to allow users to evaluate housing concepts from their point of view even at a very early phase in development. Even though neither plans, models nor photos of the planned buildings were available and these thus existed merely as narratives given by those responsible for the projects, the participants, keenly interested as they were, came to grips with the concepts relatively swiftly by means of well-targeted questions. Participants were able to name conceptual deficiencies rather quickly and to discuss suggestions for changes on the basis of their experiences. They were able to evaluate the concepts independent of personal interests, particularly since they themselves would never live in the planned buildings. The observations made during the discussions were for the most part constructive comments which could have been incorporated into the current phase of project development in an ongoing manner. Despite the plethora of results surprisingly reaped in the present case, generally it is advisable to carry out a number of focus groups (at least three) with different participants in each case.

The fact that both discussions produced an abundance of results must be attributed primarily to the participants selected in each case. Persons were intentionally selected with several years personal experience with and a special interest in environmentally friendly architecture. This procedure involved on the one hand a great deal of effort, while on the other hand it addressed residents of ecological, multi-story buildings only insufficiently.

In addition, more strongly structured moderation methods (Klebert et al., 1998) would be advantageous if focus groups are to be initiated exclusively for intervening in technological developments instead of for research purposes. Selection and evaluation results are supported by the visualisation methods constantly applied in the case of such concepts. During the entire session, results are available to all participants and also discussed in sub-groups as need arises. A further way of utilising focus groups to a greater extent for developing concrete recommendations is provided by the so-called series of focus group sessions (Dürrenberger & Behringer, 1999, p. 24ff). In this case at least three sessions are held, all on the same topic. This procedure allows participants to acquire additional expert knowledge during the process and makes for a total of more time for developing recommendations which may be put into action immediately.

References

Akrich, M. (1995). User representations: practices, methods and sociology. In A. Rip, T.J. Misa, J. Schot (Eds.), *Managing technology in society. The approach of constructive technology assessment*, pp. 167-184. London and New York.

Bijker, W. E. (1996). Democratization of Technology. Who are the Experts? Available online http://www.desk.nl/~acsi/WS/speakers/bijker2.htm

BMWV. (Ed.) (1999). *Impulsprogramm Nachhaltig Wirtschaften. Konzept [Program on technologies for sustainable development. Concept]*. Vienna: BMWV.

Bröchler, S. & Simonis, G. (1998). Konturen des Konzepts einer innovationsorientierten Technikfolgenabschätzung und Technikgestaltung [Outline of the concept for an innovation-oriented technology assessment and shaping of technology]. *TA-Datenbank-Nachrichten*, 7(1), 31-40.

Catterall, M. & Maclaran, P. (1997). Focus Group Data and Qualitative Analysis Programs: Coding the Moving Picture as Well as the Snapshots. *Sociological Research Online*, vol. 2, no. 1. Available online http://www.socresonline.org.uk/socresonline/2/1/6.html

Dienel, P. C. (1993). *Die Planungszelle. Der Bürger plant seine Umwelt. Eine Alternative zur Establishment-Demokratie [The Planning Cell. Citizens are planning there environment. An alternative to the established democracy]*. Opladen.

Dürrenberger, G. & Behringer, J. (1997). *Die Fokusgruppe in Theorie und Anwendung [Focus groups. Theory and practice]*. Stuttgart.
Gibbs, A. (1997). Focus Groups. *Social Research Update, 19*. Available online http://www.soc.surrey.ac.uk/sru/SRU19.html
Hofmann, J. (1997). Über Nutzerbilder in Textverarbeitungsprogrammen – Drei Fallbeispiele [User images in text processing programs – three case studies]. In M. Dierkes (Ed.), *Technikgenese. Befunde aus einem Forschungsprogramm [The shaping of technology. Findings from a research program]*, pp. 71-98. Berlin.
Hörning, G., Keck, G., & Lattewitz, F. (Eds.). (1999). *Die gesellschaftliche Bewertung zukunftsweisender Energieszenarien – Fokusgruppen [The social assessment of future oriented energy scenarios – Focus groups]*. Akademie für Technikfolgenabschätzung in Baden-Württemberg.
Jungk, R. & Müllert, N. (1989). *Zukunftswerkstätten [Future workshops]*. München.
Klebert, K., Schrader, E. & Straub, W. (1998). *Kurz-Moderation. Anwendung der Moderations-Methode in Betrieb, Schule und Hochschule, Kirche und Politik, Sozialbereich und Familie bei Besprechungen und Präsentationen [Short Moderation. How to use the moderation method in business, school and university, church and politics, for social and family issues, meetings and presentations]*.
Littig, B. & Wallace, C. (1998). Möglichkeiten und Grenzen von Fokus-Gruppendiskussionen für die sozialwissenschaftliche Forschung [Possibilities and limitations of focus group discussions in the field of social research]. *Österreichische Zeitschrift für Soziologie*, 23(3), 88-102
Owen, H. (1997). *Open Space Technology. A User's Guide*. San Francisco
Schot, J.W. (1998). Constructive Technology Assessment comes of age. The birth of a new politics of technology. In A. Jamison (Ed.), *Technology Policy Meets the Public*, pp. 207-231. Aalborg.
Schot, J.W., Hoogma, R. & Elzen, B. (1994). Strategies for shifting technological systems. The case of the automobile system. *Futures, 26*, 1060-1076.
Sclove, R. (1995). *Democracy and Technology*. New York/London.
Sperling, C. (Ed.). (1999). *Nachhaltige Stadtentwicklung beginnt im Quartier. Ein Praxis- und Ideenhandbuch für Stadtplaner, Baugemeinschaften, Bürgerinitiativen am Beispiel des sozial-ökologischen Modellstadtteils Freiburg-Vauban [Sustainable urban development starts in the district. A practical handbook for town planners, building cooperatives, citizens' initiatives based on the case of the sustainable model district Freiburg-Vauban]*. Freiburg.
Weyer, J., Kirchner, U., Riedl, L., & Schmidt, J. F. K. (1997). *Technik, die Gesellschaft schafft [Technology that shapes society. Social networks as places of technological development]. Soziale Netzwerke als Ort der Technikgenese*. Berlin.

IV. Urban sustainability and cultural diversity

Citydwellers' relationship networks:
Patterns of adjustment to urban constraints[1]

Gabriel Moser, Alain Legendre & Eugénia Ratiu

Université René Descartes-Paris V, France

Abstract. Urban life has been repeatedly described as being constraining. What is the impact of living in a metropolis on interpersonal relationships?

Data on the extent of the relationship networks of Parisians, suburbians, and inhabitants of a minor town and the way they relate were collected by standardized interviews. The heads of 302 families, half of them having the possibility to escape for weekends (second home, family house), described their relationship network, the origin of each relation and the modalities of frequentation.

Results show that urban constraints, and specifically the short free-time left to people living in Paris and its suburbs, result in adjusting relational behavior. Difficulties in interactions, frequently with the traditional friendship network, are compensated for by an increase in local sociability. The number of relations originated in neighborhood associations and workplaces is much higher for the Parisians and suburbians than for the inhabitants of the small town. Furthermore, urbanites meet their friends less spontaneously, even when they keep as many relationships as people living in smaller towns. Only those who have the opportunity to escape have a mean number of relations similar to the small town inhabitants, and they meet their friends preferentially on weekends out of town.

Keywords: urban constraints, interpersonal relationships, origins of relations, extend of network

---

[1] Study financed in the context of the "Urban Ecology" project submitted to the Ministère de l'environnement and the Ministère de l'Équipement, du logement,, des transports et de l'espace, contract # 92/31197002237501

Residents of Paris or the Greater Parisian region often consider that living in or around Paris implies a mediocre quality of life. Living in a large urban area is more constrictive, and among the numerous urban constraints, daily travel between home and work is particularly stressful (Stokols, 1978), as well as demanding in terms of the time available. Extension of the metropolitan Parisian region been accompanied by both diversification and lengthening of travel, especially due to the separation of where one lives from the workplace, and a progressive decentering of social life, which no longer occurs mainly within the residential neighborhood (IPSOS, 1991). Francilians (referring to all those living in the region making up the Île-de-France surrounding and including Paris) are not living at the pace of where they reside (Lacaze, 1993). Social life extends throughout the metropolitan area and is no longer tied to the local community, as is still the case in the French provinces. New needs are emerging as a reaction to the economic restructuring of space and time characteristic of the organization of urban life in postmodern society (Sèze, 1994). The development of second homes outside of the urban environment, "a place in the country," is one way of responding to these new needs: getting away from the main place of residence in the urban milieu to a house belonging to other family-members or to one's own second house provides a way of evading the numerous urban constraints.

Contact is an essential condition for social cohesion (Festinger, Schachter, & Back, 1950; Newcomb, 1961). But contact can only contribute to creating interpersonal relationships through repetition and over an extended period of time. In addition, geographic mobility tends to reduce opportunities for interpersonal activities by not leaving people enough time to establish and maintain stable relationships.

Using both experimental and survey methods, many researchers have clearly demonstrated that interpersonal relations are particularly important for coping with stressful situations (Moser, 1994). Friendship provides social and emotional support, making help available when needed and making it possible to conduct activities in common (Argyle & Henderson, 1984). The fact of being a part of a large group of friends and acquaintances allows us to overcome stressful periods of our lives more effectively (Arling, 1976), and prolonged interaction with a large number of friends is correlated with a high level of well-being (Palisi, 1985). As urban residents become more physically distant from their families, affective support in the urban setting is built principally on such friendships (Amato, 1993).

Most research studies on sociability in the urban environment have focused on anonymous contacts, typically with strangers, and support the notion of increased withdrawal, characterized by avoiding interactions with others (Moser, 1992). Few studies have, on the other hand, looked into interpersonal contacts at the level of friendships. Yet we know that individuals find it harder to get to know others in large cities (Baum & Paulus, 1987; Sundstrom, 1978). It is more difficult for students, for example, to make friends when they arrive in a large city compared to smaller city or town (Franck, 1980). Neighborly relations are weakest in the Parisian region (Héran, 1987). Those living in large collective housing units in the Parisian urban area are twice as likely to have no relationship with their neighbors, compared to those living in single-family houses in the provinces. Alongside this, one notes also more relationships involving friendship and more satisfaction with family-life in small towns compared to the large urban areas (Oppong, Ironside, & Kennedy, 1988).

To the extent that living conditions in large urban areas leave less time for being sociable, one could ask the following questions: Is the extent of relationships a function of one's place of residence? What role does the opportunity to escape from the urban environment to another second home out of the city play in relational behaviors? Can we talk about a "global sociability deficit" in interpersonal relations among those living in the greater Parisian region? In other words, are we witnessing in effect a weakening of social bonds, with consequent effects on people's health and well-being?

The purpose of the research described in this article is to study the effects of 1) the type of urban tissue (living in the center of a major city, in its suburbs, or in a small provincial town) and 2) the opportunity of getting away from this principle residence on the number and composition of city-dwellers' relational universe. Our general hypothesis is the following: The size of the urban milieu in which a person lives and the time constraints which it imposes influence the extent of interpersonal relations.

Methodology and procedure

In order to determine the mediating role of the urban tissue on different parameters of interpersonal relationships, we used the following methodological procedures: firstly, we made comparisons between two milieus, Paris and the Parisian region on the one hand, and an average-sized provincial city (Tours), on the other; secondly, we made the distinction between two subgroups within our sample according to the opportunity, or lack of opportunity, to get away from the principle residence to a home outside of the city; and, thirdly, we controlled for certain variables with respect to our survey sample, the variability of which we anticipated to influence the development of acquaintanceship- and friendship-networks.

Within the Parisian region, we retained two zones: Paris "intra-muros," which we refer in this article as "central Paris," and the outlying suburbs, relatively distant from the city center. This distinction can be justified to the extent that those living in the suburbs are likely to have to travel longer distances than the Parisians (for work, leisure activities, an evening out, shopping, etc.) and so experience more restrictions on their time.

The sample consisted of 302 respondents, all living in relatively stable family situations: traditional-type families (couples living together), with at least one child educated to the undergraduate level, and in which the head of the family household was regularly employed. The sample was further subdivided by the opportunity to get away from the daily routine of the city during the weekends and vacations. The final sample contained 89 residents of central Paris, 57 having the opportunity to get away, 122 residents of the Paris suburbs, of whom 63 had the opportunity to "escape." In addition to this were 91 inhabitants of the smaller provincial city of Tours, 40 of whom could get away to a second house, out of the city.

Data were obtained with questionnaires made up of two sections: the first contained information on specific variables concerning each respondent and the definition of his or her network of relationships; the second part was directed at determining how they encountered other people. Each respondent made a list consisting of the people with whom he or she had a relationship which they had chosen themselves and with whom there was a feeling of affinity, while excluding at the same time all people sharing the same family dwelling unit.

Results

The research context

The length of residence was similar for those living in the Parisian suburbs and the provincial town, while those living in central Paris were clearly more mobile: 51.7% (46/89) of those living in central Paris had lived there for less than five years, comparable figures for the suburban and provincial samples were 36% (45/122) and 27% (25/91) respectively. While 81% of the provincial sample (73/90) spent a maximum of 30 minutes to reach their workplace, this was true of only half of those living in the suburbs (50.4%; 61/121) and of 39.3% (35/89) for the central Parisians, differences which are significant ($\chi^2 = 35.21$, $df 2$, $p < .01$). So the time constraints were clearly greater for the Francilians (in central Paris or the suburbs) than for those in the provincial town.

Place of residence and relationship network

Respondents mentioned on average 7.4 relationships, as we had defined these, with other people, the mean varying between different locations: 7.53 in central Paris, 6.80 in the suburbs and 8.13 in the provincial city. The provincial sample thus had significantly more interpersonal relationships than those living in the suburbs (6.80/8.13; $t = 2.56$, $p < .02$). The central Paris-suburban difference was not significant.

Table 1. Origin of relationships and place of residence

	Old relationships			Recently established relationships (territorially bond relationships)			total
	Family	School	Friends	Work	Organizations	Neighborhood	
Paris	85 13.3% **195** **30.6%**	110 17.2%	**194** **30.4%**	115 18.6%	74 11.6% **249** **39.0%**	60 9.4%	638
Suburbs	87 11.4% **174** **22.9%**	87 11.4%	**220** **28.9%**	162 21.3 %	100 13.1% **367** **48.2%**	105 13.8%	76
Provincial city	90 12.6% **172** **24.0%**	82 11.4%	**309** **43.2%**	139 19.4%	53 7.4% **235** **32.8%**	43 6.0%	716
Total	262	279	723	227	416	208	2117

Note.
Friends: Paris/Suburbs $\chi^2 = 0.37$, ns; Paris/Province $\chi^2 = 23.49$, $p < .001$; Suburbs/Province $\chi^2 = 32.6$, $p < .001$
Work/organizations/neighborhood: Paris/Suburbs $\chi^2 = 11.97$, $p < .001$; Paris/Province $\chi^2 = 5.66$, $p < .02$; Suburbs/Province $\chi^2 = 36.26$, $p < .001$

What was the structure of relationships described by respondents? Relations established initially through the family or educational institutions were equivalent, independent of the place of residence. In the provincial setting, nearly half of relationships (43.2%), were developed through friendship networks, while this was the case for slightly less than a third of central Parisians and those living in the suburbs. Moreover, in the provincial city, relationships established recently and with people in close territorial proximity, that is those which develop through work, membership in various organizations (sporting clubs, political associations, religious groups, etc) and in the neighborhood, are significantly less likely to be reported than among Parisians, particularly those living in the suburbs (see Table 1).

Nearly half (48.2%) of relationships of those living in the Parisian suburbs arose from either the workplace, through contacts in the neighborhood or from organizations in which respondents were involved. If we factor out workplace relationships, which were equivalent in the three settings, we can see that, for this sample, the network was largely developed out of local contacts, without the mediation of other friends, which is consistent with the fact that in the provincial city, respondents had significantly more relations compared to the central Paris and suburban samples (Tours: 70% [473/736]; central Paris: 63% [387/606]; Paris suburbs: 59% [416/736]).

The effect of having the opportunity to get away on the relationship network

In the provincial setting, the mean number of relations did not depend on the opportunity to get away, while in Paris and the suburbs, those who had a house outside of the city reported significantly more relationships than those who were not able to get away (see Table 2 and Figure 1).

Table 2. Extent of the relational network: Mean number of relationships

	With escape possibilities nb	With escape possibilities M	Without escape possibilities nb	Without escape possibilities M	Total nb	Total M
PARIS	448/52	8.62	218/37	6.00	666/89	7.53
SUBURBS	483/63	7.87	335/59	5.68	818/122	6.80
PROVINCE	339/40	8.48	400/51	7.86	739/91	8.13
Total	1270/155	8.28	953/147	6.52	2223/302	7.42

Figure 1. Average number of relationships according to place of residence and the possibility to escape.

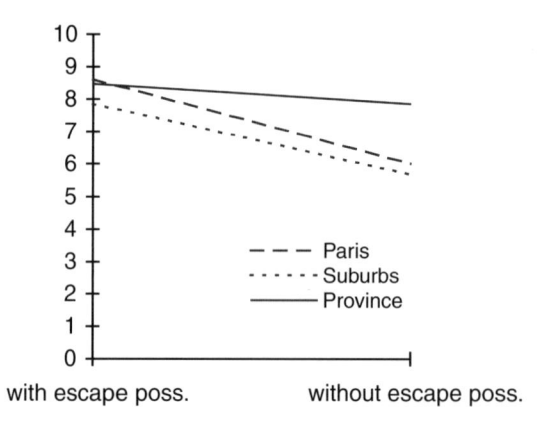

Analysis of variance:

Source of variation:	∑ of squares	DF	square means	F	significance level
Main effects:	320.207	3	106.736	7.698	.000
Paris/Suburbs/Province	100.339	2	50.169	3.618	.028
Escape/no-escape poss.	237.070	1	237.070	17.098	.000
Interaction	50.556	2	25.278	1.823	.163 ns
Explained	395.085	5	79.017	5.699	.000
Residual	4090.171	295	13.865		
Total	4485.256	300	14.951		

(301 subjects)

Patterns of relating to others

More than three-quarters of the inhabitants of the provincial town do see each other at least once a week, whereas in Paris only a little bit more than two-thirds see each other frequently. In the suburbs, there are even fewer frequent encounters with intimate friends and others (Tours: 76% [324/ 426]; central Paris: 67.2% [277/412]; Paris suburbs: 59.7% [263/440]; $\div^2 = 35.63$, d.l. 2, $p < .001$).

Table 3 shows the activity shared with friends and acquaintances when they meet together.

Just staying together, talking and drinking coffee or a glass of beer, is less frequent in Paris than in the provincial town, and even less frequent among inhabitants of the suburbs (Tours: 61.7%; central Paris: 57.5%; Parisian suburbs: 53.5% of all encounters).

When people come together with friends and acquaintances, they go out together (mostly dining out) or meet themselves during leisure or joint sport activities. Another opportunity to meet one's friends is during meetings or reunions (of clubs, vicinity-associations or political or philanthropic activities). Whereas going out together, leisure activities, and sports are equally frequent among people living in the provincial town and central Paris, the Parisian suburbs distinguishes itself by a slightly different pattern: suburbians less often go out together but share leisure and sport activities more often (Tours: 11.4% & 6.3%; central Paris: 12.7% & 5.3%; Parisian suburbs: 7.1% & 10.3% of all encounters). Furthermore, for the inhabitants of the Parisian suburbs, nearly one-third of encounters take place during meetings or reunions, while these type of encounters only constitute one-fifth for inhabitants of Tours or one-fourth for Parisians (Tours: 20.7%; central Paris: 24.6%; Parisian suburbs: 28.9%).

Table 3. Proportion of informal encounters versus other activities among all encounters

	Provincial town	Paris	Suburbs
Informal encounters	402/652 **61.7%**	302/525 **57.5%**	361/675 **53.5%**
Going out together	74/652 11.4%	67/525 12.7%	48/675 7.1%
Leisure / sports	41/652 6.3%	28/525 5.3%	68/675 10.1%
Reunions	135/652 20.7%	129/525 24.6%	198/675 28.9%
total	652 = 100%	525 = 100%	675 = 100%

Figure 2. Proportion of mere conviviality versus joint activities (all encounters)

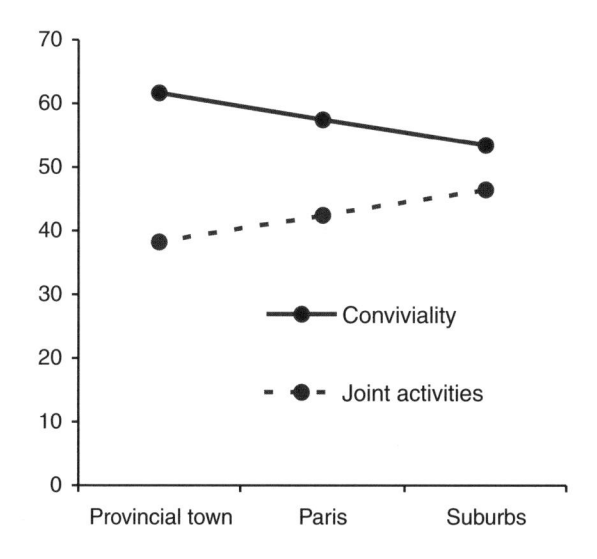

Conclusions and discussion

Parisians and those living in the suburbs of the metropolis in our sample clearly experience greater time-constraints than those living in a provincial city. One-fifth of those in the suburbs, and one-sixth of central Parisians spend more than one hour a day traveling to work. The relational network is of equivalent size between the two Paris samples, each mentioning an average of 4.7 intimate relations. This is consistent with other studies, both French (Ferrand, 1990) and English (Argyle & Henderson, 1984). Yet the composition of these relations is different when we compare the two subgroups with their provincial counterparts. In adulthood, friendship relations develop essentially around work and leisure activities (Fisher & Phillips, 1982). This was confirmed in our study for the Parisians, particularly those living in the suburbs, where friendships most often emerge from associations and organizations in which people are involved, from the work-setting and in the neighborhood. More than half of the relationships of the Parisians were accounted for by these milieus, while this was true of only one-third of the relationships in the provincial sample. The greater geographic mobility of those living in a large metropolis could account for this difference. Living in a large urban setting, people interact less with their network of friends and so make up for this deficit by forming a proximity network through neighbors, organizations and associations, and work, thus generating a specific style of life.

Despite the relative stability in the number of interpersonal relations found in many different studies (Maisonneuve, 1993), residents of the Île-de-France have significantly fewer relations if they lack the opportunity to get away from the city. It is only when such evasion is possible that the extent of relationships becomes equivalent to what is observed among those living in the provinces. This finding contradicts the results of other research, both American (Fisher, 1982) and French (Maisonneuve, 1993), which found more friends being reported in large urban areas. This difference can be accounted for by the definition of "friends": in other studies, these referred to intimate friends and not the whole network of relationships. While the number of intimate friends is not a function of geographic location, networks of more or less close friendships, as we have defined them in our study, are in fact sensitive to urban constraints.

The effects of the conditions of urban life are clearest when people do not have the opportunity to get away, and this holds true both for central Parisians and those living in the suburbs. To be confined in the urban area, without the opportunity to get away to a house out of the city, promotes spontaneous encounters with others living in the metropolis, and these are significantly more frequent, which is not the case for those living in the provincial city. It is as if the possibility of getting away allows Francilians to compensate for the impediments to conviviality which they find in large urban centers. How can these differences be explained?

For the Francilians, evasion is usually to a house belonging to the family, and the people they meet there also come from the Paris region. The number of people met in these family houses is important: one-fifth of relations are encountered during the weekends. The family house, outside of the city, provides city-dwellers with the opportunity to meet other members of the family and, in the process, to receive affective support which is lacking for those unable to escape the city and their principle residence in this way. As Amato (1993) contends, reciprocal support of friends is more important in the urban setting than in rural areas. Parallel to this, city-dwellers expect less support from their family. Amato maintains that these differences can be explained by the fact that those living in cities are further removed from their respective families than the inhabitants of smaller towns. The opportunity to get away appears to function for those living in Paris and the suburbs as an safety-valve, needed to sustain a satisfying network of relationships, to the detriment of encounters which are both less frequent and rarely spontaneous.

Patterns of relating to others show a different way of relating to others probably due to environmental constraints. In central Paris, and even more in the suburbs, people have much less occasions to meet each other frequently. They less often see their friends once a week, and when they meet their friends it is more often during one or another activity. In the Parisians suburbs, there is significantly less time for informal encounters than in Paris or the provincial town.

The possibility of getting away allows people living in the Île-de-France region to meet those with whom they have longer-lasting and more strongly anchored ties, in other words those with which the relationship is principally one of conviviality. The opportunity for evasion therefore enables city-dwellers to conserve and reinforce the nucleus of their network, and prevents the loss of friendships due to distance and less opportunity to sit together and talk.

References

Amato, P. R. (1993). Urban-rural differences in helping friends and family members. *Social Psychology Quarterly*, 56, 4, 249-262.
Argyle, M., & Henderson M. (1984). *The anatomy of relationships*. London: Heinemann.
Arling, G. (1976). The elderly widow and her family, neighbors and friends, *Journal of Marriage and the Family*, 38, 757-768.
Baum, A., & Paulus, P. (1987). Crowding. In D. Stokols & I. Altman (Eds.), *Handbook of Environmental Psychology*. New York: Wiley.
Ferrand, A. (1990). *Relations sexuelles et relations de confidence [Sexual and intimate relationships]*, p. 116. Mimeographed report. Paris: CNRS/IRESCO.
Festinger L., Schachter S., & Back K. (1950). *Social pressures in informal groups: A study of human factors in housing*. New York: Harper.
Fisher, C. S.& Phillips, S.L. (1982). Who is alone? Social characteristics of people with small networks. In L. A. Peplau & D. Perlman (Eds.), *Loneliness: A Sourcebook of Current Theory, Research and Therapy*. New York: Wiley, Interscience.
Fischer C. S. (1982). *To dwell among friends; personal network in town and city*. Chicago: University of Chicago Press.
Franck, K. A. (1980). Friends and strangers: The social experience of living in urban and non-urban settings, *Journal of Social Issues*, 36, 3, 52-71.
Héran, F. (1987). Comment les Français voisinent [How the French relate to their neighbours]. *Economie et Statistiques*, 195, 43-59.
IPSOS. (1991). *Les déplacements de Franciliens, comportements médias et urbains en Île de France [Daily commuting of the Francilians, media and urban behavior in the Isle of France]*. Mimeographed report.
Korte, C. (1980). Urban non-urban differences in social behavior and social psychological models of urban impact. *Journal of Social Issues*, 36, 29-51
Lacaze, J.P. (1993). Interview. In P. Benoit, J-M. Benoit, F. Bellanger, & B. Marzloff (Eds.), *Le Grand Desserrement. Enquête sur 11 millions de Franciliens [The great slackening. An investigation of 11 million Francilians]*. Paris: Romillat.

Maisonneuve, J., & Lamy, L. (1993). *Psycho-sociologie de l'amitié [Psychosociology of friendship]*. Paris: Presses Universitaires de France.

Moser, G. (1992). *Les stress urbains [Urban stress]*. Paris: Armand Colin.

Moser, G. (1994). *Les relations interpersonnelles [Interpersonal relationships]*. Paris: Presses Universitaires de France.

Newcomb, T.M. (1961). *The Acquaintance Process*. New York: Holt, Rinehart & Winston.

Oppong, J.R., Ironside, R.G., & Kennedy, L.W. (1988). Perceived quality of life in a center-periphery framework, *Social Indicators Research*, 20, 605-620.

Palisi, B.J. (1985). Formal and informal participation in urban areas. *Journal of Social Psychology*, 125, 4, 429-447

Sèze, C. (1994). La modification. Autrement. Confort moderne. Une nouvelle culture du bien-être [Modification. Differently. Modern comfort. A new culture of wellbeing]. *Sciences en Société*, 10, 100-124.

Stokols, D. (1978). A typology of crowding experiences. In A. Baum & Y. Epstein (Eds.), *Human response to crowding*. Hillsdale, N.J.: Erlbaum.

Sundstrom, E. (1978). Crowding as a sequential process: Review of research on the effects of population density on humans. In A. Baum & Y. Epstein (Eds.), *Human response to crowding*. Hillsdale, N.J.: Erlbaum.

Planned gentrification as a means of urban regeneration

Miriam Billig & Arza Churchman

Bar Ilan University, Israel; Technion – Israel Institute of Technology

Abstract. Neighborhood change in the guise of gentrification has been widely criticized as having negative impacts of various kinds. This paper describes a different kind of gentrification, planned gentrification, and compares its implications with those of spontaneous gentrification processes. The study examined the similarities and differences existing between this planned gentrification process and the spontaneous gentrification process described in the literature by analyzing the changes resulting from the building of six new housing developments in lower class neighborhoods in the city of Ramat Gan, Israel. The analysis is based on an ethnographic description of 240 interviews of women living in the new housing developments and in the adjacent old buildings. The results show that an advantage of the planned gentrification process is that it does not undesirably affect the original population and does not lead to their displacement. The planned gentrification process can have positive socio-physical and behavioral effects on the old neighborhood.

Keywords: neighborhood regeneration, gentrification, neighborhood planning

Introduction

Neighborhood processes of change

A number of urban mobility processes inducing change in neighborhoods have been described in the literature. Some refer to the stages of neighborhood deterioration as the process of "squatting, displacing, and inheriting" and "filtering down" (Downs, 1979;

Fisher, 1984; Goetze, 1979; Wood, 1991). Others talk of rehabilitation processes such as "filtering up" and gentrification (Carmon & Hill, 1988; Clark, 1991; Gale, 1984; London, 1980; Smith & William, 1986; Schnell & Graicer, 1993; Spain, 1992), or the process of incumbent upgrading by original blue collar residents of the neighborhood who refurbish and improve their homes (Clay, 1979), and the moving of new immigrants into downtown (Winnick, 1990).

From among the above-mentioned processes, gentrification is the most relevant to this study. The gentrification process described in the literature is characterized as an essentially spontaneous occurrence, and we therefore call it **spontaneous gentrification.** In this process middle and upper middle class people move into working class neighborhoods, particularly into houses of old architectural styles, refurbish these houses and render the area attractive also to the less daring, more established groups. As a result of pressure by the new residents, the local authority will overhaul and upgrade the neighborhood's infrastructure. This process leads to a rise in the cost of living in the area, caused by an increase in real estate values, in municipal tax rates and in prices at the local shops. All these often lead to a gradual displacement of the original residents from the neighborhood (Smith, 1996; Schnell & Graicer, 1993).

Comparisons of gentrification processes in various countries show them to differ from one place to another, in Europe versus the USA (Gale, 1984) or between different places in Europe (Carpenter & Lees, 1995; Smith, 1991). These differences are due, among other things, to the character of the place and to tastes and preferences of the local middle class, which cause an increased demand for the regeneration of certain neighborhoods over others (Gale, 1990). For example, Gonen and Cohen (1989) state that in Israel gentrification occurs in various parts of the city, not necessarily in the center or in neighborhoods with houses of old architectural styles.

The left hand column in Table 1 presents the characteristics and motivations of people engaged in spontaneous gentrification, as described in the literature.

In spite of differences in the intensity of spontaneous gentrification processes, a marked effect on the social and physical environment of the city center has often been discernable. This process has been found to counteract the ongoing decline of the centers of large cities caused by the moving of the middle class population to the suburbs (Downs, 1979). Goetze (1979) claimed that the regeneration process expressed a renewed confidence in the future of the neighborhood, which could be attributed to the emergence of spontaneous market forces, and contrasted this to the earlier urban rehabilitation processes initiated by public authorities. Many neglected areas and sites of unutilized land were found to exist in many cities; sites that could be redeveloped from less productive areas into efficient housing and commercial developments without requiring any new land. Such developments can enhance the status of the entire area (Blair et al., 1996).

However, a recent comprehensive review of the literature concluded that the majority of the research evidence on gentrification points to its detrimental effects on the neighborhood (Atkinson, 2002).

We can see that the issues raised by the literature relate to the characteristics of neighborhood change processes. Our research was designed to add another dimension, namely the question whether or not the authorities should intervene in order to bring about change processes intended to be more positive for the residents and for the city.

Unutilized land in the city and deteriorated older buildings tend to become locations of crime. The proximity of polluting industrial zones to residential neighborhoods may lead residents to move out of the area and thus to further deterioration of the neighbor-

Table 1. Characteristics and motivations of spontaneous gentrification population and of the planned gentrification population sample

Characteristics & Motivation	Spontaneous Gentrification Population as Described in the Literature	Planned Gentrification Population Described in this Study
Socio-economic status	Middle to upper middle class, higher education, professionals, both spouses employed	Middle to upper middle class, higher education, professionals, office and sales personnel, both spouses employed
Age and stage in life cycle	25 to 35, some 35 to 41, single with or without partner or married with no children	31 to 40, 41 to 60, married with mostly young children
Previous place of residence	Rented home in the area or in nearby neighborhood or in the suburbs	Own dwelling in the vicinity, in or not far from the city
Social and cultural motivations	Desire to become home owners, to live in relatively high density neighborhoods, social pluralism, proximity to shopping and entertainment, status symbol, being "in", stylish architecture	Desire to improve standard of housing, social homogeneity, quiet vicinity, proximity to relatives and to public services, new prestigious home, desirable features
Economic motivations	Affordable cost of housing, ability to buy first home, taking advantage of available housing supply, speculative acquisition, expected future profit, saving gas and commuting time	Affordable, prestigious, relatively large apartment, high quality second apartment, exploitation of available land, speculative acquisition and expected capital gain, accessibility to main roads.

hood. All of these will detrimentally affect the urban fabric. Regeneration of the urban fabric could be achieved in such cases by building new housing developments on the unutilized land or instead of the deteriorated buildings, or by moving the polluting industries elsewhere and rezoning the area for housing. This may then lead to rejuvenation of the neighborhood. Such action would be of particular importance in cities having limited resources of available land.

In Israel, as in many other countries, this is often a critical problem. The increasing shortage of available land requires maximum utilization of urban space for housing. As a result we see new multi-family buildings being built in old neighborhoods. One can notice the extent of this process by driving through neighborhoods where new white modern buildings surrounded by nice gardens have been built next to old buildings. Do these new buildings constitute a threat to the old environment, or do they actually improve the neighborhood image? This is a crucial question for urban planners and decision-makers.

The Study

We undertook analyses of the processes occurring as a result of building such housing developments in the city of Ramat Gan, Israel, and the way they have affected the old neighborhoods. This paper describes the characteristics and effects of the initiated changes occurring as a result of building new housing developments in relatively distressed urban areas. The term *planned gentrification* will be used to describe the process of change in the neighborhood resulting from the new housing development and from the moving in of a population of relatively high socio-economic status. We will compare the degree of similarity and dissimilarity in the characteristics of the planned gentrification process with that of the spontaneous gentrification described in the literature, and the social, physical, and behavioral effects of this process.

Research methodology

The study combined an anthropological approach with a person-environment approach that emphasizes the importance of physical, social, cultural, and individual variables in the study. This enabled a broad and in-depth understanding of the environments studied from the point of view of the women who live there, focusing on social, physical, and behavioral aspects[1]. No prior assumptions were made as to the type of urban regeneration process to be expected, with the intention of identifying the process through the ethnographic analysis. This allowed for the identification of frameworks other than those mentioned in the literature. The methodology enabled those interviewed to describe changes in the social and physical environment resulting from the housing development, from their subjective point of view.

The results included an ethnographic description based on 240 individual interviews of women,[2] 120 of them living in the new buildings of the various housing developments and 120 living in the adjacent older buildings. The purpose was to explore the different aspects of the changes occurring in the area, as seen by the new residents who were the cause of change and by the original population on whom this change was imposed. The study therefore included only housing developments bordering on old buildings in the neighborhood, and where the arrangement of the buildings enabled the new and the original residents to see each other.

The study was performed entirely in the city of Ramat Gan, in new housing developments built in old neighborhoods, and completed between 1990 and 1994. The city council's initiative was designed to attract middle to upper middle class residents to neglected neighborhoods inhabited by residents of low socio-economic class.[3] Unutilized plots of land were identified within the city, in troubled or distressed areas as defined by the city council. Buildings in an unacceptable state of deterioration and old industrial buildings constituting a threat to the environment that had to be demolished were also identified.

---

1 The subject of the relations between the different population groups is discussed in M. Billig & A. Churchman (2003).
2 The research was based on interviewing women only, because they tend to be more often present in and involved with the neighborhood.
3 In Israel, about 70 % of the dwellings are in multi-family buildings, with the apartments owned by the residents in a condominium-type arrangement.

Planned gentrification as a means of urban regeneration

The city of Ramat Gan issued calls for proposals to private investors to build modern prestigious housing developments on these sites. This study focused on six such developments (see Figures 1 and 2 for two examples), similar in size, and within or adjacent to old residential neighborhoods.

Figure 1. Yahalom. In the foreground are the old two-story buildings, some of which have been refurbished and enlarged. In the background are the five new buildings arranged in a star shape, with a wall separating them from the old buildings.

Figure 2. Leshem. To the left of the street are the new buildings. Opposite these are old buildings, similar in height and in parking spaces.

Table 2 describes the main characteristics of the housing developments studied and their environments. This description is provided so as to give the reader some idea of the particular physical environments being discussed.

Table 2. Characteristics of housing developments studied

Name of housing development	Number of buildings		Height of buildings (No. of storeys)	
	New	Adjacent old	New	Adjacent old
Sapir	5	9	3–5	2
Yahalom	5	9	7–13	2
Odem	5	9	7–15	1–4
Leshem	5	6	7–8	3–7
Tarshish	5	11	7–15	3–6
Bareket	7	13	7	3–4

The older buildings that were part of the study were those along the streets adjacent to the new developments, and located in such a way that the new and the original residents could see each other. This area was assumed to be the most likely to be affected by the existence of the housing development and the resultant changes in the area. The housing development's surroundings also included the public services belonging to the neighborhood or located near the housing development.

The sample

Table 3 presents a summary of the most relevant characteristics of the sample. In two of the cases (Tarshish and Bareket), we found that new residents had also moved into the older buildings, and they were younger and better educated than the long-term residents. As a result we divided the interviewees into three groups: 1) New residents in the new buildings; 2) Long-term residents in the old buildings; 3) Residents (long-term and new) in older buildings that are partly inhabited by new residents.

Age differences between the groups were such that the new residents were younger than the long-term residents: 55% of the former were aged up to 40 years, while only 32% of the latter were in this age group. The third group was more evenly divided between the age groups, reflecting a combination of younger, new residents and long-term, older residents. Among the new residents there were very few families with no children, while there were more of these in the other groups; a reflection of the differences in age. The difference in education level is evident in the relatively low percentage of women among the long-term residents who have a higher education and the large percentage of those with less than a full high school education. This is also the group with considerably fewer employed women than in the other groups.

Table 3. Sample Characteristics According to Groups (in %)

Characteristic		Residents in the new developments $N = 123$	Residents in existing buildings with no new residents $N = 78$	Residents (long-term *and* new) in existing buildings with new residents* $N = 39$	Statistical significance
Age	21–30	14	18	23	$\chi^2 = 30.25$
	31–40	41	14	21	$df = 8$
	41–50	25	26	21	$p < .000$
	51–60	12	14	26	
	61–90	8	27	10	
Number of children	0	18	31	36	$\chi^2 = 22.23$
	1–3	74	54	59	$df = 4$
	4 or more	8	15	5	$p < .016$
Education	11 or less	9	49	17	$\chi^2 = 46.65$
	12–15	56	42	64	$df = 4$
	16 or more	35	10	18	$p < .000$
Occupation	Professional	50	29	49	$\chi^2 = 15.97$
	Sales	23	20	15	$df = 6$
	Pensioners	10	13	15	$p < .014$
	Unemployed	17	38	21	

* This group exists in two cases, where new residents moved in to the older buildings after the new developments were built.

Results

Planned gentrification and its effects

The planned gentrification process was characterized by middle to upper middle class people moving into new housing developments built through the initiative of the local authority. The right hand column in Table 1 summarizes the characteristics and motivations found to be typical of this population, and Table 4 summarizes the characteristics of the planned gentrification process, as compared with the spontaneous gentrification one.

Besides serving the needs of the residents of the new housing developments, the planned gentrification process was found to have a significant impact on the adjacent areas in the old neighborhoods. It was found to have improved the image of the old neighborhood and its residents, leading them to take better care of their living environment and to refurbish their homes. It led to many of their adult children remaining in the neighborhood, and even triggered spontaneous gentrification in a number of old buildings. The new housing developments have also brought about investments in infrastructure and public services, and a perceived increase in real estate value and tax revenues. These findings will be described below:

Table 4. Comparison of spontaneous and planned gentrification processes

Spontaneous Gentrification Process According to the Literature	**Planned Gentrification Process According to this Study**
Group of middle to upper middle class people moving into traditional working class area	Group of middle to upper middle class people moving into traditional working class area
Refurbishment of old buildings by the new residents	*New buildings built in open areas or in the place of demolished old buildings or factories*
Area becomes "in demand"	Area becomes "in demand"
Value of homes increases	Value of homes increases
Mainly in center of city	*In any part of the city*
Mainly buildings of old architectural style	*New buildings of modern style*
Upgrading by local authority of infrastructure in the neighborhood	Upgrading by local authority of infra- structure, in the part of the neighborhood in the vicinity of the new buildings

Note: Italics indicate difference between spontaneous and planned gentrification, plain type indicates similarities betwen the two processes.

Increased care for the housing environment

The general awareness of and care for the environment was found to have increased, even among people living in buildings that had not been refurbished. People started to tend their gardens or improve the appearance of their buildings. One of the women interviewed reported:

"If I take a look at our street it seems funny: On one side you see a 50 year old building like in Harlem, and next to it a gorgeous new house. We tried to improve the appearance of our building by ourselves, but a number of residents are older people and some others are government-subsidized tenants, so we couldn't do it. I would be ashamed if my daughter had a friend come to our home. The building is full of graffiti, the garden is totally neglected and the mailboxes are ruined. When I look through the window I see those green gardens of the new housing development. It doesn't really bother me, no envy from my part. I just like everything to be nice and neat, it's like in paradise."

Incumbent refurbishment of homes

A process of incumbent refurbishment of homes by their original residents was found to occur in old buildings adjacent to the Sapir, Yahalom, and Odem developments. This happened in addition to a general refurbishment of these buildings initiated by the city council, which had taken place a number of years earlier. These longterm residents now began to enlarge and further refurbish their homes. They mostly belonged to the younger generation who grew up in the neighborhood. They converted 2½-room apartments into 4- to 5-room ones, and regarded them as even superior to the flats in the new 8 to 16 story buildings that had no balconies. A young woman living in one of the old buildings adjacent to the Yahalom development said: "When I look at these new buildings I don't think I would want to live there. Our homes all have a little garden or a rooftop, which theirs don't. They probably envy us when they smell our barbecues."

Remaining of the second generation in the neighborhood

Another indication was the desire among the second generation who grew up in the old neighborhood to buy an apartment in their neighborhood as they got married. One of the young women described this as follows:

> "Before, young people wanted to get away from here. They did not want to carry the negative stigma of living in the neighborhood, so anybody who could manage, left. Today things are just the other way round. They all want to stay, because where else could you find such a luxury home at such a low price."

Table 3 shows a considerable number of local young residents still living in the old buildings (18% up to the age of 30 years and another 14% between 31 and 40 years). A number of women interviewed talked about the competition among mothers living in the neighborhood to buy any apartment becoming available for their children.

Stimulation of spontaneous gentrification

The process of planned gentrification was also found to stimulate spontaneous gentrification in older buildings adjacent to the new developments. This was particularly so for the Tarshish and Bareket developments, which are located near the city center. The women interviewed reported that any apartment becoming available in the area was being bought and refurbished by young couples or singles with a higher education. A young woman living in one of the old buildings described this as follows:

> "Maybe we are aware of them because of the very obvious differences between the older and the younger people. We have people who have been living in this neighborhood for a long time. Though we aren't very much in touch with them, they certainly don't bother us. The younger population is increasing all the time. The old people are gradually dying and are being replaced by young people. Out of nine apartments in this building, young people inhabit three and two additional ones are presently empty. Two residents in this building passed away this year and their apartments were taken over by young people."

A resident living in another building in the same street described a similar process:

> "Out of 12 apartments in our building, five have been rented by young people and that's the way it is throughout the street. Whoever buys an apartment does so as a

first investment. The apartments here are very small and people will therefore leave as soon as their families become larger.

Young people who moved into the old buildings refurbished their apartments, as described by one of those interviewed: "Walking on the street and looking at the old buildings, you can immediately tell by the new windows, blinds and laundry, where young people have moved in."

The spontaneous gentrification process in old buildings adjacent to the new housing developments differed from the planned gentrification process in the new buildings, though a more educated population of higher socio-economic status brought both about. Besides differences in the type of buildings, their residents were also different. The people moving into the old buildings were younger, many of them single or married without children, or with small children. Many were renting their apartment and for others this was their first home ownership situation. They were not economically "settled" in most cases, and were open-minded towards social heterogeneity. The people who moved into the new buildings on the other hand, were economically settled and preferred social homogeneity.

Investment by the city council in infrastructure and public services

As a result of a demand by the new residents living in the new housing developments, the city began to overhaul and upgrade the urban infrastructure, including sidewalks, lighting, reserved parking places and public services in the vicinity of the developments. Schools were refurbished and new shops opened. Public gardens and tree-lined promenades were provided near some developments, to be used by new as well as long-term residents. A resident of the new Bareket development described this as follows:

> "Our buildings are regarded as prestigious, while their buildings are very old and their standard of living very low. It was very objectionable: We had nice new lampposts in the development, while their lampposts were old wooden ones. As a result of our demand, the city later replaced them all by new ones. We new residents do have some leverage; therefore everything is nice and well maintained on our side, including lots of tropical plants. On their side you will only find a few shabby looking trees planted by the city. We will act to change the city council's attitude, since their buildings and gardens are also part of our landscape."

In many cases there were marked improvements in the public services provided as a result of the new housing developments and the new residents, whether due to the initiative of the city, the development project management or the new residents.

Increase in perceived real estate value

An immediate effect of building the new housing developments in place of polluting factories was the increase in value of apartments in the adjacent old buildings, as reported by the women interviewed. Local real estate agents reported an increase in demand for old apartments, which until then could not be sold because of air pollution, bad smell and noise emanating from the factories. A long-term resident living in an old building until recently bordering on the industrial zone, described the difference as follows:

> "Now the neighborhood is OK, but for 35 years it was horrible. There used to be a factory, it was very unpleasant, a smell of sugar, workers hanging around, warehouses and trucks. Now it's wonderful, a nice and clean environment, good people, nice view. I am an old woman sitting in front of the window and watching it all."

A resident of one of the old buildings adjacent to the Sapir development remarked: "Those new buildings raised the status and prestige of the neighborhood. Prices of apartments went up because of the new tall buildings."

Increased tax revenues

A further advantage of the planned gentrification process is the possible increase in municipal tax revenues, obtained by charging higher municipal tax rates for the new buildings inhabited by the new population than for areas inhabited by the long-term residents. Thus the new population of higher socio-economic status increases the tax revenues of the city from the neighborhood, without penalizing the original lower income population. A resident of one of the old buildings near the Sapir development describes this as follows:

> "We are not a high-class neighborhood. Whoever lives in the new development is regarded as living in a high-class neighborhood, and therefore pays higher tax rates. The place where we live is listed as a low rate area, for low-income families. We pay much less, although we live just across the street."

Discussion and conclusions

This study has found the planned gentrification process to have a number of advantages over the spontaneous gentrification process. One such advantage is the beneficial effect of planned gentrification on the social and physical characteristics of the neighborhood. The presence of high standard multi-family buildings for the new population can improve the neighborhood's image and serve as an example for the original population of the neighborhood. It tends to stimulate the refurbishment of old buildings by young people who move into the neighborhood, and by young people of the second generation who grew up in the neighborhood; similar to the process of incumbent upgrading described by Clay (1979). This clearly represents a process of revitalization of the old neighborhood. Additional improvement of the neighborhood can also be brought about by leverage exercised by the new population on the city council to invest more in upgrading of the existing infrastructure in the neighborhood.

The change in social structure of the neighborhood can also have beneficial effects. The original population not being displaced, the second generation remaining in the neighborhood, young educated people moving into the old buildings, middle to upper middle class families moving into the new buildings, all can contribute to regeneration of the neighborhood without penalizing the original population.

The experience gained in the city of Ramat Gan, by building housing developments of five to six buildings with 180 to 250 apartments per development, has shown that relatively small developments can have a very desirable effect on the old environment. The building of such small size housing developments is therefore recommended as a possible strategy for the regeneration of distressed neighborhoods of the city. As shown in this study, the beneficial effect may initially be local, limited primarily to the population living in the adjacent buildings. However, in the course of time this process of change may well transfer to more distant areas of the neighborhood. This aspect could be the subject of further investigation in a follow-up study.

A more in-depth study of the planned gentrification process is justified. Would any new housing development built in any existing neighborhood generate such a process? Does it require the existence of particular conditions? The hypothesis was raised in the course of the study that the planned gentrification process might also serve the needs of people previously involved in spontaneous gentrification, but who wished to move because their families had grown to include several children. It would be worthwhile to investigate the relation between the two processes and whether they involve the same population. In two cases in this study, spontaneous gentrification was found to occur in buildings adjacent to the new housing developments, as a result of the planned gentrification process. The question as to whether this is a coincidence, or whether this can be predicted to occur under certain circumstances, should be further investigated.

Marked beneficial effects have been noted in the behavior and attitudinal changes among the original population. This was an unexpected outcome of the planned gentrification process. Many of these residents did not remain unaffected by the examples created by the new buildings and the new population of higher socio-economic status. In some cases there seemed to be an almost direct influence, such as refurbishing the façade of their buildings or enlarging their homes. Sometimes there was a more indirect influence, such as a change in the concept of the desirable façades for their buildings. One may conclude from this that an intercultural social discourse may take place, in spite of the absence of much social interaction, provided the two populations live in dwelling proximity and see each other, as occurred in this case.

Attempting to impose encounters on the two populations is counter-productive and unnecessary, since mutual influence occurs in any case. On the contrary, spontaneously initiated emulation is more desirable. The new population makes the long-term residents aware of the existence of a variety of alternative role models of behavior and attitudes. The advantage is that no particular behavior is imposed, but it is left to the person concerned to choose between various role models. This increases the probability of achieving a beneficial influence. The gentrifying population thus provides role models previously not available in the neighborhood, which can serve as stimulants for rejuvenation and change of image of the old environment.

As in the case of spontaneous gentrification, an increase in real estate values of old buildings resulting from the planned gentrification process was reported by those interviewed. In this case of planned gentrification, however, the original population remained in the neighborhood, and their second generation also wanted to remain there. This may be explained by the fact that no competition was caused between the new and the old populations in the acquisition of old buildings in the neighborhood. The higher-class population was attracted by the availability of apartments in the new buildings, which in turn stimulated the second generation of long-term residents to buy apartments in the old buildings. The resulting increase in property value of the old buildings thus enabled a lessening of the socio-economic differences between the new and original populations.

Since the new buildings are built as a unit, the city council could differentiate between the municipal taxes, charging higher rates for the new population living in the new buildings, while retaining the lower rates for the original population. This too contributed to the remaining in the neighborhood of the original residents; whereas in the case of spontaneous gentrification, across the board increases in tax rates are among the causes for displacement of the original population from the neighborhood.

Above all, planned gentrification enables the rehabilitation and regeneration of old neighborhoods at relatively little cost to the public. Financing of such housing developments is done mainly by private investors and by the new residents, while the original long-term residents share the resulting benefits of enhanced status of the neighborhood and increased value of their property.

The similarity between the planned gentrification process as described in this study and the spontaneous gentrification process as described in the literature (Atkinson, 2002; Carmon & Hill, 1988; Clark, 1991; London, 1980; Smith & William, 1986; Smith, 1996; Gale, 1984; Schnell & Graicer, 1993; Spain, 1992;) is primarily in the moving in of a relatively high class population into a neighborhood of lower class residents. This process can create a renewed demand for the area, which in turn can cause an increase in the value of real estate in the old neighborhood. A process of "filtering up" may be created, reflecting a renewed confidence in the future of the neighborhood (Goetze, 1979). The most significant difference between the two processes is that in the planned gentrification process, moving in of the new population does not lead to displacement of the original population. Planned gentrification also may enable efficient utilization of non-productive land inside the existing neighborhoods, in accordance with the findings of Blair (1996).

Thus, planned gentrification by building new housing developments inside distressed neighborhoods was found to be a suitable method for upgrading the status of such neighborhoods. This study has shown that local authorities can initiate planned gentrification of distressed areas with the help of private investors, thereby simultaneously solving a number of problems: a) increasing the supply of homes in the city by building housing developments on unutilized land, b) diversifying the population of socially homogeneous neighborhoods by attracting people of higher socio-economic status, c) improving public services in these areas, and d) enhancing the image of these neighborhoods.

References

Atkinson, R. (2002). Does gentrification help or harm urban neighborhoods? An assessment of the evidence-base in the context of the new urban agenda. Paper presented at the 32nd meeting of the Urban Affairs Association, Boston.

Billig, M., & Churchman, A. (2003). Building walls of brick and breaching walls of separation. *Environment and Behavior*, 35(2), 227-249.

Blair, J. P., Staley, S. R., & Zhang, Z. (1996). The central city elasticity hypothesis – A critical appraisal of Rusk's theory of urban development. *APA Journal*. Summer, pp. 345-353.

Carmon, N., & Hill, M. (1988). Neighborhood rehabilitation without relocation or gentrification. *Journal of the American Planning Association*, 54(4), 470-481.

Carpenter, J., & Lees, L. (1995). Gentrification in New York, London and Paris: An

international comparison, *International Journal of Urban and Regional Research,* 19, 286-303.
Clay, P.L. (1979). *Neighborhood Renewal: Middle Class Resettlement and Incumbent Upgrading in American Neighborhoods.* Lexington, MA: Lexington Books.
Clark, E. (1991). On gaps in gentrification theory, *Housing Studies,* 7(1), 16-26.
Downs, A. (1979). Key relations between urban development and neighborhood change, *Journal of the American Planning Association,* 45(4), 462-472.
Fischer, C.S. (1984). *The Urban Experience,* San Diego, CA.: Harcourt.
Gale, D. (1984). *Neighborhood Revitalization and the Post Industrial City.* Lexington, MA: Lexington Books.
Gale, D. (1990). Conceptual issues in neighborhood decline and revitalization. In N. Carmon (Ed.), *Neighborhood Policy and Programs: Past and Present.* London: Macmillan.
Goetze, R. (1979*).* *Understanding Neighborhood Change.* Cambridge: Ballinger.
Gonen, A., & Cohen, G. (1989). Multi-Facetted Gentrification Process in Neighborhoods of Jerusalem. *City and Region,* 19-20, 9-27. (Hebrew).
London, B. (1980). Gentrification as urban reinvasion: Some preliminary definitional and theoretical considerations. In S.B. Laska & D. Spain (Eds.), *Back to the City.* New York: Pergamon Press.
Shnell, I., & Graicer I. (1993). Causes of in-migration to Tel-Aviv inner city, *Urban Studies,* 30, 1187-1207.
Smith, N., & Williams P. (1986). *Gentrification of the City.* Boston: Allen & Unwin Press.
Smith, N. (1991). On gaps in our knowledge of gentrification. In van Weesep and Musterd (Eds.), *Urban Housing for the Better-Off: Gentrification in Europe.* Utrecht: Stedelijke Netwerken.
Smith, N. (1996). *The New Urban Frontier. Gentrification and the Revanchist City.* New York: Routledge.
Spain, D. (1992). A gentrification research agenda for the 1990s. *Journal of Urban Affairs,* 14(2), 125-134.
Winnick, L. (1990) *New People in Old Neighborhoods.* New York: Russell Sage Foundation.
Wood, C. (1991) Urban renewal: The British experience. In R. Alterman & G. Cars (Eds), *Neighborhood Regeneration: An International Evaluation.* London and New York: Manssel.

Ubiquitous technology, the media age, and the ideal of sustainability

Andrew D. Seidel

Texas A&M University, College Station, TX, USA

> **Abstract.** As technology proliferates and decreases the conceptual distance between people and countries, another gap widens, further separating the information-rich and the information-poor and affecting the physical environment. While the Internet, for example, was once touted to be a liberating and decentralizing force, the results, thus far, have been the opposite. What technology has become is a centralizing force that is one of many factors leading to the centralization of decision-making, wealth, and the world of work. In turn, these changes have affected other areas. To explain this phenomenon, this paper briefly examines the changes occurring in how wealth and work are distributed around the world; and, then, how changes in technology (i.e., the Internet), media, and the ideal of sustainability may be following a similarly centralizing pattern.
>
> **Keywords:** sustainability, technology, environment, media, Internet, community

Introduction

Gary Gumpert and Susan Drucker often do us the favor of organizing a session on a topic that is both extremely timely and highly interesting. Yet, these topics are also often thought-provoking and highly controversial. This is the case here. This paper, while a form of thinking out loud – a speculative think piece – also provides the opportunity to explore connections that have been too little explored. In this case, it is a brief discussion of changes that are occurring worldwide concerning the centralization of wealth and the world of work and how similar changes have occurred in ubiquitous technology, the media age, and the ideal of sustainability.

The digitalization of the globe (the technology under consideration here) has affected the way in which people coexist with one another and with the natural and built environments. No longer are people isolated by distance, but, instead, they utilize technology to coexist, communicate, and span these distances. However, while some view the world as shrinking, others are becoming more secluded by the modernization of the human lifestyle.

One could argue that many factors shape perceptions concerning the interconnectedness of today's world. Those who live in the midst of technological progress have seen its far-reaching effects and would be the first supporters of a digitized society. However, those who live in areas where technology is in a primitive or infantile state cannot yet embrace this modernization; not because they are not willing, but more so because they are unable. Thus, it can be argued that there is an unequal distribution of resources around the globe, which has a profound effect on how people live and work in relation to modernization.

While originally intended to be a decentralizing force, the Internet and technology have achieved the opposite effect by centralizing wealth and the labor market, in general. Robert Reich defined four categories of work that can be ranked in order of demand or order of importance. The knowledge brokers, or those on the upper end of the scale, are seeing their incomes and their opportunities increase. Those who hold jobs at the lower end of the scale are seeing their incomes and workload decrease because it is being contracted out to those elsewhere on the globe who will do it for less money. Therefore, it has become commonplace in today's world that work always goes to the cheapest labor, no matter what the location, labor and environmental regulation, or other context of that work. Even work once regarded as "thought work" (or "knowledge brokering" in Reich's terms) is now being exported for pennies on the dollar, and, consequently, that knowledge work becomes redefined to be more routinized.

Similarly, the built environment has been affected by this centralization of wealth. One prominent example is the growth of call-centers as a building type in post-industrial countries, such as Australia, Ireland, and the United States. We once relied on door-to-door sales. However, the exportation of labor soon facilitated the rise of department stores and, then, discount stores. Now, call-centers are emerging because contracting cheap labor in countries worldwide proves less expensive than the international sales phone calls. Workers in Australia are trained in an American accent so the calls can appear to be local. In a way, we are automating our society and depending less and less on human interaction.

As Reich's work-type pattern predicted, wealth is increasing among the wealthy, and the poor are forced to settle for even less in order to maintain work. Thus, this overall pattern helps to manage the issues of technology, the media, and sustainability, as well as the affect they each have on the increasing centralization of work and wealth. To help explain the correlation between these topics, we must first address them separately and, then, as interrelated subjects.

Ubiquitous technology

In modern life, technology is seemingly ubiquitous. Businesses, academic institutions, and individuals all rely heavily on the use of computers, as well as the Internet. However, while many view technology as essential to today's way of life, in some areas of the world, this is not so. In fact, it can be seen that technology is positively associated with the amount of wealth in a particular area.

The Internet is perhaps one of the best technological indicators. Therefore, in order to measure the prevalence of technology, many projects have been aimed at finding Internet penetration rates in countries around the world. One such example is a project by the International Telecommunications Union in which they are conducting a series of case studies on the diffusion of the Internet in countries at different stages of development. "The aim of the project is to seek to understand the factors, which accelerate or retard the development of the Internet in different environments and, through comparative analysis, to advise policy makers and regulatory agencies on appropriate courses of action" (Minges, 2001).

The first series in the study concentrated on Bolivia, Egypt, Hungary, Nepal, Singapore, and Uganda, all of which vary widely in socioeconomic status and levels of telecommunications development. Nepal and Uganda are among the least developed nations in the world and have extremely low telephone densities. On the other end of the spectrum is Singapore, the eighth richest country in the world on a GDP per capita scale, where there is nearly one telephone per household (Minges, 2001).

The study focused on the GDP per capita and the amount of Internet penetration in each country. Thus, it was found that in countries with substantially higher GDP per capita, the Internet was more prevalent (Figure 1). As of January 2000, the GDP per capita in Nepal is $1,219, and the number of Internet users was at 35,000. However, this number accounts for only 0.15% of the population. In contrast, Singapore, dubbed the "e-city," had a much higher rate of Internet penetration. With a GDP per capita of $27,024 and a population of 4 million, Singapore had 1.8 million Internet subscribers in January 2000 (Minges, 2001).

Figure 1. 1999 fiscal year. Basic indicators.

	Populations (millions)	GDP per capita	Telephone Density	Mobile phone Density	PC Density	Internet Density
Bolivia	8.14	$2,193	6.17	5.16	1.23	0.96
Egypt	62.43	$3,302	7.51	0.77	1.20	0.32
Hungary	10.04	$10,479	37.09	16.21	7.47	5.97
Nepal	22.37	$1,219	1.13	0.02	0.27	0.16
Singapore	3.89	$27,024	48.20	41.88	43.66	24.40
Uganda	21.62	$1,136	0.26	0.26	0.25	0.12

Source: ITU World Telecommunication Indicators Database, as cited in Minges.

In addition to individual earnings, the governing bodies also had an impact on the prevalence and diffusion of the Internet in the countries of study. Nepal, for example, was once monopolized by a single phone service, which hindered the development of the Internet. Conversely, Singapore's government implemented Singapore One, a country-wide network, and promoted high speed Internet access and the development of broadband and infrastructure services (Minges, 2001). Therefore, wealth, not only of the individual but also of the state, influences how quickly and the extent to which technological advances are available.

Figure 2. Where are the world's poor?

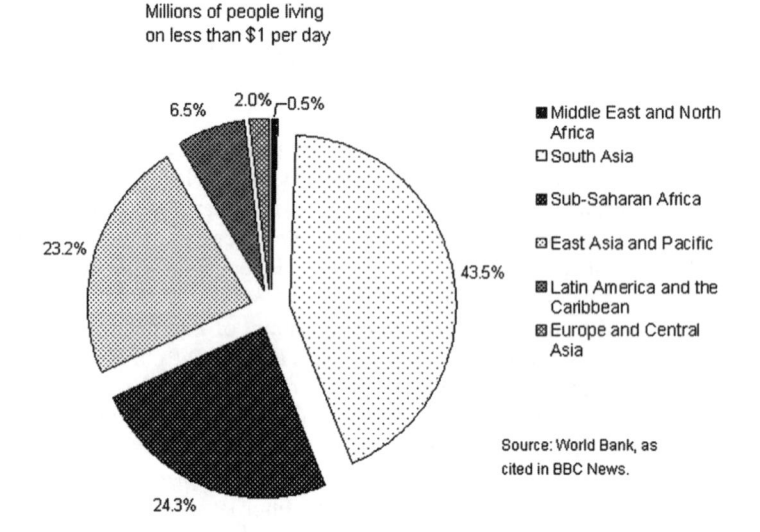

Millions of people living on less than $1 per day

- Middle East and North Africa: 2.0%
- South Asia: 43.5%
- Sub-Saharan Africa: 24.3%
- East Asia and Pacific: 23.2%
- Latin America and the Caribbean: 0.5%
- Europe and Central Asia: 6.5%

Source: World Bank, as cited in BBC News.

The pie chart in Figure 2 gives the percentage of the world's poor according to the region in which they live. As the chart indicates, the majority of the world's poor live in South Asia. Ironically, the region of southern Asia houses both Nepal, a technologically deprived city, and Singapore, a technologically rich city, further emphasizing the relationship between technology and wealth. Just as wealth is a distinctive feature of the countries in which the Internet, and technology in general, are prevalent, an opposite relationship is also true. Just as the centralization of wealth is made possible by technology, technology helps countries build upon existing wealth, as well as accumulate more wealth.

The graph in Figure 3 shows the number of Internet hosts per 1,000 people. With reference to Figure 2, it is evident that technology is lacking in poor countries but is proliferating in countries of great wealth, such as the United States.

As indicated, technology is not as ubiquitous as some would like to believe. In fact, is seems to be ubiquitous only in wealthy regions of the world. However, even in countries where wealth is centralized, a digital divide still exists. In U.S. households earning $75,000 or more per year, 86.3% have Internet access. However, in U.S. households earning less than $15,000 per year, only 12.7% had Internet access (Digital Divide Network, 2000). The graph in Figure 4 shows levels of income compared to the percentage of Internet usage at home, outside the home, and at any location.

It can be seen that in almost all categories (the exception being the under $5,000 at home and outside home categories), as income level increases, Internet usage in all places increased. Thus, this, in conjunction with the other findings, indicates that wealth and technology are positively associated – an increase in one leads to an increase in the other.

Figure 3. The digital divide.

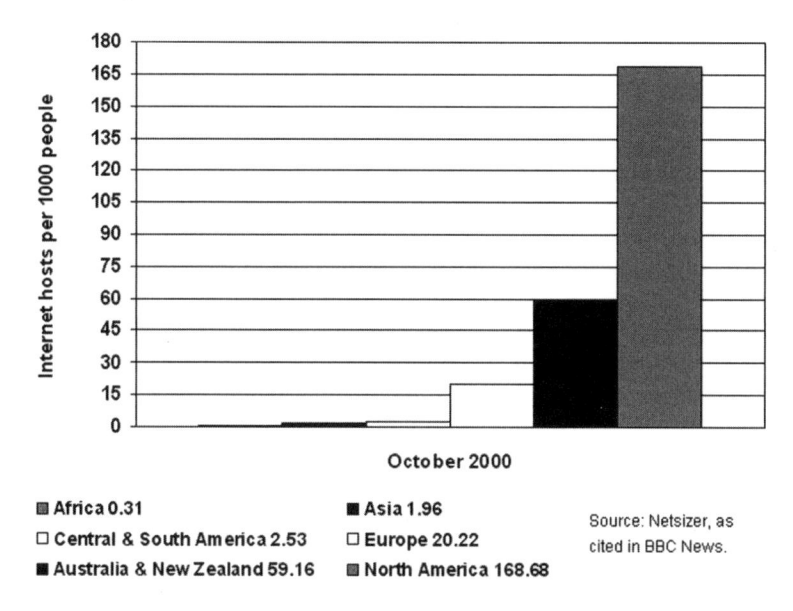

- Africa 0.31
- Asia 1.96
- Central & South America 2.53
- Europe 20.22
- Australia & New Zealand 59.16
- North America 168.68

Source: Netsizer, as cited in BBC News.

Figure 4. Levels of income versus percentage of Internet usage in different locations.

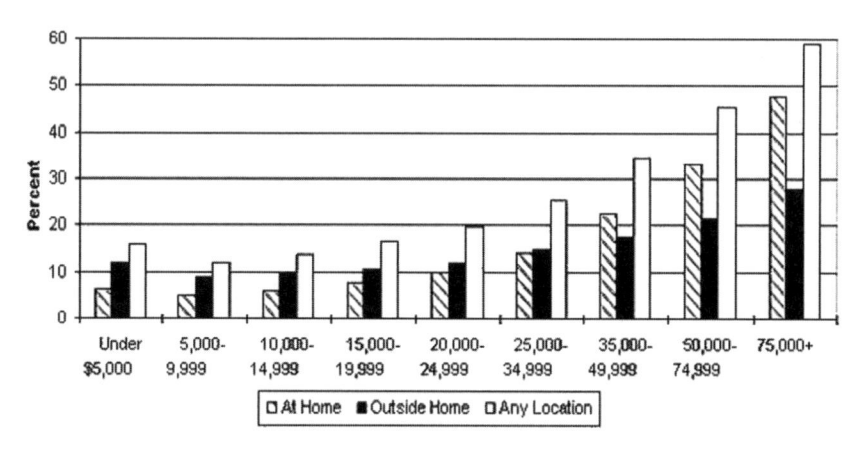

Source: Digital Divide Network.

In addition, those who are technologically poor cannot achieve the status of knowledge brokers, as described by Reich. In order to make a living for themselves, they must increasingly hire themselves out for menial amounts of money. Even the middle class, as Reich notes, is seeing former knowledge work made routine and pushed down in value. Therefore, where technology is far from ubiquitous, it can be seen that wealth is also far from ubiquitous, further emphasizing the relationship between the two.

The media age

Information is always desirable. Whether it is in the form of grounded facts or mere opinion, people are always knowledge hungry. As technological advancements make a constant stream of data possible, from continuous media coverage, on-line sources, or otherwise, many have become information connoisseurs. "Humankind appears to be on the verge of achieving mastery over information, turning a scarce resource, knowledge, into an abundant one. But sometimes the worst that can happen is to get what one wants" (Noam, 1995). As with anything, excess is not always desirable. While the benefits reaped from the mass of information available are a valuable resource, we should be wary of information overload. We are able to attain information of any kind, but we do not know exactly how to evaluate, assimilate, and use all of it. Thus, we are on the verge of what could be considered the Paradox of Information Technology: "The more information technology we have and the more knowledge we produce, the further behind we are in coping with information" (Noam, 1995).

In order to deal with this paradox, we seek help, and, in essence, we seek editors. However, we blindly place our trust in the media gatekeepers, paying little attention to the editing process the information we receive has gone through. Thus, we run the risk of receiving messages tailored to communicate a specific point of view.

As many continue to grapple with the onslaught of numerous media channels, what goes nearly unnoticed is the shrinking world of the mass media, and the consequences may be affecting democracy and consumers (*NOW* In-depth, 2002). North America boasts a wide variety of media outlets, including 1,800 daily newspapers; 11,000 magazines; 11,000 radio stations; 2,000 television stations; and 3,000 book publishers (*NOW* In-depth, 2002). However, despite the variety and seemingly democratic state of the mass media, the number of companies who hold a controlling interest in the aforementioned media has been shrinking since 1984. In 1984, 50 companies held an interest in the book publishing media, but the number has since dwindled to 6 in 2002 (*NOW* In-depth, 2002). Fourteen web sites (13 of which are affiliated with traditional brick-and-mortar sellers) currently dominate consumer sales on the Internet. As the media becomes more centralized, the forum for free speech and democracy is slowly dissipating, as is the content of the information provided. Instead, we are bombarded with visual themes or eye-candy toting, "feel good" messages, while the simple stories provide audiences with just enough information to comprehend the issues at hand. Thus, a danger lies in the pseudo-facts much of the public may accept without question or simply because no other information is readily available, whether on the Internet or not.

People have been trained to trust what they observe, and, thus, they accept the messages delivered by the media at face value. Such is the case with issue ads. These TV messages promote ideals that people are quick to support. However, what people do not realize is that the self-proclaimed noble cause they favor is actually an advertising mechanism for an undisclosed source. On the U.S. public television program *NOW with Bill Moyers*, guest Kathleen Hall Jamieson, Dean of the Annenburg School for Communication, addressed this topic of issue advertising and its relations to Medicare.

The particular issue advertisement addressed was "Citizens for Better Medicare," which ran on television during the 2000 Congressional campaign. The premise of the ad was to convince people not to allow Congress to import Canadian governmental regulations regarding medicine and health care. While the ad seemed to promote a cause most citizens would deem important, it failed to reveal the source behind the message (*NOW* transcript, 2002).

"'Citizens for Better Medicare' is an issue advocacy group set up largely by the pharmaceutical industry to advance its interests in the debate. The problem is, there is nothing in the law that requires that it tell us that it's the pharmaceutical industry or that it tells how much it's paying for the campaign," Jamieson said (*NOW* transcript, 2002).

Thus, people endorse the ad's cause but do not realize that its purpose supports interests other than their own. Instead, the ad acts as a facade behind which the pharmaceutical industry wields its powerful influence on the unsuspecting public. People back issues like "Citizens for Better Medicare" to achieve the "feel good" factor, for, in the public's mind, the ad promotes what they accept as a noble cause. They are willingly persuaded because, after all, what senior, or any person, does not want the best medical care and technology?

The role of the media has taken on a new dimension and some have shed their objectivity in favor of profit. While advertising has and will continue to be an effective means of persuasion, current trends hold different implications. With the emergence of pseudo-facts and an increasing public dependency on the media, many large corporations implore the use of the media to conjure socially desirable images that will enhance profits.

The ideal of sustainability

The industrialization and modernization of the world has led to increased initiatives regarding sustainability. Gro Harlem Bruntland defined sustainability in terms of sustainable development. She described it as "development that meets the needs of the present without compromising the ability of future generations to meet their own needs" (Bartlett, 1997). However, despite efforts to fuse the needs of the environment with those of humans, many have found that these needs remain in conflict (Bartlett, 1997/1994).

Thus, it appears that sustainability is always a compromise. The article "Reflections on Sustainability, Population Growth and the Environment" examines the question of compromise and finds that in a series of ten compromises, each planning to affect a 70% reduction in environmental damage, only 3% of the damage is actually reduced under the planned eventual agreement (Bartlett, 1997/1994).

These kinds of compromises span many types of industry. In a corporate world in which the bottom line drives profits and decision-making processes, wealth and prestige often supercede environmental concern. Such is the case with tradable pollution permits, which allow companies to buy and sell permits allocating specific amounts of pollution in order to maintain maximum production capacity. Other such compromises exist in the automobile industry.

As suburbia continues to lure more people from the city, the types of transportation needed are changing. When leaving the city, many choose to leave behind the efficient and environmentally friendly solutions of mass transit, and, because of the widespread nature of the suburban landscape, an increased dependency on the automobile has emerged. However, people are not opting for the fuel-efficient electric hybrids. Instead, they are purchasing an environmental demon – the sport-utility vehicle.

Not only do consumers realize that SUVs are detrimental to the environment, but manufacturers and recent reports are willing to admit the same. In the book *Connect-*

ing with Society, Ford Motor Company admits that the SUVs "burn more gas and emit more pollution than cars and can pose a danger to smaller vehicles in crashes" (Hyde, 2000). The Sierra Club has even affectionately renamed the Ford Excursion "the Ford Valdez."

Despite pressure from both sides of the environmental community, automobile manufacturers have another factor to consider – the business of catering to the consumer. In an interview with SustainAbility Chairman John Elkington, Bill Ford, Chairman of the Board of Ford Motor Company, spoke of Ford's attempt to be environmentally conscious. However, he continued to emphasize the importance of catering to the consumer, regardless of the threat posed by SUVs:

> "The real issue, though, is that we can't be in the business of dictating what the customer wants to buy. What we can do is make sure that whatever customers buy is done as responsibly as it possibly can... You could buy the Excursion or you could buy its competitor. If you buy the Excursion, its 43% cleaner in terms of its tailpipe emission – and it's made at the cleanest paint shop in the industry. If we were to withdraw from the market, customers would be forced to buy the competitive vehicle, with a truly negative impact on the environment." (Elkington, 2000)

Ford believes it is exercising a socially responsible role by catering to the consumers' desires, while still remaining somewhat environmentally conscious. However, this further proves that wealth and profit play a distinctive role in the promotion of sustainability. Once again, it is evident that the accumulation of wealth dictates the social agenda, despite possible negative impacts.

How are these movements in support of each other?

Many believe technology, the media, and sustainability are cohesive forces that aid in the progress of society and culture. They reject the idea that the modernization of the world is negatively affecting the social and built environments and continue to push for technological advancements in light of what they would deem "minor setbacks" experienced in society. Others feel that these things work in opposition to one another and, therefore, cannot be successfully intertwined. They would most likely agree with Reich in that technology is decreasing standards of living among the poor while aiding the wealthy. In addition, they would claim that, by computerizing society, we are merely emphasizing the gaps technology has created. In reference to ideology and sustainability, John McCarthy, Professor Emeritus of Computer Science at Stanford University, states, "For many, prophesying doom if we don't change our ways is a signal of virtue. Others are irritated by doom-saying and have an immediate favorable reaction" (1995). In response to the three issues, the debate of the curmudgeon's view versus the Pollyanna view surfaces.

The curmudgeon view

In its early stages, the Internet was touted as a democratizing force. However, in a document entitled "The Seven Priorities for the U.S. Department of Education," it was stated that, "just as technology can be a great democratizing force, it can also exacerbate inequities in our society" (n.d.). While the Internet can serve as a connecting tool, it can also serve as a dividing force. If the freedom and power provided by technology can only be utilized in wealthy countries, as wealth continues to centralize, a growing majority will be denied the positive aspects of technology.

With the onset of technology, an increase in consumerism has been observed. As the accumulation of material things continues to be equated with success, people are driven to constantly raise the level of technology. This ultimate fixation on elevating technology has become so intense that the benefits previously attained are lost in the midst of chaos. Thus, this consumerism is altering the way business is conducted and the way in which society, as a whole, operates.

In an article about his book, *The Future of Success*, Robert Reich addresses the idea that the intensity of consumerism has been matched by retailers. Reich states, "The easier it is for us as buyers to switch to something better, the harder we as sellers have to scramble in order to keep every customer, hold every client, seize every opportunity, get every contract..." (2001).

As the world strives to become more advanced, both technologically and otherwise, society must adapt to meet the demands placed on it. However, because everyone is not afforded the luxuries of technology or that acclaimed status of a "knowledge broker," society will continue to see a centralization of wealth, as well as a centralization of the world's poor. In addition, the labor market will become increasingly competitive and sparse as those on the lower end of the work-type spectrum are forced to lower the living wage in order to attract some type of work. The curmudgeon says, in order to benefit everyone, technology must first be made available to everyone.

The Pollyanna view

While some remain skeptical, others serve as the world's Pollyannas and equate technology with progress, despite any setbacks that may be encountered. This sector sees technology as the driving force behind the progress of the world and is quick to tout economic successes, especially those of the United States, as products of innovation.

Dr. Cheryl L. Shavers is one such proponent of technology. In a testimony before the U.S. House of Representatives, she argued that the technological innovations experienced in the past half-century could be correlated with the changing marketplace. She went on to state that "... information technology has defined the recent past. Even our common language has been affected, with Americans 'surfing' the 'Net and the prevalence of 'dot com' advertising" (2000). According to Shavers, technology is the driving force behind all progress, for she believes, "... innovation is at the heart of productivity and economic growth" (2000).

Despite setbacks and technological "flops," the true essence of progress depends on the continuing expansions of the technology behind it. The optimist, like Shavers, feel that the economic force unleashed by the Internet will drive the business market by rectifying past problems and promoting timeliness and availability of information (Shavers, 2000). Thus, even with the income disparities and the mechanization of the world, both socially and physically, the Pollyannas feel the world would be in a much different and much darker place, had it not been for technology.

Conclusion

In light of all the information presented, one might speculate on the relationship between technology, the media, and sustainability. However, they are all connected by an underlying factor – the centralization of wealth and the world of work. It is my belief that these three issues are greatly affected by the unequal distribution and increasing centralization of wealth across the globe, and, therefore, they affect people differently.

The recent accounting scandals in the U.S. have brought issues of wealth and corporate greed into the public's eye. What should have been a mathematical method of bookkeeping was quickly transformed into a creative game of disguise. Financial statements no longer stated the facts but, instead, reported information companies wanted to believe was factual.

However, it seems as though these practices are not as outlandish as once suspected. In fact, shuffling the books and shifting papers have become norms, but these practices are not a recent product of the consumer-driven, wealth-crazed society we live in. In fact, some, like Thomas Donlan, editor of *Barron's*, insist that this method of doing business has been the norm for years. Donlan stated:

> "The level of professional shock that was expressed last week at this was itself shocking. Sure, it was a violation of 'Accounting 101,' as the experts and professors told the TV audience, but after the chicanery, misrepresentation and proforma numbers that became standard procedure in press releases and annual reports, capitalization of expenses might not look so indefensible to a person running a company in crisis. Wall Street had lapped up their excuses for years." (2002)

Just as accounting scandals have been the norm, so has the bribery of politicians. However, bribes now have a new, legal name and can be disguised as "campaign contributions." While politicians cannot accept direct donations, corporations can hope to persuade politicians by providing them with unlimited "soft-money" contributions. Politicians are even allowed to establish "527 groups," which allow them to personally accept soft-money bribes (*Legal Bribes*, 2002). "... 527 groups are a mechanism for legalized bribery," Public Citizen Joan Claybrook states. "This corporate investment is for one purpose only: to shape the laws that Congress votes on and ultimately approves" (*Legal Bribes*, 2002).

Ultimately, the wealthy control the leaders of democracy, especially as described in the United States. With the increasing centralization of wealth, the ideals of democracy, technology, and progress can be demoted to mere hype.

In response to today's consumption-driven mentality, we have seen movements like voluntary simplicity and co-housing surface. Commercialization has led some to revert to a more simplistic lifestyle and, in response to the overspending of the 1980s, a slight resurgence in the voluntary simplicity movement has been observed (Edwards, 1998). Co-housing emerged in response to environmental concerns, but the number of active co-housing communities (20) in North America pales in comparison to its population (The Cohousing Network, 2002). Thus, the effects of both, thus far, have been marginal at best. We simply embrace idealistic lifestyles to relieve, temporarily, the burdens we, as a society, have created.

At a time when freedom of speech and thought seems supreme, the opposite is actually true. As discussed earlier, the number of companies controlling the media is shrinking, and the media is becoming more selective about the information distributed to the general public.

In "Reporting the News on a 'Need to Know' Basis," Norman Solomon discusses the centralization of the media. "We've all heard that knowledge is power. But the ultimate power can flow from being a gatekeeper," Solomon states. "In practice, few media companies determine what most Americans "need to know" on a daily basis" (n.d.). Robert Reich calls these knowledge brokers the future's most economically profitable forms of work. Those with the money exert the power on what the public is told.

Just as technology and the media are ultimately controlled by wealth, so is the ideal of sustainability. In theory, everyone is an environmental advocate because to feel otherwise would be unethical. However, as the old adage goes, "Actions speak louder than words." If society harbors an environmental passion, then why does it continue to destroy the globe?

This destruction comes in the form of consumer choices and consumption and is perhaps best exemplified in the vehicle of choice – the sport-utility vehicle. Despite proof that they damage the environment, consumers and motor companies continue to hail SUVs as the rulers of the road. In a study conducted by Polk that surveyed the percentage of people who considered gas mileage critically or very important, the results revealed that SUV buyers had a much lower concern for the fuel efficiency of their vehicle, when compared to all other new vehicle buyers (2001). This goes to support a point made earlier: sustainability is always a compromise. As the world becomes more affluent, it must identify with a status symbol to exemplify this success. Unfortunately, SUVs have become this status symbol, despite the drastic threats they pose to the environment.

While ubiquitous technology, the media age, and the ideal of sustainability are all interrelated subjects, the unifying link between them seems to be that of wealth. The wealthy have the resources to purchase and dictate the kinds of information they want disseminated, and they are constantly contradicting the ideal of sustainability through the choices they make. Thus, as wealth becomes more centralized, we move farther away from the universality of technology, widespread information dissemination, and saving the environment.

Robert Reich discusses three types of work that will begin to dominate society: routine production services, in-person services, and symbolic analytic services (which include the knowledge brokers mentioned earlier) (Guppy, 1994). He realized the centralization of wealth would create very few wealthy people, pushing the middle class and everyone else down to the lowest worldwide labor wage. This, of course, includes many professors, if I may say so. With the division of the work force, Reich has created a pattern that molds and confines the world. This pattern of increasing the gap between the rich and the poor through centralization of work, wealth, and resources will continue to transform the economic, physical, and social environments to the benefit of some and the detriment of others.

Therefore, as the world becomes wealthier, and as wealth becomes more centralized, the gaps between workers will remain wide, as will the digital divide. If our true desire is to interconnect the world, we must recognize the role of wealth in this process.

I want to end with one, perhaps hopeful thought. Has technology always been predominately available first only to the wealthy? Probably, yes. Yet has it always tended to reinforce the status quo or "make the rich richer?" The answer to that is less clear. Some technologies have produced significant redistribution of wealth. But does the ubiquitous technology of the Internet, the "feel good" effect of the current media age, and the "anything you want it to be" definition of sustainability appear to have the prospect of, as politicians like to state, all boats rising with the tide? I think it will be anything but that result. All three of these topics, as Reich might have predicted, are having the effect of further centralizing wealth.

Individuals, it would appear, are recognizing this and, rather than seeking to protect the common good, may be seeking primarily to focus on their own accumulation to take care of "them and theirs."

References

Bartlett, A. (1997/1994). Reflections on sustainability, population growth and the environment. *Population & Environment,* 16(1), 5-35.

Bartlett, A. (1997, August). *Environmental Sustainability*. Paper presented at the Annual Meeting of the American Association of Physics Teachers. Denver, Colorado.

BBC News. (2002, June 18). *World Inequality.* Retrieved June 2002 from http://www.news.bbc.co.uk

Digital Divide Network, Benton Foundation. (2000). *United States Digital Divide.* Retrieved July 2002 from http://www.digitaldividenetwork.org/content/stories/index.cfm?key=168.

Donlan, T. (2002, July 1). Does Everybody Do It? *Barron's*, 35.

Edwards, K. (1998). *An overview of the voluntary simplicity movement.* Retrieved July 2002 from http://www.stretcher.com/stories/960415c.html.

Guppy, M. (1994). Abstract of Dissertation on Robert Reich's *Work of Nations*. Retrieved July 2002 from http://jcomm.uoregon.edu/~robinson/j649/viii1.html.

Hyde, J. (2000). *Wheel Trouble.* Retrieved July 2002 from http://abcnews.go.com/sections/living/DailyNews/Ford00512.html.

Elkington, J. (2000). [Interview with Bill Ford, Chairman of the Board of Ford Motor Company]. Retrieved July 2002 from http://ford.com/en/ourCompany/environmentalInitiatives/environmentalActions/elkingotnOfSustainAbilityInterviews-BillFord.html.

Legal Bribes in Politicians Pockets. (2002). Retrieved July 2002 from http://www.socialistworker.org/2002-1/397/397_02_LegalBribes.shtml.

McCarthy, J. (1995). *Ideology and Sustainability*. Retrieved July 2002 from http://www-formal.stanford.edu/jmc/progress/ideology.html.

Minges, M. (2001). *Internet Around the World.* Retrieved July 2002 from http://isoc.org/isoc/conferences/inet/01/CD_proceedings/G54/Inet2001_100501.html.

Noam, E. (1995, June). *Visions of the Media Age: Taming the Information Monster.* Paper presented at the Third Annual Colloquium at the Alfred Herrhausen Society for International Dialogue. Frankfurt am Main, Germany.

NOW In-Depth. (2002). Retrieved July 2002 from http://www.pbs.org/now/politics/media.html.

NOW Transcript. (2002). Retrieved July 2002 from http://www.pbs.org/now/transcript/transcript123.full.html.

Polk study shows SUV sales strong despite rising fuel costs. (2001). Retrieved July 2002 from http://www.polk.com/news/releases/2001_0104d.asp.
Reich, R. (2001). *Robert Reich On 'The Future of Success'*. Retrieved July 2002 from http://www.jobsletter.org.nz/jb13810.html.
Shavers, C. (2000) [Testimony of Dr. Cheryl L. Shavers before U.S. House of Representatives]. Retrieved July 2002 from http://www.ogc.doc.gov/ogc/legreg/testimony/106s/shavers0309.html.
Solomon, N. (n.d.). *Reporting The News On A "Need To Know" Basis*. Retrieved July 2002 from www.fair.org/media-beat/990304.html.
The Cohousing Network. (2002). Community Status Key. Retrieved October 2002 from http://www.cohousing.org/cmty/status.
The Seven Priorities of the U.S. Department of Education. (n.d.). Priority Six. Retrieved July 2002 from http://www.ed.gov/updates/7priorities/part8.html.

Acknowledgement:
The author wishes to thank Drs. G. Gumpert and S. Drucker for organizing the session at which this paper was presented.

EMILY CARR UNIVERSITY OF ART + DESIGN
LIBRARY
1399 JOHNSTON STREET
VANCOUVER, BC V6H 3R9
TELEPHONE : (604) 844-3840

Applying urban indicators to clarify the urban development of Taipei

Yung-Jaan Lee

Chinese Culture University, Taipei, Taiwan

Abstract. A dialogue must be established comparing cities from different perspectives and in different areas. The Urban Indicators Programme (UIP) of the United Nations Center for Human Settlements (UNCHS) provides an excellent basis for comparing urban development in different cities. The UIP of UNCHS not only performs data collection, but also integrates policy assessment and strategy development. Additionally, the UIP focuses on providing national and local governments with the ability to select indicators and employ them for integrating national and local policies on urban development. Consequently, this work adopts the UIP of UNCHS as a framework to explore the sustainable development of Taipei and compare the urban development of Taipei with 237 cities included in the UIP of UNCHS. Furthermore, this work applies the Human Development Index (HDI) to compare Taipei with other cities.

This paper is organized as follows. First, Section One contains an introduction outlining objectives and contents. Section Two then investigates the implications of globalization and global urban development. Next, Section Three considers the context, content, and policy implications of the UIP, as well as its relations with other cities around the world. Subsequently, Section Four adopts the UIP of UNCHS to explore the urban indicators program applied in Taipei, with needed justifications. Finally, Section Five employs the UIP of UNCHS as a framework for examining the urban development of Taipei, and compares analytical results with those for other global cities from the UIP of UNCHS, using the HDI concept. Conclusions and suggestions for future research are finally made in Section Six.

Keywords: Urban Indicators Programme (UIP), Human Development Index (HDI), globalization, sustainability, Taipei

Introduction

Regardless of the controversy surrounding *globalization*, it continues at an accelerating pace, drawing more countries, cities, and people into increasing interdependent relationships. Globalization can be considered "*complex connectivity*" (Tomlinson, 1999) or "the intensification of global interconnectedness" (McGrew, 1992). The process of globalization is multidimensional, and can be defined as the global interconnections and interdependences of goods, capital, services, ideas, beliefs, and culture. Globalization is characterized by reducing distances owing to a significant reduction in the time required to cross them, both physically and representationally (Tomlinson, 1999). On another level, connectivity shifts gradually towards spatial proximity through the "stretching" of social relations across distance (Giddens, 1994).

Rather than describing a concrete object, globalization interprets a social process, thus making it difficult to define. While some see globalization as indicating Westernization or Americanization, others interpret it as the growing importance of the world market, and others see it as a cultural or ideological reality: "globalization as the victory of market plus democracy" (Lubbers & Koorevaar, 1998, p. 1). For most commentators, globalization is a complex concept involving political, economic, ecological, social, and cultural changes (for example, Giddens, 2000; Tomlinson, 1999; Lubbers, 1999). For example, Giddens identifies five basic areas in which global institutions require strengthening or further development, namely: "the governance of the world economy, global ecological management, the regulation of corporate power, the control of warfare and the fostering of transnational democracy" (Giddens, 2000, p. 124). Notably, Tomlinson criticized Giddens for neglecting "the *concept* of culture" (Tomlinson, 1999, p. 21, emphasis original). Generally, globalization also describes its subsequent consequences. Therefore, globalization is frequently considered a multifaceted dialectal dynamic.

Controversy regarding the nature of globalization has generated numerous debates, for instance on world trade development. The income gap between rich and poor countries has reached unprecedented levels over the past decade. For example, the income gap between the richest and poorest fifths of the world has increased from 60 to 1 to 74 to 1. Moreover, 80 countries have a lower per capita GDP today than a decade previously, and the assets of the world's 200 richest individuals now exceed the total assets of 41 percent of the world's population (UNDP, 1999). Globalization not only has spread new technologies, products, lifestyles, and living standards around the globe, but also has created a global culture.

Taking Taiwan as an example, the post-war "Taiwan Miracle" was based on adopting a policy of export-oriented industrialization to achieve economic development vis-à-vis global economic division. The international economic division was the main reason for Taiwan, South Korea, Singapore, and Hong Kong becoming the "Four Little Dragons". Internationalization appears to be the only path to development for island economies. However, owing to the transformation of the structure of the international economic system, dependent production systems based on cheap labor, low-tech products, and high pollution need to be changed. Taipei, as the capital of Taiwan, has begun economic and social restructuring to meet the economic needs of globalization, but the results remain incomplete. The problem is that Taipei faces competition from other cities, and its own advantages are being eroded by environmental degradation. Consequently, a new direction for urban development must be established because of globalization, the generation of new tools, and the continuation and expansion of old problems.

Urban lives are inevitably characterized by tension, and globalization implies the generation of a new spatial order for urban areas. Urbanization symbolizes not only wealth creation, domination, and opportunity, but also conflicts among pollution, poverty, and living standards. Importantly, the urban order problem is related to political economy, which is the product of globalization. The endogenous complexity and tension are connected to the heterogeneous diversity of the outside world. Clearly, the tension of contemporary urban areas is embedded in globalization. Social relations are extended through "time-space compression", while simultaneously local contacts are intensified and accelerated. Consequently, to enhance understanding and management of modern cities and monitor their future development, the relationships and connections between urban and global development must be explored. Accordingly, this work employs the Global Urban Observation (GUO) of the United Nations Center for Human Settlements (UNCHS, 1997) as basis for exploring urban development in Taipei, Taiwan.

A common baseline is required to make comparisons and connections among different cities. The Urban Indicators Programme (UIP) of the UNCHS provides an excellent foundation for developing a common knowledge base. Notably, numerous cities worldwide have adopted the UIP. Therefore, this work employs UIP to discuss the feasibility of transferring UIP to Taipei City. This work then investigates the status of the 237 cities listed in the UIP of UNCHS. Additionally, the Human Development Index (HDI) is adopted to compare Taipei with other cities in terms of regional spatial structure and development status.

Development of globalization and global cities

Globalization and global cities

The affluent lifestyles enjoyed in northern hemisphere cities are based on intensifying unequal and unsustainable capitalist relations among cities globally (Blowers & Pain, 1999). The notion of sustainable development (SD) has to some extent become a rhetorical device that avoids challenging current patterns of consumption and pollution or the need to redistribute wealth and resources. The most negative radical viewpoint, that of Ulrich Beck (1992, 1995), forecasts environmental catastrophe owing to what he terms the "risk society". The "risk society" refers to a state in which we "are confronted by the challenges of the self-created possibility, initially hidden and then increasingly apparent, of the self-destruction of all life on this earth" (Beck, 1995, p. 67). The risk society is an outcome of high technology, which has the potential to create devastating changes and which is ultimately beyond our control. While Beck analyzes the dangers, he does little to develop defensive measures in the context of urban development patterns (Blowers & Pain, 1999).

The drawing effect of cities provides a certain level of opportunity, freedom, choice, and quality of life, which contrasts with the limited, closed, and conservative nature of rural societies. If current trends continue, over 50 percent of the world's population ultimately will live in urban areas (Sustainable London Trust, 1996). During the late 1990s, between 20 and 30 million people left rural areas in favor of urban areas annually (Massey, 1999). This trend is unprecedented in human history, and the world is rapidly urbanizing as a result.

Owing to population migration, urban areas are growing rapidly. Some cities have become control centers, while others have become control concentration centers. For some scholars (such as Sassen, 1991), the above phenomenon is important evidence of the development of *global cities*. Consequently, another important issue requiring clarification is the relationship between the size and power of urban areas. Manuel Castells observed that the size and power of urban centers are not directly and simply related (Castells, 1996). The population of Sao Paulo (19.235 million in 1992) exceeds that of New York (16.158 million), but Sao Paulo clearly is not as influential than New York. Nor is Mumbai (13.322 million) more influential than Los Angeles (11.853 million). Thus, some cities can be enormous, yet still lack international power. Furthermore, small cities can also be powerful as exemplified by Washington, DC, which is very influential despite its small size (Massey, 1999). Defining the power of cities thus remains controversial. Saskia Sassen, in her book *The Global City*, stated the following:

> "... the combination of spatial dispersal and global integration has created a new strategic role for major cities. Beyond their long history as centers for international trade and banking, these cities now function in four new ways: first, as highly concentrated command points in the organization of the world economy; second, as key locations for finance and for specialized service firms, which have replaced manufacturing as the leading economic sectors; third, as sites of production, including the production of innovations, in these leading industries; and fourth, as markets for the products and innovations produced. These changes in the functioning of cities have had a massive impact upon both international economic activity and urban form: Cities concentrate control over vast resources, while finance and specialized service industries have restructured the urban social and economic order. Thus a new type of city has appeared. It is the global city. Leading examples now are New York, London, and Tokyo." (Sassen, 1991, pp. 3-4)

In fact, banks or financial centers are not always the most powerful institutions linking cities, and other forms of networks also control or influence inter-city relationships. Religious centers continue to occupy an important position today (a good example is Rome and the global power and influence of the Vatican). Global cultural networks among cities also exist, and examples of cultural centers on these networks include Los Angels (Hollywood), Paris, and London, or even the gay and lesbian cultural centers in San Francisco and Sydney (Massey, 1999).

Finally, "networks of ideas" (Massey, 1999) can also exist. The construction of the "modern" buildings in Mexico City's Plaza not only reflected the spread of particular dominant cultural styles (the International Style), but also embodied the spread of ideas about what was necessary to become "modern". This spread of dominant ideas and of particular dominant theories is an element of globalization. For survival, signing up to these dominant ideas and theories may be necessary for a city; their very dominance may create difficulties for imagining and pursuing an alternative future.

Network and impact of global trends

The global community is experiencing various threats, including poverty, the population explosion, financial crises, health scares like mad-cow disease, and so on. This situation clearly reflects the fact that global society does not simply affect the world overseas, but also affects individuals (Beck, 1992).

Global integration is accelerating rapidly and with far reaching consequences. However, the process is uneven, with different countries and individuals participating unequally in the process – including the global economy, global technology, global spread of culture, and global governance. The new rules of globalization, as well as the players writing them, focus on integrating global markets, neglecting popular needs concentrating power in their hands and marginalizing the poor, including both countries and people (UNDP, 1999). For instance, consider the issue of population: fears about the future of Western cities, and about disorder within them, often invoke concerns of population loss – the so-called "doughnut-effect" (Mooney, 1999).

Surprisingly, globalization produces urban diversity rather than uniformity. Given multiple global processes, cities react differently, and thus adopt different developmental strategies. While urban problems tend to be analogous (such as, traffic congestion, overcrowding, pollution, and management crises), the responses of cities are varied. Even where successful policies are transferred geographically, they are localized, thus changing their character. Carefully investigating the overall impacts of global trends on urban areas reveals three general trends (UNCHS, 1999): (a) increased speed, diversity and extent of the processes of change affecting cities, (b) shortcomings of urban management in a globalizing world, and (c) innovative reactions to crisis situations. Despite severe economic problems, cities are gradually becoming more efficient as centers of productivity, knowledge generation and technological innovation.

Consequently, some changing global trends appear to be producing a new group of negative factors], including increasing social exclusion, growing unemployment, increasing urban poverty, and environmental exploitation. Considering these trends further complicates the development of a global vision. UIP aims to increase knowledge of cities and thus provide a basis for quantifiably analyzing urban development.

UIP and the global city

Since the early 1990s, numerous discussions and conferences have focused on the development of sustainability and/or environmental indicators. Notably, the United Nations has challenged countries to establish sustainability indicators and initiate national policy developments. Most research on indicators can be classified based on spatial units, with the following three groupings generally being adopted: geographical, administrative, and ecological areas. Geographic groupings include global, continental, regional, and local, while administrative areas include nation, state, province, and town/city, and ecological system areas include rivers, deserts, slopeland, and rainforest. However, the most common groupings are the spatial structures based on administrative/jurisdictional boundaries. These groupings are the most common because of both the availability of statistical data and physical operational issues. Meanwhile, another reason for their popularity relates to hierarchical construction. Presently, systems of SD indicators can be divided into economic-based models, three-context-models and subject models (including environmental, social, and economic fields), pressure-response models (including the Pressure-State-Response model of OECD, 1998), connecting human/ecological system welfare models (such as, sustainability barometers), and compound capital models (including natural capital, human-made capital, human capital, and social capital) (Hardi et al., 1997).

Model selection depends on the desired goals, and the required backup effects vary among indicator structures. The UIP of UNCHS belongs to the subject indicator category, and essentially focuses on monitoring social, economic, and environmental issues. Meanwhile, related fields include national, metropolitan and urban areas.

Background of the UIP of UNCHS

Phase I (1994-1996)

The UIP of UNCHS (Habitat) was created to address the urgent global need to improve basic knowledge of urban areas by helping countries and cities to design, collect and apply data on policy-oriented indicators. The program began in 1988 as the Housing Urban Indicators Programme (HUIP), in response to the objectives of the *Global Strategy for Shelter to the Year 2000*. The program collected policy-sensitive housing indicators for the major cities of 53 countries in the period 1991–1992.

The HUIP was used to design effective methods for monitoring the implementation of the *Global Strategy for Shelter*, which called for establishing a legislative and regulatory environment to promote housing development. A set of indicators that were policy-sensitive and easy to regularly collect and update was proposed to create a framework for monitoring the performance of the housing sector from various perspectives. These indicators were designed to provide a management approach for the key stakeholders, namely housing consumers, housing producers, finance institutions, local governments, and central governments, to identify policy imperatives for addressing pressing housing problems. Closely examining the links between housing policies and outcomes shows that poor housing outcomes frequently result more from inadequate policies than from income or expenditure levels.

Phase II (1997-2001)

Phase One of the global UIP demonstrated a double demand on the UIP that can be successfully addressed through an integrated set of regional networks. On the one hand, the UIP was conceived as a global means of gathering indicator data to permit comparisons among cities, countries, and regions. On the other hand, national and local participants have expressed a need for indicator data that reflects their individual circumstances. Global and local expectations can be reconciled in two ways: 1) adopting two sets of indicators, namely universal key indicators and locally developed indicators; and 2) establishing a global network of local and national UIP that supports increasingly refined global analysis (Table 1).

Continuing to consider this duality, as well as the lessons from Phase One and the capacity building goal of the program, four objectives have been adopted for Phase Two of the UIP. These objectives closely resemble those of Phase One, and suggest a strong coherence and continuity between the two phases (UNCHS, 1997).

OBJECTIVE A: develop networks for information exchange and capacity building.
OBJECTIVE B: develop policy-oriented urban indicators and indices.
OBJECTIVE C: develop tools for collecting and analyzing of indicator data.
OBJECTIVE D: analyze and disseminate global indicator data.

Table 1. List of UIP key indicators

Background data

D1: Land use	D6: Household formation rate
D2: City population	D7: Income distribution
D3: Population growth rate	D8: City product per person
D4: Woman headed households	D9: Tenure type
D5: Average household size	

1. Socioeconomic Development	**4. Environmental Management**
1: Households below poverty line	18: Wastewater treated
2: Informal employment	19: Solid waste generated
3: Hospital beds	20: Disposal methods for solid waste
4: Child mortality	21: Regular solid-waste collection
5: Life expectancy at birth	22: Housing destroyed
6: Adult literacy rate	
7: School enrollment rates	
8: School classrooms	
9: Crime rates	
2. Infrastructure	**5. Local Government**
10: Household connection levels	23: Major sources of income
11: Access to potable water	24: Per-capita capital expenditure
12: Consumption of water	25: Debt service charge
13: Median price of water	26: Local government employees
	27: Wages in the budget
	28: Contracted recurrent expenditure ratio
	29: Government level providing services
	30: Control by higher levels of government
3. Transport	**6. Housing**
14: Modal split	31: House price to income ratio
15: Travel time	32: House rent to income ratio
16: Expenditure on road infrastructure	33: Floor area per person
17: Automobile ownership	34: Permanent structures
	35: Housing in compliance
	36: Land development multiplier
	37: Infrastructure expenditure
	38: Mortgage to credit ratio
	39: Housing production
	40: Housing investment

Source: UNCHS, 1997

Relationship between the UIP and global city

SD requires efficient urban management. New mobile patterns also forecast a new governance pattern. Moreover, GUO is designed to establish an open global resource network to form an urban knowledge infrastructure. GUO achieves this by providing a knowledge base structure and locally appropriate data collection methods and developments to adjust and maintain data banks and information systems. However, UIP is not merely a data collection system, but also is a plan for policy development, strategic development, and technological cooperation. UIP aims to develop the ability of governments to collect and implement indicators nationally and locally to form a necessary part and development network for a comprehensive policy for national and local administrations. Nevertheless, UIP should be used to measure the extent to which the social goals of all actors are achieved, rather than to measure narrow governmental activities. Government activities are measured to stress the goals of sustainability and efficiency, rather than the simple goal of production stressed by traditional government. Indicators simplify complex systems, making them easily understood by decision-makers and the general public. Indicators are designed mainly to provide a clear and accurate basis for spatial comparison among various cities, and to clearly display changes in urban conditions to provide a reference for improved decision-making. However, UIP aims not only to provide guidance for administrators, but also to design a monitoring process open to participation by all urban stakeholders. UIP can achieve these goals by placing policies in their political contexts, focusing on housing for the poor, or expanding the housing market. Indicators can be used to examine the regular context of urban development.

Determining UIP for Taipei

The UIP of Taipei aims to increase the capacity of the city to gather and use policy-orientated indicators for sustainable city development. It follows the UIP of UNCHS to measure Taipei's sustainability. However, different SD definitions and environmental problems develop from different time-space backgrounds and requirements. Accordingly, the UIP of UNCHS should be revised based on the characteristics and capacity of Taipei to gauge the problems and sustainability of the city (Lee, 2002).

Indicator development

More than a data collection program, the UIP of Taipei is a policy and strategy development and technical co-operation program, which attempts to establish the capacity for collecting indicators as an integral part of the national and local policy and development framework. Wherever possible, the indicators are expected to be part of an empowering process, measuring the progress of all actors toward social goals, rather than narrowly measuring government activities. Indicators of governmental activity stress sustainability and efficiency over simple production.

Indicators should be:
- linked to the four aspects of environmental, social, economic, and institutional sustainability;
- easily understood by the public and private sectors;
- related to participant interests;
- measurable using immediately available data at the city level; and
- clearly related to urban policy goals and changeable using policy instruments.

The indicators should be readily available, easily collected or estimated, and should not require special surveys or studies for quantification.

Criteria for indicator selection

Establishing indicators for Taipei City has largely depended on developing criteria to assess alternative indicators. Although indicators have been selected based on policy requirements, specific indicators are frequently adopted for many other reasons. This work uses the following criteria to determine the preferred set of indicators (UNCHS, 1997).

- *Importance for policy:* Indicators should be directly relevant to existing or proposed Taipei urban policy, and should directly measure outcomes.
- *Comprehensiveness:* The indicator "package" should be able to offer an immediate, broad overview of the environmental, social, and economic "health" of Taipei city that is accessible to residents and mainly based on existing data sources.
- *Priority:* Two levels of priority exist. Key indicators are based on available data, and the Taipei city government is encouraged to provide this data. Secondary indicators are less important to policy or require difficult to collect data.
- *Easily understood:* Simple indicators that can be understood by individuals without specialist knowledge are likely to attract more interest and be more accurately and readily adopted.
- *Cost-effective and timely:* Indicator data should be able to be gathered both economically and regularly, reflecting the rate at which the indicators are expected to vary. The level of detail and comprehensiveness of the data required should always remain within the capabilities of the collecting agency.
- *Measurable:* Indicators should display the magnitude of problems, and should be measurable using a dimensionless and time-independent scale.
- *Inclusion of the most disadvantaged:* Where equity is important, indicators should focus on the most disadvantaged.
- *Reliable:* Indicators should convincingly show that objectives are being met, based on sound observation, and should not be subject to statistical "noise".
- *Unambiguous:* Indicators should be clearly defined and refer to specific objectives and goals.
- *Independence:* Different indicators should measure different outcomes.

Comparison of UIP of Taipei with UIP of UNCHS

First, this work explores Taipei's the urban characteristics and development trends of Taipei and compares them to the international situation from the perspective of regional spatial structure and urban development. From the regional spatial perspective, from the UN data, until 2000, 237 cities from 110 countries participated in UIP planning and construction. Divided regionally, 37 of the countries were in Africa, the continent with the highest rate of participation in the UIP, and there were also Arab states, Asian-Pacific states, highly industrialized countries, Latin-American countries, and transitional countries (such as, Europe's Eastern Bloc). The proportion of cities in a region participating in the program, was highest in Africa, with 87 cities participating, or approximately 36.7%. Africa was followed in terms of participation rate by the Asia-Pacific region, highly industrialized countries, Latin America, transitional countries, and the Arab region. Presently Taipei is not participating in the UIP of UNCHS. However, if Taipei was to participate in the UIP it would do so in the category of Asian-Pacific states (Table 2).

Table 2. Countries and cities contributing to the UNCHS Global Urban Indicators Database by region.

Region	Countries	Cities
Africa	37	87
Arab States	8	11
Asian-Pacific	14	42
Highly industrialized	12	33
Latin American countries	17	32
Transitional countries	22	32
Total	110	237

Source: UNCHS, 1999

In terms of extent of development, UNCHS did not establish annual goals regarding UIP. However, UNCHS categorizes the 237 cities into different groups based on the Human Development Indicators (HDI) for each country.[1] For example, 60 cities are classified as being in highly developed countries; 78 cities as being in medium developed countries, and 99 cities as being in under developed countries, providing a standard for comparing various regions and cities around the world. The highly developed areas are concentrated in Western Europe, North America, Australia and Japan, while the medium developed areas are mostly located in the Asia-Pacific and Latin-American regions, and the under developed areas are mostly located in African countries (see Table 3). The so called highly, medium, and less developed areas differed as follows: the coefficient of the highly developed HDI was between 1.00 and 0.80; that of the medium HDI was between 0.79 and 0.50; and that of the low HDI was between 0.49 and 0.00 (UNDP, 1999). From statistical data for Taiwan, the HDI of Taiwan is 0.87, placing it in the group of high HDI countries. Moreover, using the UNDP (1999), Taiwan ranks 26 among 46 countries. Taiwan's HDI was adopted as the classification criteria for Taipei city's HDI grouping. Given the lack of assessment criteria, this work uses the classification of the HDI to reflect the value of the UIP and also as a reference point for goal setting using Taipei's UIP.

The HDI was also used as evaluation criteria, much as GNP has been adopted to assess development degree of an area or country. In this regard, GNP has been criticized for being too simple and abstract. Consequently, to avoid assessing developments purely from an economic perspective, this study recommends adopting HDI to assess the degree of urban development.

The original statistical data of the UIP does not all originate from the same unit, for example household income distribution is based on five income categories. Therefore, in assessing the context and the efficiency of the evaluation of the indicators, this work attempts to select indicators that better reflect the meaning of urban development.

Background data

Five indicators for Taipei are compatible with the UIP of UNCHS: city population, population density, population growth rate, average household size, and income distribution

(Table 3). These indicators provide information on important demographic and economic characteristics of cities. Taipei resembles other Asian cities in terms of population and population density, and also has a high population density. However, Taipei differs from other Asian cities in population growth rate. Furthermore, regarding income distribution, the income gap between the richest 20% and the poorest 20% of people in Taipei is just 4.31 times, compared to a world average of 11.3 times; an Asian average of ten times, and an average of 10.1 times in industrialized countries. Additionally, from the HDI, besides income distribution, most of the indicators for Taipei City are similar to those for medium developed countries, partly reflecting certain cultural and geographical characteristics.

Table 3. Background data analysis by region

Region	Population average (thousands)	Population density per annum	Population growth	Household size	Household formation rate (%)	Income distribution (times)
Africa	716.0	146.0	5.2	6.04	4.59	11.7
Arab States	2177.6	252.0	4.35	5.12	5.42	10.2
Asian-Pacific	3104.3	247.2	3.21	4.91	4.32	10.0
Highly industrialized	1582.7	82.4	0.59	2.45	1.50	10.1
Latin American countries	1312.6	149.8	2.32	4.18	2.78	17.5
Transitional countries	1192.0	126.2	-0.03	3.02	0.79	7.8
All cities	1497.3	154.0	2.92	4.59	3.39	11.3
Taipei City	2639.9	–	1.36	3.00	2.95	4.31
Development level (HDI)	Population average (thousands)	Population density per annum	Population growth	Household size	Household formation rate (%)	Income distribution (times)
Low	1127.0	190.4	5.16	6.00	4.73	10.6
Medium	2416.9	158.5	2.11	4.17	2.89	13.4
Highly	1823.1	98.8	0.69	2.74	1.52	9.4

Source: UNCHS, 1999; excluding Taipei City

Socioeconomic development

Six indicators are compatible with the UIP of UNCHS, including hospital beds, adult literacy rate, school enrollment rates, school classrooms, life expectancy at birth, and households below the poverty line. However, since the UN listed no information on adult literacy rates, school enrollment rates, and life expectancy at birth, this work will not make any comparisons (Table 4). Regarding households below the poverty line, Taipei has fewer than anywhere else (even more industrialized countries). Meanwhile, regarding hospital beds, the average number of residents per hospital bed for Taipei (133) is better than for almost all other cities globally. However, this indicator simply measures the average ratio of urban population to hospital beds, and does not consider those from

Table 4. Socioeconomic development analysis by region

Region	Household below poverty line (%)	Hospital beds (persons/ bed)	School classrooms (person/ unit)	Crime rates (% per thousand) Murder	Crime rates (% per thousand) Theft
Africa	38.96	954	56.6	1.49	12.3
Arab States	34.33	410	41.1	0.03	2.5
Asian-Pacific	20.53	566	43.2	0.70	4.2
Highly industrialized	14.02	132	23.5	0.15	54.1
Latin American countries	37.96	288	36.3	0.63	5.3
Transitional countries	10.57	31	11.6	0.09	6.3
All cities	30.25	502	42	0.70	17.1
Taipei City	8.00	133	33	0.17	12.9
Development level	Household below poverty line (%)	Hospital beds (persons/ bed)	School classrooms (person/ unit)	Crime rates (% per thousand) Murder	Crime rates (% per thousand) Theft
Low	36.1	930	58.5	1.50	9.8
Medium	31.4	291	37.6	0.34	7.9
High	15.8	147	24.0	0.17	37.5

Source: UNCHS, 1999; excluding Taipei City

outside Taipei who use Taipei's hospitals. Therefore, this indicator is inevitably inaccurate. Regarding the HDI, perspectives differ significantly according to development status. Taipei has better than average development status in terms of the above indicators.

Infrastructure

Regarding infrastructure indicators, Taipei is compatible with the UIP of UNCHS in only one indicator, household connection levels. This category is dominated by water related indicators, including access to potable water, water consumption, and median water prices. Differences in these indicator items clearly display difference among different development states. While no data is available for Taipei City, the level should exceed the global average (84.4%). As for water consumption, water consumption in Taipei (359 liter per person per day) is twice the global average (161.3 liter per person per day), and is also higher than the average in highly developed countries (96.7 liter per person per day). This comparison is quite alarming for an island country like Taiwan. As for wastewater con-

Table 5. Infrastructure analysis by region

Region	Household connection levels (%)	Access to potable water (%)	Consumption of water (liter/day/person)	Median price of water (cubic meter/USD)
Africa	12.7	69.1	53.6	1.302
Arab States	58.9	88.2	157.9	0.645
Asian-Pacific	38.4	87.5	160.7	0.536
Highly industrialized	97.8	99.6	262.3	2.240
Latin American countries	62.5	86.9	182.8	0.908
Transitional countries	88.8	99.1	306.6	0.409
All cities	51.8	84.4	161.3	1.076
Taipei City	47.78	99.4	373.7	–
Development level (HDI)	Household connection levels (%)	Access to potable water (%)	Consumption of water (liter/day/person)	Median price of water (cubic meter/USD)
Low	14.3	68.9	66.1	0.998
Medium	63.7	92.4	211.0	0.645
High	91.2	99.3	271.6	1.776

Source: UNCHS, 1999; excluding Taipei City

nection levels they are only around 47.78% in Taipei, considerably lower than in the highly developed countries, and even below the world average (Table 5). Overall, the water resource pattern of Taipei City is characterized by "high usage and low treatment". In terms of the HDI, the water consumption of Taipei City significantly exceeds the average for highly developed countries. Moreover, Taipei scores much lower than less developed countries in terms of wastewater connection levels. Consequently, moves to improve Taipei's infrastructure should focus first on wastewater treatment and water consumption.

Transport

Two indicators for Taipei correspond to the UIP of UNCHS – expenditure on road infrastructure and automobile ownership. Transportation is crucial to urban efficiency. Poor transportation plans invariably cause symptoms of urban dysfunction, including traffic congestion, increased transportation time, careless operation of public transportation,

Table 6. Transport analysis by region

Region	Modal split connection levels (%)	Travel time potable water (%)	Expenditure on road infrastructure	Automobile ownership meter/USD)
Arab States	58.9	88.2	157.9	0.645
Africa	11.8	37	6	32.4
Arab States	27.4	32	33	63.3
Asian-Pacific	9.0	32	2	78.9
Highly industrialized	54.6	25	116	423.5
Latin American countries	25.2	37	15	101.1
Transitional countries	18.4	36	55	177.6
All cities	21.9	34	33	144.6
Taipei City	–	–	23.28	258
Development level (HDI)	Modal split connection levels (%)	Travel time potable water (%)	Expenditure on road infrastructure	Automobile ownership meter/USD)
Low	10.0	34	4	41.1
Medium	18.4	39	32	78.6
High	45.1	28	81	354.4

Source: UNCHS, 1999; excluding Taipei City

poor local traffic control, accidents, air pollution, noise pollution, and costly domestic cargo transportation. Levels of automobile ownership are one of the most effective indicators for monitoring the above symptoms. Private car use and ownership are closely related to local income levels, and automobile ownership currently is highest in cities in industrial countries, which have 424 cars for every thousand residents, 13 times the level of car ownership in African cities. The level of automobile ownership in Taipei City is 258 cars per thousand residents (see Table 6), which essentially places Taipei in the high automobile ownership category. Other factors also influence choice of transportation tools, for example transportation efficiency, urban density, and the distribution of road networks, fuel prices, social conditions, and so on. Automobile ownership in Taipei City is not necessary related directly to urban transportation volume. However, this indicator indicates the lack of a regional transportation system.

Environmental management

Two of Taipei's indicators in this category adhere to the UIP of UNCHS: wastewater treated and solid waste generated. Comparing wastewater treatment with wastewater connection levels demonstrates that connection levels appear to have reached a high stan-

Table 7. Environmental management analysis by region

Region	Wastewater treated (%)	Solid waste generated (tons/person/year)
Africa	15.3	0.27
Arab States	54.1	0.28
Asian-Pacific	26.0	0.29
Highly industrialized	86.8	0.51
Latin American countries	18.1	0.60
Transitional countries	64.2	0.49
All cities	38.3	0.39
Taipei City	8.25	0.52
Development level (HDI)	Wastewater treated (%)	Solid waste generated (tons/person/year)
Low	14.1	0.26
Medium	37.6	0.46
High	80.1	0.50

Source: UNCHS, 1999; excluding Taipei City

dard, implying that continuing stagnation of wastewater treatment will significantly and negatively influence the urban environment. In Taipei City, although wastewater connection levels are 47.78%, wastewater treatment levels are just 8.25%, meaning that 91.75% of wastewater is simply dumped untreated into rivers or the sea. Taipei lags cities in highly developed countries by a long way in terms of wastewater treatment. Even in African countries, wastewater treatment levels are higher than in Taipei City (reaching 15.3%), an issue that deserves serious attention from the Taipei City Government. Regarding solid waste generation, Taipei generates 0.52 tons per person annually, a figure that almost equals that of highly developed countries (see Table 7). Finally, in terms of HDI levels, the consumption of various resources of Taipei City has reached the level of highly developed countries, yet Taipei lags highly developed countries in terms of pollution treatment.

Local government

The UIP of UNCHS focuses mainly on local government because most urban construction depends on various local government services. Consequently, a measurement tool is needed to examine governmental functions. Moreover, contemporary management systems have shifted from central governments to local jurisdictions, making increasing local government efficiency a priority. The indicators in this category mostly relate to financial and personnel management, a phenomenon that occurs mainly because local financial and personnel situations are the basis for implementing various infrastructures. In terms of local government indicators, five indicators are compatible with the UIP of UNCHS: major sources of income, per capita capital expenditure, debt servicing charges, number of local government employees, and wages in the budget. Table 8 lists that the major source of income for Taipei City is around 2,146 USD, ranking close behind that in the cities of highly industrialized countries (2,763 USD/person), and around 8.6 times that for cities in Asian countries. Regarding per capita capital expenditure, that in Taipei is approximately 1,726 USD per person, the highest level among all the regions. This figure not only exceeds that of the Asian countries, but also is also 583 USD higher than that of the highly industrialized countries. This high value reflects the significant recent capital investment in Taipei City, a positive indication that the city can meet the needs of its industries and residents.

Housing

Regarding housing policy, two implementation approaches are available: (a) adopting housing subsidies to support low-income households, such as social housing; and (b) relying on market mechanisms to provide appropriate and affordable housing to residents. Housing indicators provide cities with a clear baseline for analyzing their performance in the area of housing and helping in policy development. Four of Taipei's indicators adhere to the UIP of UNCHS, including the ratio of rents to incomes, housing floor area per person, infrastructure expenditure, and housing investment. Regarding the ratio of rents to incomes, although house rents are approximately one-fifth (21.3%) of income levels, equal to the world average (20.5%), they are still expensive compared to highly

Table 8. Local government analysis by region

Region	Major sources of income	Per capita capital expenditure	Debt service charge (%)	Local government employees (person/per thousand)
Africa	15	10	6.90	5.2
Arab States	1682	32	1.89	35.5
Asian-Pacific	249	234	5.88	10.0
Highly industrialized	2763	1133	7.77	23.6
Latin American countries	252	100	5.78	19.3
Transitional countries	237	77	3.01	4.7
All cities	649	245	5.95	11.9
Taipei City	2146	1726	25.55	6.1
Development level (HDI)	Major sources of income	Per capita capital expenditure	Debt service charge (%)	Local government employees (person/per thousand)
Low	12	4	–	5.1
Medium	178	70	–	14.6
High	2356	928	–	21.2

Source: UNCHS, 1999; excluding Taipei City

and medium developed countries, at around 5%. Regarding total residential floor area per person, Taipei has roughly 28.7 square meters, about 15.0 square meters higher than the Asian average. Regarding infrastructure expenditure, the highly developed countries lead with 589 USD, while Taipei has about 327 USD, comparable to the level in highly industrialized countries (see Table 9). Overall, the above indicators are intended to measure the contribution of housing to economic development, and to reflect the input of the improvement of living conditions. Assessing all the indicators reveals that housing rent is rather expensive, indicating that the ratio of house rent to income is too high. This finding also indicates not only that Taipei's citizens must bear a heavy burden, but also that demand for rental houses exceeds supply. Finally, infrastructure expenditure displays a strong interaction between development and construction, and also demonstrates that the living environment in Taipei has improved.

Table 9. Housing analysis by region

Region	House rent to income (%)	Floor area per person (square meters)	Infrastructure expenditure (USD/person)	Housing production (thousand persons)	Housing investment (%)
Africa	25.3	8.4	23	7.5	8.4
Arab States	19.5	13.7	71	5.6	7.3
Asian-Pacific	22.7	10.2	21	9.6	9.3
Highly industrialized	20.7	35.8	589	4.6	2.8
Latin American countries	20.2	14.7	138	7.3	6.8
Transitional countries	4.5	17.6	82	2.5	3.9
All cities	20.5	13.8	87	6.6	7.1
Taipei City	21.3	28.7	327	—	2.4
Development level (HDI)	House rent to income (%)	Floor area per person (square meters)	Infrastructure expenditure (USD/person)	Housing production (thousand persons)	Housing investment (%)
Low	26.7	7.6	16	7.0	10.3
Medium	16.36	14.8	67	7.4	5.9
High	16.1	28.3	304	4.5	3.6

Source: UNCHS, 1999; excluding Taipei City

Conclusions

While contemporary social developments and related inequalities can prevent cities from achieving sustainability, the SD concept still provides a possible path to sustainability. Restated, sustainability can be achieved under various social conditions, indicating that the spatial form and social process of sustainability can be achieved by paying increased attention to ecological constraints and efficient governance. Creating a sustainable global city requires a new way of thinking, which reconsiders the basic principles used in constructing human settlements. Occasionally a comprehensive procedure must be implemented to examine the context of global cities. The UIP can provide a comparison basis, not only for exploring the endogenous problems of cities, but also for comparing the development status of global cities.

The above analysis shows that cities in densely populated island countries, such as Taipei in Taiwan, have a relatively high population concentration, generating considerable physical and non-physical environmental pressures, for example in wastewater treatment, water consumption, solid waste production and so on. This phenomenon shows that Taipei has high levels of resource consumption, yet poor waste treatment and disposal abilities.

Taipei is a highly developed city based on HDI analysis. However, these indicators still show that Taipei has significant room for improvement. From the perspective of highly industrialized countries, Taipei still reflects a preliminary pattern for future urban development. The analytical results presented in this work provide a valuable reference for Taipei. On the other hand, for the Taipei city government, the above analysis indicates that the government should no longer seek to ascribe simple causes to problems because problems are always inter-related. Neither should the Taipei city government adopt a single attitude and single role in deciding the future of the city's development. Instead, future development directions should be decided by community residents, urban citizens, and island residents. The above approach can be used to apply the UIP to effectively gauge and monitor the urban development and sustainability of Taipei City.

Note: The Human Development Index (HDI) was first developed in 1990. The HDI is a comprehensive indicator; including life expectancy, education (adult literacy rate and gross enrollment rate), and economic factors (per capita GNP). The index is measured on a scale between 0 and 1, with a higher number indicating a higher level of development. One hundred and seventy-four countries now assess their cities using the HDI criteria.

References

Beck, U. (1992). *Risk Society: Towards a New Modernity*. London: Sage.
Beck, U. (1995). *Ecological Politics in an Age of Risk*. Cambridge: Polity Press.
Blowers, A. & Pain, K. (1999). The Unsustainable City? In S. Pile, C. Brook, & G. Mooney (Eds.), *Unruly Cities? Order/Disorder* (pp. 247-298). London: Routledge.
Castells, M. (1996). *The Information Age: Economy, Society and Culture*. Volume I: *The Rise of the Network Society*. Oxford: Blackwell.
Giddens, A. (1994). *Beyond Left and Right*. Cambridge: Polity Press.
Giddens, A. (2000). *The Third Way and Its Critics*. Cambridge: Polity Press.
Hardi, P., Barg, S., Hodge, T., & L. Pinter. (1997). *Measuring Sustainable Development: Review of Current Practice*, Occasional Paper Number 17, Industry Canada.
King, A. & Schneider, B. (1991). *The First Global Revolution: A Report by the Council of the Club of Rome*. London: Simon & Schuster.
Lee, Y.-J. (2002). Adopting the Urban Indicators Prgramme to Measure Taipei's Sustainable Development. *Asian Pacific Planning Review,* 1(1), 111-127.
Lubbers, R.F.M.& Koorevaar, J.G.. (1998). *The Dynamic of Globalization*. (//cwis.kub.nl/globus/Lubpdfs/Globaliz/Global07.pdf)
Massey, D. (1999). City in the World. In D. Massey, A. John, & P. Steve (Eds.) *City Worlds* (pp. 99-156). London: Routledge.
McGrew, A. (1992). A Global Society. In S. Hall, D. Held, & A. McGrew (Eds.), *Modernity and its Futures* (pp. 61-102). Cambridge: Polity Press.
Mooney, G.. (1999). Urban 'Disorders'. In S. Pile, C. Brook & G. Mooney (Eds.), *Unruly Cities? Order/Disorder* (pp. 7-52), London: Routledge.
OECD. (1998). *Towards Sustainable Development: Environmental Indicators*. OECD.

Sassen, S. (1991). *The Global City: New York, London, Tokyo*. Princeton, NJ: Princeton University Press.
Sustainable London Trust. (1996). *Creating a Sustainable London*. Available online http://www.greenchannel.com/forums.htm
UNCHS. (1999). *The State of the World's City: 1999-Cities in a Globalizing World*. Available online http://www.urbanobservatory.org/swc1999/cities.html
UNCHS. (1997). *Monitoring Human Settlements with Urban Indicators (Draft)*. United Nations Center Human Settlements (Habitat), Nairobi, Kenya.
UNDP. (1999). *Human Development Report 1999*. New York: Oxford University Press.

Choosing sustainability:
The persistence of non-motorized transport in Chinese cities

John Zacharias
Concordia University, Montreal, Canada

Abstract. Non-motorized transport expanded rapidly in most Chinese cities following market liberalization in 1978. The bicycle and pedestrian share of intra-city trips continued to grow until the mid-1990s while incomes continued to rise and cities invested in new highways and public transportation. The affection for the bicycle in particular is closely related to extensive urban infrastructure and the predominance of bicycles on many urban thoroughfares. The commercial and service structure of cities has developed around non-motorized modes, further reinforcing these forms of transportation. In spite of central government policies to promote the use of motorized modes and the automobile industry in particular, street culture remains closely allied with non-motorized modes. Personal characteristics, income, and household composition have relatively little to do with the persistence of sustainable transport in China. While the future of non-motorized modes is unclear, the use of bicycle, tricycle, and pedestrian modes remain a significant advantage for Chinese cities facing substantial challenges to a viable and sustainable transport future.

Keywords: sustainable transport, bicycles, transport policy, China

This paper is concerned with the factors underlying the continuing widespread use of non-motorized modes – in particular, the bicycle – in major Chinese cities. It is argued that the tenacious hold of non-motorization on the daily living patterns of urban-dwellers in China today can be found primarily in the environment. The extensive use of the bicycle today in many Chinese cities was not expected. After 1978 and the "opening up" of the Chinese economy, it was thought that the bicycle would be abandoned in favor of public transportation, in keeping with a widely held view that rising income accompanies motorization (Steininger, 2002); but in the 1980s, just the opposite happened. Public transportation declined while use of the bicycle boomed (Yang, 1985). The boom continued into the 1990s even with massive investment in public transportation, highway development, promotion of the car, and restrictions on bicycle use in some cities. The dominant discourse in China with regard to use of the bicycle centers on economic and social factors and hardly at all with the physical structure of the city and the way that people interact with the environment.

The paper begins with a brief summary of the push for motorization since 1994 in China (Gakenheimer, 1995). Factors in support of the bicycle are then considered with illustrations from Shanghai, Tianjin, Shaoxing, Guangzhou, and Shenzhen. The paper concludes with notes and questions for Western societies increasingly concerned with sustainable practices in planning and environmental quality. The argument presented here for the China case, largely supported by developments in parts of Europe (Evans et al., 2001; OECD, 2000) is that if environmental conditions favor sustainable transport, behavior may follow.

The drive for motorization

A variety of policies militate against non-motorized travel but have to a limited extent transformed the physical environment of cities. In 1994 the central government declared the automotive industry one of the major motors for economic growth, in keeping with the abundant evidence from Western societies that the car industry has been closely associated with rapid economic growth (Figure 1). Following this decision, substantial national resources were marshaled in support of highway systems within and between cities, while the passenger and freight rail system languished or declined. Civic leaders often refer publicly to the bicycle as a symbol of backwardness and economic deprivation associated with the lean years of doctrinaire socialism. Planning and transport studies invariably refer to use of the bicycle as an irritant in the transport system. For example, "the rapid growth of bicycles, which will slow down the speed of the whole traffic system and increase the accident rate, will lead to a worse traffic situation" (Shanghai, 1990). Local governors and mayors often refer to the omnipresence of bicycles in their city as a sign of backwardness. Anxious to show they have made significant change in their city during a term of office that may only be for three or four years, they often focus on traffic management. These approaches to the problem are closely connected with the top-down policy of promoting the automobile industry, itself closely connected with a particular vision of the modern city (Gan, 2003). The linkage made between personal betterment and individual motor transport is certainly not novel, although looking increasingly doubtful in literature on sustainable transport in the West (Stead & Banister, 2001). While the prevailing vision for urban development in China includes a transport policy that seems dated and fraught with problems, examples of similar motorization outcomes in other developing contexts, and precious few examples of rapid economic growth without motorization make it even more difficult to provide a persuasive argument for a policy review (see for example, Bae & Suthiranart, 2003 and Whitelegg & Williams, 2000).

Recent studies in sustainable transport have emphasized the system properties of policy formulation in local contexts (Rosen, 2001). Actors and decision-makers are constrained within a local context of accumulated decisions, commitments, and sustained relationships that may pose significant barriers to innovation. In China, the presence of the Communist party at all levels of public administration has supported a

Figure 1. A civic symbol erected at the centre of a sub-centre in Shanghai known as "Five Corners."

monolithic policy structure with appointed decision-makers whose careers are at least partly tied to their performance within the party decision-making structure. Thus it becomes difficult for local adminstrators to veer sharply from a pattern that is all too visible in other cities. The positive side of this is the rapidity with which enlightened policy can result in physical transformation in cities. The administrative system is described in detail by Sun (2000).

Substantial private resources both within the country and in research labs in the West are working to lower the cost of car production in the effort to make private motorization available to more Chinese. Privately owned automobiles were a tiny part of the vehicle fleet just a few years ago but are now growing at the rate of thousands per month in major cities such as Guangzhou, Shanghai, and Beijing (Hook & Replogle, 1996). At an annual rate of growth of 7% or more in GDP overall in China, somewhat higher in the cities, it is predictable that a rapidly growing proportion of China's 1.3 billion people will be able to purchase a car.

Social pressures also work against popular adherence to the bicycle. In contemporary China where economic growth is a nearly universal top priority, use or possession of a car denotes both economic and social standing. The use of a car is often conferred as a privilege on valued employees in the public and private sectors. It is also a matter of convention, though not of law, that car-drivers have the right of way on and off streets, with respect to cyclists and pedestrians. Such practices are widespread in the developing world. American popular imagery increasingly available in China displays a car culture without traffic jams, car-jacking, air pollution, accident fatalities, and urban sprawl. Advertising, following Western patterns, uses car-based imagery to promote a variety of products and associated lifestyles. There remains a disconnection between this imagery and the daily reality of the Chinese city, however. The disconnection has long been identified in the West, where constantly rising loads on the road system have eroded the promised experience of personal freedom and thrill.

Cities remain anchored in a post-liberation physical structure that favors the bicycle. In the Socialist era, housing was tied to workplace in a city that consisted mostly of large, undifferentiated estates (Zhu, 2002). These large estates with narrow internal streets given over to pedestrians and bicycles have virtually no space for car parking. With developable land now at a premium, it is difficult to see just where all the new cars can be parked. Because most intra-urban travel was tied to this local environment, roads were widely spaced but were designed sufficiently large to accommodate separate bicycle and motor vehicle traffic. This system is ideally suited to bicycle travel since the broad dedicated paths can be found on all major streets. On narrow, traditional streets where mixed traffic prevails, the bicycle generally dominates, leading to multiple conflicts or near-conflict situations. With a handful of exceptions, cities have been reluctant to undertake radical changes in the structure of the public environment and in transport choices.

Bicycle travel remains the fastest way to get to destinations within the city for the majority of trips and users (Zacharias, 2001), a finding not inconsistent with data for some European cities where the bicycle is a viable transport option (ECMT, 1994). For trips up to 5 km in length, it is faster to bicycle than to take a bus. Increasingly because of street congestion, it is also faster to cycle than to drive or take a taxi the same distance. Other questions concern relative comfort, convenience, and peace of mind, and of course a land use structure already built up around bicycle travel.

The effect of vast numbers

The explosion in the number of bicycles during the 1980s was followed by a period of sustained growth (Yang, 1985). Today, many large cities have one or even more bicycles per person in use. Bicycles offer a quick, inexpensive, and comfortable means to access destinations at up to several kilometres distance from home (Shen, 1997; Xin, 1996). As a consequence, cities that have the environmental structure described above and a bicycle fleet to exploit it, experience daily flows in the tens of thousands of bicycles on important traffic streets. In this environment, the bicycle then becomes the fastest, most comfortable, and most flexible means of transportation for local travel.

In the last ten years the motor vehicle fleet has been growing faster than the bicycle fleet but remains very small in comparison, even in very large cities such as Shanghai. In that city, 300 thousand automobiles compare with 7 million bicycles (Shen Qing, 1997). Even in Beijing, with nearly one million passenger cars, the bicycle fleet remains several times larger. The daily presence of cyclists on the streets makes radical change in the use of street space difficult for local leaders wishing to meet the demands of the motoring public. Efforts to encourage cyclists to switch to public transportation have met with resistance in some cities. For example, in Shanghai, the share of trips on buses has remained stagnant at 15% from 1995 to 1998 (Shanghai, 2001). The number of bus riders in most cities has still not attained the levels reached in the early 1980s, before massive defection to the bicycle. The city of Shenzhen resorted to the elimination of bicycle paths and a massive reduction in transit fares to move cyclists onto buses in 1999. The bicycle nearly disappeared overnight from the road environment but still has 13% of all trips over 500 m. in that new city of 7.5 million (Shenzhen 2001). Guangzhou is implementing a system of bridges and a reduction in the number of bicycle streets as part of an effort to reduce the bicycle share of local travel from 26% to 16% (Guangzhou, 1997). Shenyang has recently eliminated some major bicycle routes for the same reasons.

By contrast, Tianjin, the third largest city in China, still has over half of all trips of over 500 m by bicycle – 52.5% in an O-D study conducted by the Tianjin Planning Bureau in 2000. In that particular case, bicycles dominate the traffic environment of the whole central area. The new pedestrian zone, also the largest in China, depends heavily on visitors arriving by bicycle. The parking lots and spaces surrounding the pedestrian streets are lined with bicycle parking facilities, in this case an economic necessity.

Sheer numbers of bicycles and bicycle users make for a big difference in the traffic environment and the experience of cycling (Kubota & Kodokoro, 1994). Qualitative assessment suggests that there is some minimum threshold of bicycle users above which many more people will be willing to adopt the bicycle mode. No study on this particular question appears to exist.

City planning and bicycle infrastructure

Bicycle system infrastructure was integrated into the plans of all cities after 1949 and strongly marks most of the cities of the central and eastern plains today (Xin, 1996; Yan and Zheng, 1994). Such infrastructure may be regarded as fundamental to the support of cycling. While major cities undergo massive rebuilding, transforming the look and places of sociability, street infrastructure remains hardly touched (Ning, 1995). This section provides some typical examples and how they support the bicycle mode.

Bicycle lanes are typically 3 to 4 m wide and are mandatory on all major streets. Separated from motor traffic by a concrete barrier, they are typically generous in dimension when compared with the volume; that is, the carrying capacity is substantially higher than the actual volume, even in the highest volume situations (Figure 2). When such dedicated lanes connect with shared streets the sheer volume of bicycles ensures that cycling rules of the road prevail and cars conform to the speed and trajectory of the cyclists. In other words, the comfort level provided by such a system is a major consideration in personal decisions about travel mode. One has only to observe buses, crowded to capacity and mired in jams to appreciate the relative freedom, speed and flexibility of the bicycle.

Bicycle parking facilities have sprung up wherever users require them. Sidewalk space, plazas in front of buildings and many residual spaces close to major destinations have been given over to parking. A simple lock integrated with the rear wheel is sufficient since a parking attendant typically manages the parking. Such parking facilities are operated as micro-enterprises and are quite lucrative for the operators. Nevertheless, for the users the cost is relatively low, typically 0.2 RMB ($0.025US). The proliferation of such parking facilities on sidewalks has led to pedestrian use of the street, in effect displacing bicycle movement toward the centre of the street and narrowing the lanes for motor vehicles.

Figure 2. Traffic on the ring road in Tianjin.

Flats do occur in China although the streets are remarkably free of sharp debris when compared with the typical North American city street. When they do occur, or when some other mechanical problem arises, a bicycle repair facility is never far. Such repair shops are micro-enterprises on the street and offer a wide range of repair services at very reasonable cost. No one carries the usual tune-up and repair kit virtually mandatory in Europe and North America, since the bicycle repair people are better equipped and more skilled than the typical rider.

Finally, simple one-speed bicycles are available at about $30US, an affordable price for the great majority of Chinese. In fact, the cost is increasingly a relatively minor consideration in the purchase of a new bicycle. While growth in the vehicle fleet during the 1980s was closely tied to the ability to pay the purchase price, today considerations other than price typically enter into the purchase decision. However, because of the rapid growth of the bicycle fleet through the 1980s and 1990s, the market is saturated in central cities. People continue to buy bicycles, including electrically assisted ones.

Overall then, it will be seen that the bicycle infrastructure system strongly supports the widespread and intensive use of the bicycle as a daily means of travel within Chinese cities.

Street culture and non-motorization

Market liberalization starting in 1978 led quickly to the establishment of street-oriented commercial activity in many cities. The increasing flow of bicycles on major streets supported enterprises operating out of ground floor apartments and converted factories. The symbiotic relationship between non-motorized traffic and street-level commercial activity has continued today. Many cyclists then combine shopping trips with trips to work or study, since they need not travel out of their way to access most daily services.

A tradition of street use for recreation and communication is then augmented with commercial activity. Bicycle culture easily fits this model of street life (Baudon, 2002). The space between the shop and moving traffic is remarkably different from that in Western cities. The line between the commercial operator's shop space and the sidewalk is already blurred with commercial activity and services spilling into the nominally public space. Micro-enterprises and bicycle parking occupy the residual space such that the remaining walking space is only suitable for movement between shop and street or between the enterprises. Pedestrian movement is usually relegated to the street itself, often in an informal shared arrangement with moving bicycles. Least important and often completely obscured by carts and parked bicycles is the street curb. This street feature, while apparently replicating a Western convention, has little function in the typical Chinese urban context and might even be considered an irritant, allowing detritus to accumulate where it cannot easily be removed.

Shopping centres have emerged in recent years in new urban areas but also at key intersections in long-developed areas. The layout of such centres, often with the building itself isolated from the street by vast areas devoted to parking, are car-ready but are seldom actually used by large numbers of car-drivers. They are, of course, popular attractions in the city since they provide air-conditioned comfort in the summer, ambiance, and distractions. They have not themselves had a significant effect on travel mode, however.

The restoration of a street-based popular culture is then both a product of bicycle culture and supportive of bicycle culture. While the public bus system also supports commercial development, stops tend to be much farther apart than in a Western city, sometimes two or three times farther apart. In effect, bus movement generates nodes of commercial development while bicycles support strip commercial development. Moreover, since bicycles can penetrate narrow side streets and residential lanes, they also support a much vaster commercial network, closely allied with local pedestrian circulation.

Radical changes in this system locally would have a major perturbing effect on the quality of streets and the quality of daily life for many people. As has been witnessed in many places worldwide, the disturbance of local social life also reduces the opportunity for local enterprise.

Personal and household characteristics and cycling behavior

It is often taken as axiomatic that rising income will lead to more costly and, it is understood, more motorized travel. Certainly evidence worldwide suggests that higher income individuals demand more and more varied transportation services than do lower income people. In the context of the physical environment, local culture, and the cumulative behavior of millions of people, does this trend hold?

Thus far in China, it would appear that the relationship between personal characteristics and travel behavior is rather weak. In two studies conducted by this author (Zacharias, 2001; Zacharias & Pan, under review) such a transformation of behavior is not axiomatic.

In a study conducted at two bicycle-parking destinations in Shanghai, 250 cyclists were interviewed. Their home locations were plotted along with their chosen paths to the two destinations – Shanghai Central Library and a major parking facility on the Nanjing Road. As can be seen in figure 3, cyclists traveling to one destination, in this case the parking facility associated with office buildings in the vicinity, chose a great variety of different paths through the city. It is not apparent that all these paths could be grouped along fewer major dedicated bicycle routes. These cyclists not only follow the path of least resistance but also the shortest possible route through the city. This inevitably leads to considerable distribution of cyclists across the city and not the extreme concentration of movement experienced with car traffic.

These cyclists were also interviewed with regard to policies or factors that might lead them to switch to bus transportation, reported in the above-cited article. In general, little dissuaded them from their chosen mode and behavior. While less crowding on the buses and free bus rides were attractive to a minority of users, the great majority were not persuaded that bus transportation offered them significant advantages.

Figure 3. Bicycle trips made from home to a parking facility in central Shanghai.

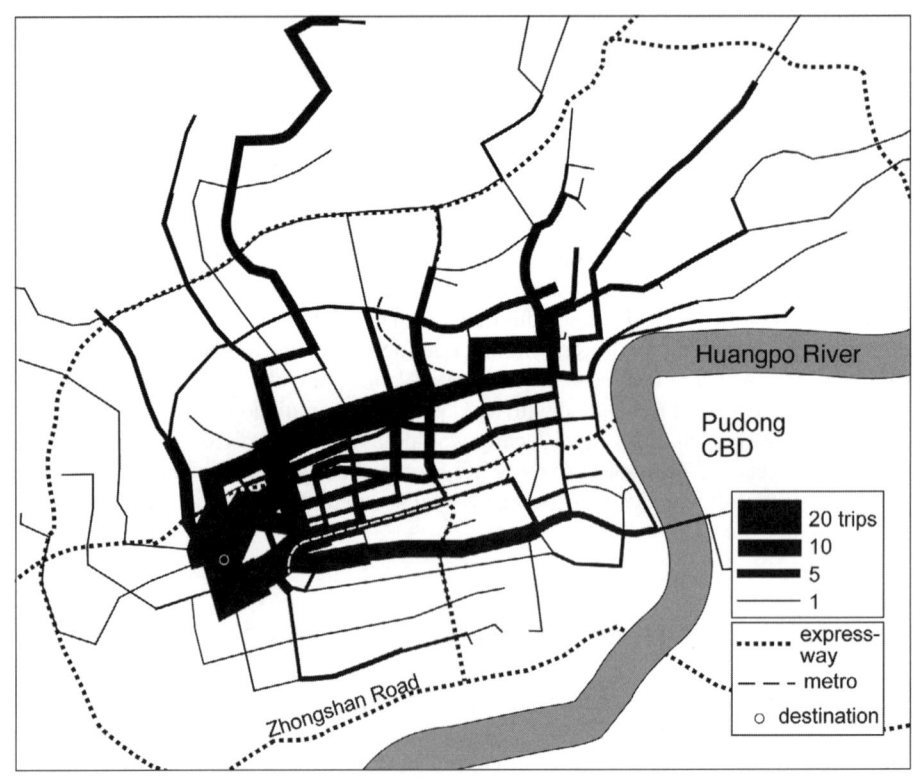

In the second study, 1,900 individuals were interviewed in four neighborhoods in Shanghai. The differences between the neighborhoods in the article in review. We found that there was little difference in the frequency of travel by non-motorized modes according to income. On the other hand, higher income individuals tended to travel greater distances by bicycle or by other means, presumably to better-paying jobs farther away. Age also made no difference in travel distances and frequency by non-motorized modes, although elderly people traveled much less and began to give up the bicycle at about age 60. There were also no differences detectable between the sexes. Overall, we might conclude that the reported personal characteristics had no significant effect on the use of bicycles in local transportation.

A small proportion of the respondents did have access to a car. As expected, they tended to use it. On the other hand, their travel to nearby destinations tended to follow the pattern of those who did not have access to a car, while destinations farther away, several kilometers or more, were accessed using the car.

It is not clear, then, that changing household composition, an aging society and rising income have much effect on the rate of travel by non-motorized modes in China. Much more important is variation in transport behavior between sharply different urban districts. The conclusions of that study, as indeed those suggested throughout this article, strongly support the importance of the environment in sustaining public affection for bicycle use in China.

Conclusions

It is increasingly suggested in the West that environmental policies in support of sustainable transport are essential (Westermark, 2001). This has, however, proven to be difficult to implement in a uniform way across Europe, because environmental issues are closely tied to those of local autonomy. Nevertheless, it has been observed that public response follows local environmental measures, including the provision of bicycling facilities and reformed rules of the road (Pucher, 1997).

It is possible to see a similar phenomenon in China, where longstanding environmental conditions have made it possible and generally supported substantial local resistance to planned alternatives to the bicycle. In an increasingly complex urban world of multiple choices and varied personal situations, the reasons for sustainable transport choices are complex and cannot be tied to a single factor or even several factors individually. The vast volumes of bicycle traffic – as many 6,000 bicycles per hour on certain major streets in Shanghai and Tianjin – suggest that it is more than simply the provision of bicycle paths that accounts for the scale of popular choice. The environment as a whole has supported and reinforced a pattern of use and behavior that has developed over one generation but little more than that. Urban development and the re-emergence of a street-based popular culture have been closely intertwined with non-motorized transport today.

The future of non-motorized transport in China is unclear. As discussed above, some local environmental policies have had a major disturbing effect on the use of the bicycle in some cities. The major disruptive effects of such policies have led other cities, to support multiple transport networks. Nevertheless, all cities continue to invest in road networks, traffic engineering and expressway systems in an effort to slow or halt the decline in traffic speed. It is widely believed by planners, however, that such investment is simply a stop-gap while sustained growth in the motor vehicle fleet will eventually outstrip investment. At that point or before, re-allocation of road space might well occur. The ability to restructure the transport network of Chinese cities is limited. The cost of acquiring the right to build infrastructure is enormous in the cities experiencing the fastest growth, where coincidentally urban development is largely led by real estate (Wu, 2000). A powerful local leadership and central government direction remain major factors in the eventual effects on cities. The planning system itself remains a powerful instrument of local control. Such control is very evident in the most cursory examination of Chinese cities today. The emerging connections between sustainability, non-motorization and environmental enhancement may provide the framework for a new urban transport policy direction in China. Preliminary evidence from across countries does suggest that behavior is responsive to environmental conditions in very similar ways. The massive behavioral response of cyclists in some Chinese cities provides suggestive evidence that the benefits of environmental improvements are cumulative and can be sustained during rapid economic growth.

References

Bae, C., & Suthiranart, Y. (2003). Options towards a sustainable urban transportation strategy for Bangkok. *International Development Planning Review*, 25(1), 31-51.

Baudon, L. (2002). Mutations de l'espace urbain a Shanghai: une mégapole entre ville globale et culture locale? *Géographies*, 4, 375-388.

European Conference of ministers of transport (ECMT) (1994). *Report of the Ninety-sixth Round table on transport economics*, 10-11 June 1993. Paris: ECMT.

Evans, R., Guy, S., & Marvin, S. (2001). Views of the city: multiple pathways to sustainable transport futures. *Local Environment*, 6(2), 121-133.

Gakenheimer, R. (1995). Motorization in China. Discussion paper. Massachusetts Institute of Technology, Cambridge MA: Department of Urban Studies and Planning.

Gan, L. (2003). Globalization of the automobile industry in China: Dynamics and barriers in greening of the road transportation. *Energy Policy*, 31(6), 537-551.

Gaubatz, P. (1999). China's urban transformation: patterns and processes of morphological change in Beijing, Shanghai and Guangzhou. *Urban Studies*, 36 (9), 1495-1521.

Guangzhou Transport Planning Research Institute. (1997). *Guangzhou Transport*. Guangzhou: city of Guangzhou.

Hook, W. & Replogle, M. (1996). Motorization and non-motorized transport in Asia. *Land use policy*, 13(1), 69-84.

Kubota, H. & Kodokoro, T. (1994). Analysis of bicycle-dependent transport systems in China: case study in a medium-sized city. *Transportation Research Record*, 1441, 11-15.

Ning, Y. & Yan, Z. (1995). The changing industrial and spatial structure in Shanghai. *Urban Geography*, 16(7), 577-594.

Organization for Economic Cooperation and Development. (2000). *Sustainable transport policies*. Paris: OECD 2000(8), 1-38.

Pucher, J. (1997). Bicycling boom in Germany: a revival engineered by public policy. *Transportation Quarterly*, 51, 31-46.

Shanghai Comprehensive Traffic Research Institute. (1990). Study of land use and traffic development in large Chinese cities, Shanghai, China (in Chinese).

Shanghai Comprehensive Traffic Research Institute. (1999). Traffic data statistics (in Chinese).

Shanghai Comprehensive Traffic Research Institute (2001). *Transportation changes in Shanghai 1995 to 1998* (in Chinese).

Shen Qing (1997). Urban transportation in Shanghai, China: problems and planning implications. *International Journal of Urban and Regional Research*, 21 (4), 589-606.

Shenzhen Urban Planning Bureau. (2001). *Origin-destination study and the transportation situation of Shenzhen* (in Chinese).
Stead, D. & Banister, D. (2001). Influencing mobility outside transport policy. *Innovation: The European Journal of Social Sciences*, 14(4), 315-330.
Steininger, K. (2002). Transport, access and economic growth. *World Economics*, 3(2), 75-91.
Sun, S. H. (2000). Shanghai between state and market in urban transformation. *Urban Studies*, 17(11), 2091-2112.
Whitelegg, J. & Williams, N. (2000). Non-motorised transport and sustainable development: evidence from Calcutta. *Local Environment*, 5(1), 7-18.
Wu, F. (2000). The global and local dimensions of place-making: Remaking Shanghai as a world city. *Urban Studies*, 37(8), 1359-1377.
Xin, C.J. (1996). Bicycle transportation in Shanghai: Status and prospects. *Transportation Research Record*, 1563, 8-15.
Yan, K. & Zheng, J. (1994). Study of bicycle parking in Central Business District of Shanghai. *Transportation Research Record*, 1441, 27-35.
Yang, J.-M. (1985). Bicycle traffic in China. *Transportation Quarterly*, 39(1), 93-107.
Zacharias, J. (2002). The bicycle in Shanghai: movement patterns, cyclist attitudes and the impact of traffic separation. *Transport Reviews*, 22(3), 309-322.
Zhu, J. (2002). Urban development under ambiguous property rights: a case of China's transition economy. *International Journal of Urban and Regional Research*, 26(1), 41-57.

A cultural comparative analysis of two villages in Storm Valley, Rize, Turkey

Fitnat Cimşit, Erincik Edgü and Alper Ünlü
Istanbul Technical University, Turkey

Abstract. This paper aims to explore the relationship between concepts such as culture, ecology, life style, built environment, and resource utilization. The research area of the paper concentrates on the context of two villages located in Storm Valley of the Black Sea region of northern Turkey. Due to cultural differences, the villages reflect distinctive physical and social choices along with different utilization of the available natural resources.

In this research, it is emphasized that regardless of its scale, the physical environment is independent from neither the socio-cultural group of behaviors and choices nor the ecological milieu of the habitat. In this sense, the inhabitants of the two villages of our research have altered the physical settlement of the environment. Hence, this research shows that the two different cultures living in the Storm Valley present different identities, life-styles, habits, choices, and ecological adaptation strategies. Spatial organizations of the villages are also differentiated in semantic, syntactic, and pragmatic ways. Consequently, the aim of this paper is to search for possible answers to the questions indicated below:

- Which socio-cultural parameters characterize the physical structure of both villages?
- Which parameters determine natural resource use and ecologic adaptation?
- How can the economic necessities determine the social compositions and house forms?

Keywords: ecological adaptation, resource utilization, culture, life-style, built environment

Ecological adaptation as a key to the built environment

When the relations of all behavioral variables within ecology and culture are explored, two basic contexts are found. The first one is ecological context; that is the milieu, which balances the relationship between humans and the settlement. The second is cultural context, which enables the use of adaptation strategies using common choices from possible alternatives (Berry, 1980). Cultural norms of a society may be the aspects of the ecological needs and expectations of the same society. Considering ecological adaptation and the cultural aspects of the environmental design, the determined parameters of this research are life style, utilization of the natural resources, and the built environment.

Adaptation is a term that has been transferred from biology to psychology. In general, adaptation is the natural sense of organizing the behavior for survival. As it is defined by Lawton and Nahemov's (1973) model, the stress felt in a foreign environment can be dealt with in time. The behavior of the individuals, their assessments, and cognitive evaluations all reflect adaptation modes conducted against environmental distress. The environmental adaptation of the individual may occur in three ways: the individual may adapt to the environment with full congruence, the individual may react to the environment trying to change it according to his/her needs, or the individual may withdraw him/herself from the environment.

On the other hand, ecological adaptation generally means the development of strategies in order to cope with the changes that take place in the environment. This study relies on the hypothesis that suggests that adaptation is directly related with the way resources are used (Bennett, 1980). As one of the parameters of ecological adaptation, achievements can be improved or transformed by the ecological context. This system suggests that adaptation is the structural appearance of the habitat and the society (Berry, 1980). The choices of using the natural resources in order to feed oneself organize the social classifications of the community. On the other hand, economic development is closely related with urbanization by means of the character of the built environment and the life styles of the residents (Drakakis-Smith, 1990). Thus, culture is an aspect of adaptation strategies as a result of resource use. Cultural aspects of the environmental adaptation organize the variety of choices that are acceptable to that culture.

The core elements of culture define the norms and criteria of communal life, and they are the unchangeable rules of continuity. All these common choices have a role in the meaning, behavior, space, and time organization; and the choice that the group makes is a function of what is possible in that milieu (Rapaport, 1980). Since cultural characteristics are resistant against time, minor changes in the living environment do not affect the lifestyles of the people. Traditional settlements are places where one can be aware of the concrete outputs of the cultural connections. Comprehension of the communal behavior modes is necessary in order to understand the social structure of a traditional environment. As Cooper et al. (1994) and Hummon (1988) suggested, in traditional communities, the physical pattern of the environment reflects the symbolic implications of the social structure, and people tend to consider the social control of the environment more important than the existing physical pattern.

A comparative case study

The research aims to compare two villages of Rize by means of cultural differences and choices of environmental adaptation strategies. These two villages, called Konaklar and Muratköy, have populations of nearly 600 and 400 respectively. The villages share the same ecological environment of the research area called Storm Valley, which is located in the Black Sea region of northern Turkey.

Being a natural site itself, Storm Valley hosts a number of villages where the inhabitants are descendants of two different cultures called as Hemsin and Laz. Having emigrated from Caucasus during the 11th century, Hemsin settlements were located at the upper parts of the Storm Valley where the plain land is scarce. Although Hemsin people had originally a different language, clothing, eating, and dancing habits, most of the Hemsin settlements have stopped speaking their language and lost many of their traditions.

The settlement of Laz culture in the area traces back to the 5th century B.C. They chose to settle on the lower regions, which are closer to the sea. Lazs also initially emigrated from Caucasus, as did the Hemsins, however, contrary to the Hemsins, it would not be wrong to say that Laz culture clings to their original language and traditional values of their cultural background.

In order to understand the relationship between and comparison of cultural differences and choices of environmental adaptation strategies of these two villages, a comparative case study of a two staged data analysis was conducted. The first data collection was a face-to-face inquiry-based technique that gives direct information about the life styles, ways of survival, social preferences, habitual choices, relationships among the community, and attachment to cultural traditions.

The second data collection is based on the spatial qualities of the built environment. The built environment possesses cues about the semantic, syntactic, and pragmatic expressions of socio-cultural differences, such as the meaning of the behavioral modes, juxtaposition of spaces, or utilization of the resources. By operating the spatial analysis method both in dwellings and on the settlement scale, the relations between the space and the socio-cultural structure were compared. In both cultures, house typologies, connections of the spaces, allocation of the buildings, the organization of the private and public spaces, give us implications about the functional uses, social status, privacy regulations, spatial priorities, and life styles.

Data analyses

First data analysis: Social characteristics

The face-to-face inquiry included a questionnaire comparing the indoor and outdoor activities, ecological adaptation strategies, and worldviews of the inhabitants of the two villages. The inquiry was done randomly with a total of 20 people from the Hemsin village Konaklar, and 24 people from the Laz village Muratköy. The questionnaire included questions to obtain data for structural parameters such as the ecology, socio-cultural background, and the semiology of the spaces. In this sense, the questions are grouped into three categories that define the socio-demographic structure, socio-cultural structure of the environment, and the properties of the built environment.

In this research, the interviewed people were randomly selected and they are composed of different genders and age groups. The retired population of Konaklar made it possible for us to interview an approximately equal amount of men and women. However, the case was different in Muratköy because of the men working outside of the village. The age distribution of the villages were also different; as the people aged over 65 years were in the majority in Konaklar, while Muratköy village had many middle-aged people along with a number of children.

The distributions of the household number and number of children per family in both villages were also examined. Ninety percent of the occupants of Konaklar live in houses composing of 5 persons or less, while this percentage decreases to 71.00 in Muratköy. This result is interesting to show that although unlike Konaklar, there is no patriarchal family structure in Muratköy, and there is always a new house to be built when a marriage takes place in the family; there are families composed of 10-12 persons. This situation is directly linked with the economic use of the natural resources and number of children in the family. In Muratköy, families need to have enough work power in order to cultivate land; however, Konaklar people choose to work abroad and only after retirement do they return home. In this case the population and the families are smaller in number in comparison to Muratköy.

Improvements in the education of women have been showing a slowly increasing curve in the area. However, the increasing means of communication helps to change traditional mentality opposing the education of women. The amount of illiteracy among Muratköy women is extremely high at 79.17%. In Konaklar, 55.00% of women have an elementary school degree; while there is no university degree available in the village, the only high school degree available is also from Konaklar. However the education of the new generation is considered as a status symbol among Konaklar people, and regardless of the gender, the children are encouraged to get a higher education.

Both Konaklar (75.00%) and Muratköy (83.00%) inhabitants have left their home for bigger cities to settle and work for good. However, the reasons of migration among the youngsters show different tendencies in both villages. In Konaklar, youngsters leave home for education while in Muratköy they leave for temporary work.

Both villages use the means of natural resources as effectively as they can. The additions to the Laz houses are convenient for storing both agricultural and fishery products along with related tools. In Konaklar, fishing is possible in the Storm River; however, Hemsins do not own a boat or have any relation to sea fishery while the people of Muratköy sail on the sea for fishing. Forestry is the main energy resource for both villages. The important factor of the natural resource use is that, in both villages, fishing and forestry is done to meet the family needs and is not utilized as a source of trade and income.

Table 1 gives us indications about family income. The occupational structure of Konaklar shows a dispersion slightly concentrating on retired workers and tradesmen. Muratköy on the other hand, concentrates on husbandry and animal husbandry. In Muratköy, the majority of the housewives also work in their own fields and they are busy with grazing sheep, goats, and breeding poultry. Animal husbandry is the most important means of livelihood for Muratköy people (83.33%). Muratköy people use the Storm Valley slopes as a natural resource for grazing the animals, while Konaklar people consider the place to be a recreation area. The income of many families in Konaklar comes from trading and other sources, such as a retirement salary. Thus, animal husbandry is not a concept for Konaklar people; the village also has long left agricultural business unlike Muratköy. In this sense Laz people use all sorts of natural resources for their

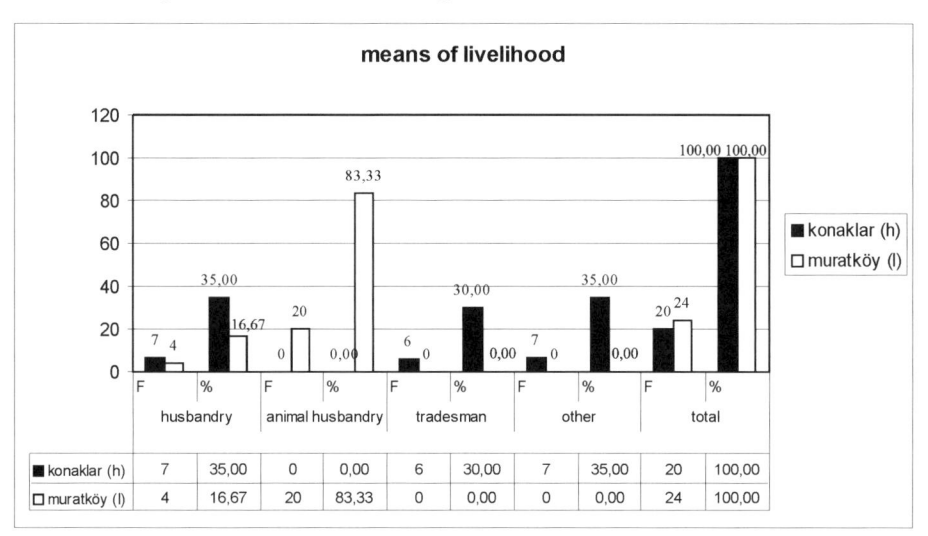

A cultural comparative analysis of two villages in Storm Valley, Rize, Turkey 235

Table 1. Comparison of means of earning a livelihood

	F	%	F	%	F	%	F	%	F	%
	husbandry		animal husbandry		tradesman		other		total	
konaklar (h)	7	35,00	0	0,00	6	30,00	7	35,00	20	100,00
muratköy (l)	4	16,67	20	83,33	0	0,00	0	0,00	24	100,00

Table 2. Comparison of neighboring villages

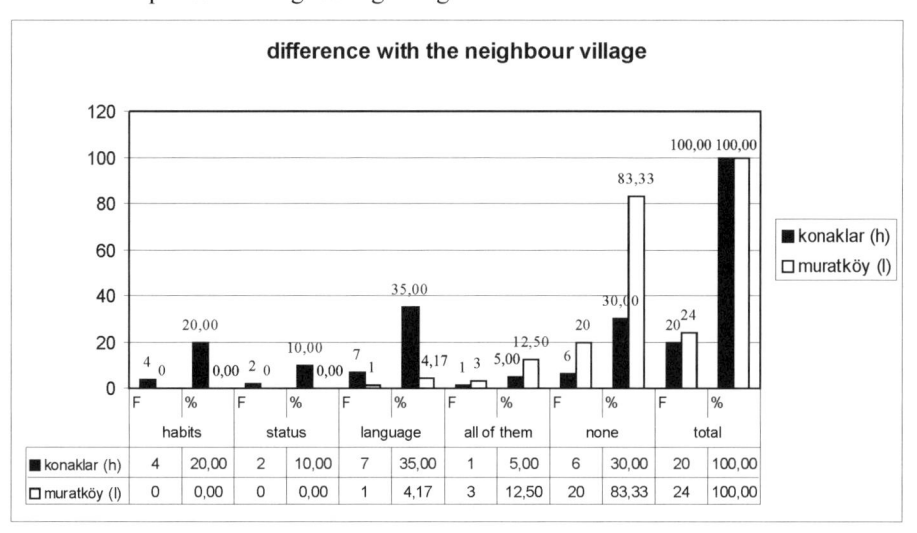

	F	%	F	%	F	%	F	%	F	%	F	%
	habits		status		language		all of them		none		total	
konaklar (h)	4	20,00	2	10,00	7	35,00	1	5,00	6	30,00	20	100,00
muratköy (l)	0	0,00	0	0,00	1	4,17	3	12,50	20	83,33	24	100,00

livelihood, while Hemsin people traditionally have always gone abroad for work; i.e., until the closing of borders with Russia in 1938, Hemsin men were working at the bakeries as they do nowadays in the bigger cities of Turkey.

Both villages have certain images of the neighboring villages, and consider them somehow different to themselves. As can be seen in Table 2, Konaklar people believe that neighboring villages are different than theirs in habitual ways, social status, and spoken language. Laz people prefer to speak the traditional language among themselves; however, Hemsins do not speak any language other than Turkish. Only 30.00% of Konaklar people believe that the neighboring village has the same social structure as theirs; however, this percentage is as high as 83.33% in Muratköy.

As for the social relations among neighbors, both villages share a similar temporal schema of mutual visits. The amount of neighbors visiting each other everyday is 100.00% in Konaklar, and 91.67% in Muratköy. The major type of relations with neighbors has a friendship basis, 90.00% in Konaklar and 91.97% in Muratköy. However, the concept of friendship shows some differences among villagers; in Konaklar shopping is considered as a form of friendship, while in Muratköy collective working for husbandry or grazing is. Sixty-five percent of the inhabitants of Konaklar and 75.00% of the inhabitants of Muratköy define weddings as the most important means of communication among villagers. Twenty-five percent of Konaklar people prefer to visit neighboring villages on a friendship basis, while in Muratköy this percentage decreases to 4.17. On the other hand, in Muratköy 16.66% of the families do not wish to communicate with other villages. This approach clearly shows us that Konaklar people have tendencies for more social interaction with others compared to the Muratköy people. However, although Hemsins are socially more active compared to the Laz people, the Hemsin villages are dispersed throughout the valleys while Laz settlements are united together in groups.

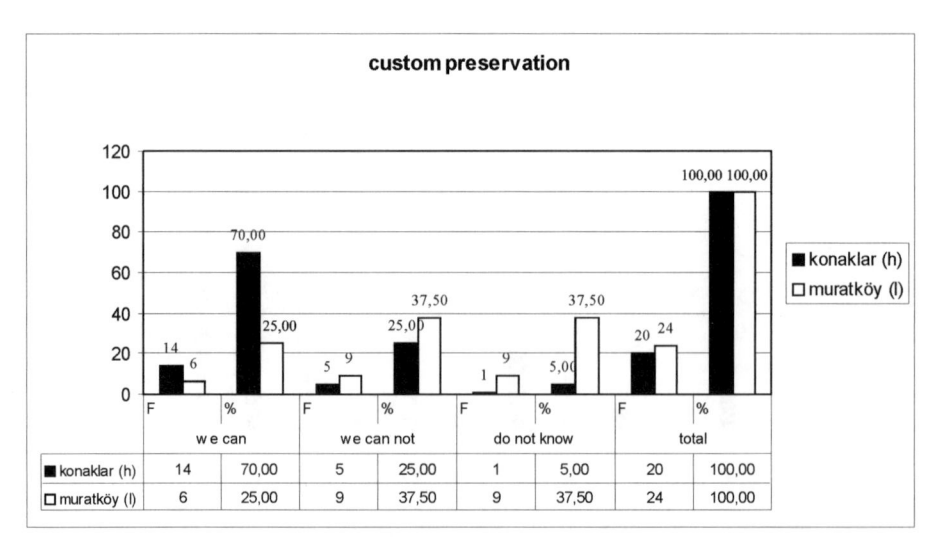

Table 3. Comparison of custom preservation

	F	%	F	%	F	%	F	%
	we can		we can not		do not know		total	
konaklar (h)	14	70,00	5	25,00	1	5,00	20	100,00
muratköy (l)	6	25,00	9	37,50	9	37,50	24	100,00

Table 3 indicates the preferences of these two cultures regarding the preservation of customs. Seventy percent of Konaklar people can preserve their customs, while this percentage decreases to 25.00% in Muratköy. It is also interesting to learn that although the Laz community are more introverted comparing to the Hemsin community, in Muratköy, 37.5% of the families do not have any information of their traditional characteristics.

Child raising may also be regarded among the parameters of family compositions and social structures of the villages. In 63.16% of the families of Konaklar, parents share the responsibility of their children. However in Muratköy, child raising is usually the duty of the mother. In this village, 83.30% of the families state that mothers have the full responsibility of their children. This situation is significant because in Muratköy women usually work in the family field and cannot find many chances to leave the village. On the other hand, men in Muratköy usually work outside the village for temporary jobs, which prevent them equally sharing the responsibility of their children.

Due to the Laz tradition of building a new house when there is a marriage in the family, the age of Hemsin houses are older compared to Laz houses. Although Hemsins admit that new houses are easier to clean, 90.00% of Hemsin people do not approve of new house typologies and do not wish to live in one of them. The house is usually considered solely as a shelter by both cultures. However, 15.00% of Konaklar people also attribute a social status to the houses, while there is no such concept of evaluation in Muratköy. There is also another interesting point here; the people of Konaklar usually own more than one house and these are embellished compared to the one-storey high Laz houses.

Second data analysis: Spatial characteristics

In order to analyze the spatial qualities of the built environment, the village spaces are examined both with settlement and house scales.

The settlement scale

As we have seen in the face-to-face inquiry data, the two cultures have different effects on the daily life of the villagers, while the reflections of cultural structure can be examined in many details. The analyses of settlement structure of these villages also show us clear differences of choices. Figure 1 shows the settlement patterns of Konaklar and Muratköy. The analyses of the data related to the settlement scale based on this map.

Figure 2 shows the collective comparison of the settlement properties of Konaklar and Muratköy. In this figure, comparisons of properties such as the settlement pattern, centre of villages, house compositions, silhouette formation, house-road connection, and spatial privacy are shown. As is seen in Figure 2, the scarcity of land for settlement in Konaklar has led the inhabitants to compose a dispersed settlement on the banks of the Storm River and the village has developed a branch-like pattern. On the other hand, in Muratköy the settlement has formed a linear pattern while concentrating alongside the main road.

The centre of the village is the place where the inhabitants prefer to gather for common purposes (Figure 2). However, the preference for these areas differs in the functional uses of the area, rather than the actual place that they are located in the village. In this sense, Konaklar people mostly prefer to gather in the coffeehouse and grocery, which are located face-to-face at the entrance of the village, while the Muratköy people prefer to gather at the mosque which is located at the midpoint of the main road. The noteworthy difference here is that Konaklar people attach great importance to their privacy; so that the centre point of the village is actually located at the entrance of the village, symbolizing a sense of control and keeping the strangers away. In the case of Muratköy, however, the centre being located at the midpoint of the main road implies a symbolic approach, suggesting that the villagers prefer to act as a community. On the other hand, patios of the houses can also be considered as places of gathering, which enables more social interaction especially in Muratköy.

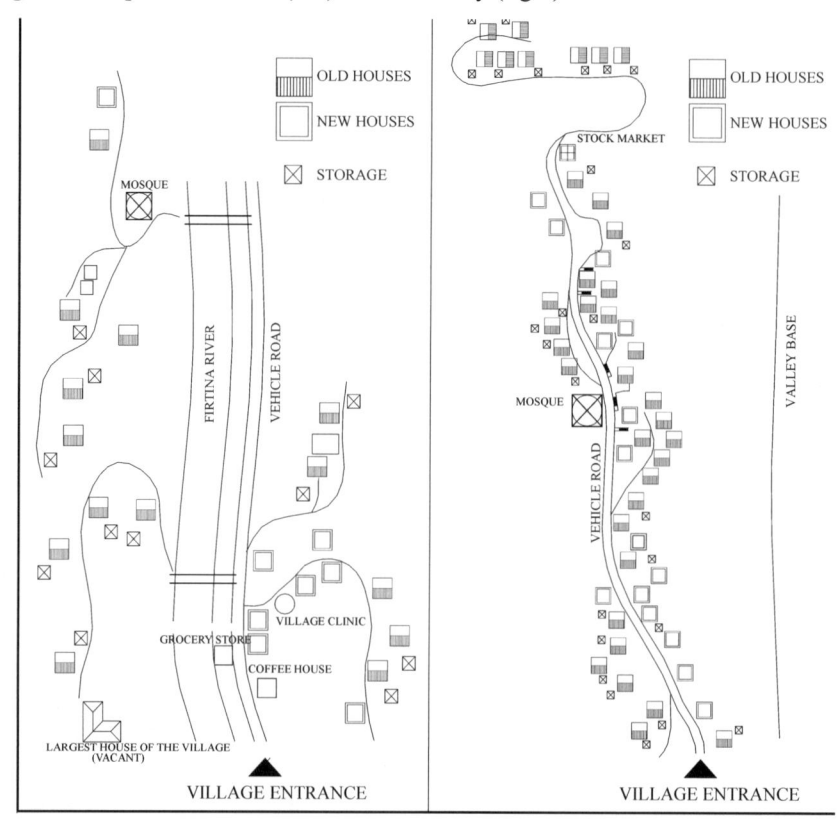

Figure 1. Maps of Konaklar (left) and Muratköy (right)

Figure 2 also shows the house patterns in the villages. In Konaklar, the physical form of the valleys and the privacy preferences of the inhabitants have caused the village to develop a dispersed structure. The transition axis of the houses does not allow a continuous circulation between the buildings; transition is limited to the boundaries of the family property. In the case of Muratköy, however, the houses follow a linear path that allows a continuous transition circulation. The patios and backyards of the houses face each other, and they allow a semi-private/semi-public area, which can also be used by neighbors. It is also necessary to remember at this point that in Muratköy a new marriage means a new house, so that the connections of the houses through the patios also symbolize a connection among the families.

The silhouette pattern of the settlements is shown also in Figure 2. In Konaklar, the dispersed settlement in the valley and the height of the houses enable an alternating visual order. In Muratköy, however, the linear structure of the settlement and the one storey high houses show the similar tendency in the third dimension, as they do in plans. However, linear planning does not necessarily force a monotonous silhouette for Muratköy.

Figure 2 also emphasizes the relationship between the main road and the entrance of the house. As it was mentioned before, the transition from a public space to a private space is directly oriented in Muratköy houses; however, the transition is somehow hidden

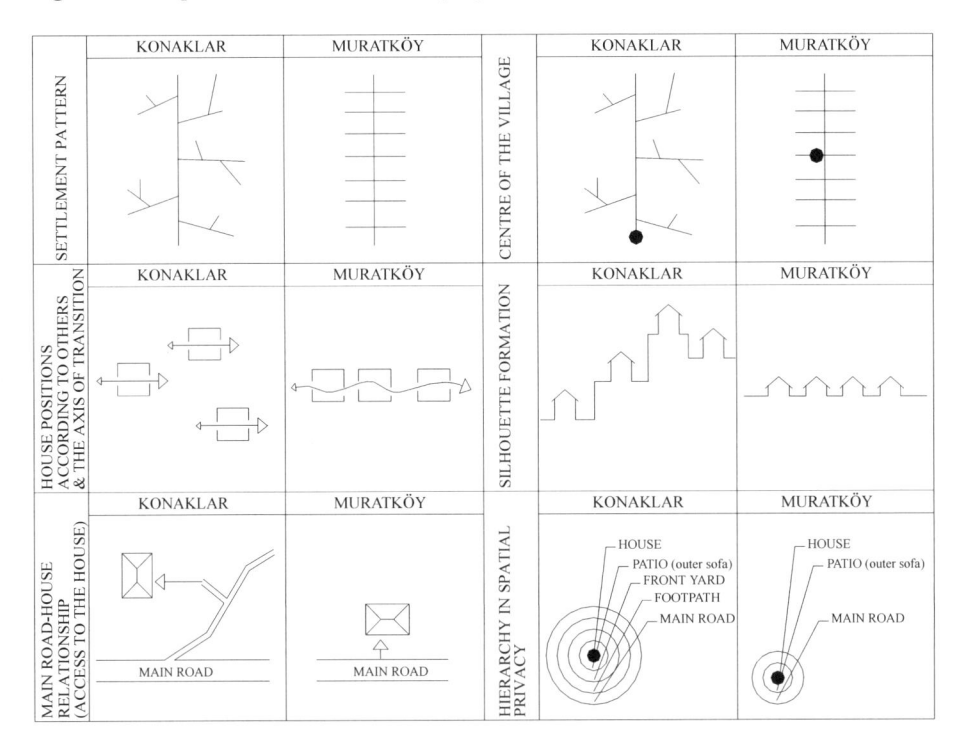

A cultural comparative analysis of two villages in Storm Valley, Rize, Turkey   239

Figure 2. Comparison of the settlement properties

and indirect in Konaklar houses. It should also be mentioned that, in settlement scale, the facades of the houses face the main road in Muratköy, while in Konaklar the entrance cannot be seen from the road.

The barriers or the stages of the transition axis from the main road to the houses in both villages can also be seen in Figure 2. With regard the many barriers preventing direct communication, in Konaklar, reaching the private space from a public space is relatively harder. As for Muratköy, however, there is direct access to the private space, indicating the importance of circulation between house and the main road. In Muratköy, house patios can be used as a semi-public space that enables even easier access to the private space.

As it was mentioned before, husbandry is an important means of livelihood for Muratköy. This situation attaches great importance to the connection of the family field to the entrance of the house. The women of the family who work on the field also are responsible for the housework. Thus, they need to have easy access to the main road and to the circulation area between the patio and the field. In this case, Muratköy people prefer a direct transition from the house to the field. On the other hand, in Konaklar husbandry is not the major source of income; in this case the field is located farther from the main road. In Hemsin culture, fields are perceived as property, rather than a production area.

The house scale

The second scale in the spatial data analyses is connected with the actual spaces of the home environment, with the exploration and comparison of the differences between the two cultures. A marriage in the Laz family leads to the building a new house; the Hemsins, on the other hand, have patriarchal family types, and a new married couple can also live in one part of the family house. In this case, Hemsin houses are older and have more rooms compared to Laz houses. In our research, it has been determined that 50.00% of the Hemsin houses are older than 50 years and there is no house newer than 5 years. Although there are common similarities in the house typologies, the symbolic meaning and the functional use of the spaces reveal the differences between the cultures. It is common to see 10–12 rooms and few storeys in a Hemsin house, while a Laz house has a maximum of 5 rooms and one storey. As we have mentioned before, Hemsin houses are more embellished compared to Laz houses. One can directly reach the main room, called *salomani*, of Laz houses, as is the case with reaching the house itself from the main road. *Salomani* is located right across the entrance door. However, in Hemsin houses, the main room, called *hayatI*, is rather hidden in the house and can only be reached after a series of transition spaces

Figure 3 shows the collective comparison of the spatial properties of typical Konaklar and Muratköy houses. In this figure, comparisons of properties such as the temporal uses of spaces, service areas, and bathroom spaces are shown.

Figure 3 shows the temporal use of the spaces in both house typologies. As is seen in Hemsin houses, functional use of the spaces is more widespread through the house. Having a patriarchal family structure, Hemsin houses have separate rooms for each married couple and these are located around the main room of the house, without having a transition space. This is interesting because, as we have seen in the settlement scale, Konaklar houses present barriers from the main road to the house in order to maintain privacy. However, the privacy barriers, which are present at the outside of the house, do not seem to be dominant inside the house. The guests are also entertained in the main room that is surrounded by private bedrooms, and there is no partition that can prevent their easy access. On the other hand, in Muratköy, daytime and nighttime uses of the spaces are completely separated from each other. Even though the neighbors share common patios on the settlement scale, house spaces clearly pronounce their functions.

As we see in Figure 3, service areas, where cooking, eating, and such take place, are the places related to the diurnal use. In this analysis, we see that, although Laz culture is very strict about the allocation of diurnal vs. nocturnal use of spaces, other than sleeping, they prefer to handle every function of house in the same space. In Muratköy houses, cooking, eating, sitting, and guest entertaining all take place in the living room. In Hemsin culture, however, there is a place used only for service facilities and related purposes. This space is located along the circulation axis from the main entrance; it may be considered similar to a large kitchen for it also has storage functions. We may conclude that the functional use of spaces in Konaklar is more definite compared to Laz houses.

Figure 3 also gives us data about the location of bathrooms. Sanitary necessities have traditionally forced people to locate their bathroom spaces near, but outside the houses. We see this situation in Laz houses where the bathroom is located under the patios; the doors of these bathrooms are located also outside the house, where it is made difficult especially for bathing facilities. As for the Hemsin houses, bathroom spaces are located inside the house attached to the service area. We may conclude that although Hemsin

Figure 3. House patterns

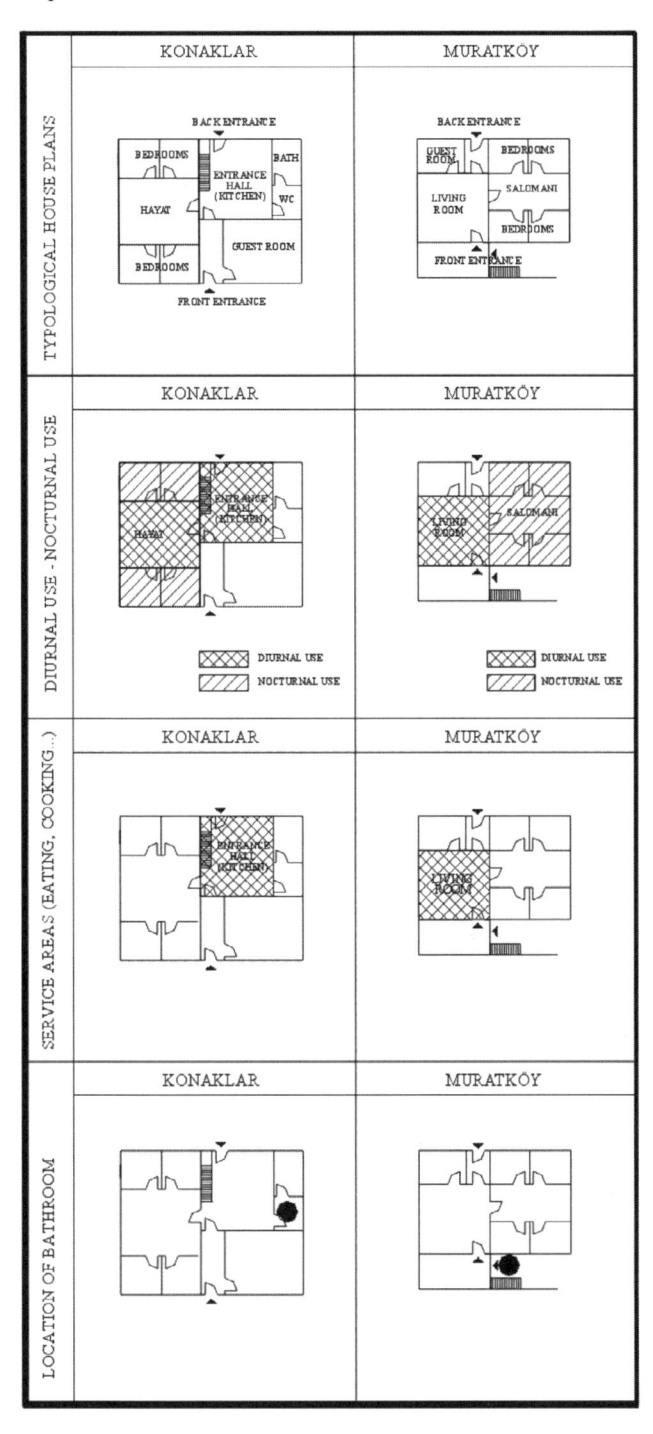

people prefer to keep on preserving the traditional house typologies, their spatial solutions are more convenient for contemporary preferences of life styles and standards, whereas Laz houses resemble the traditional preferences.

Conclusion

When we deal with any scale of a settlement, it is impossible to consider human beings independent from parameters such as behaviors, preferences, community attachment, and ability of adaptation to a certain ecological environment. Other than its physical appearance, space gives us clues about hidden dimensions. The functional and temporal use of spaces, social behavior, and the symbolic meanings are attributed to spaces in a way that their reflections can be considered as cultural specifications.

The data, acquired by the case study of this research, clearly show that the formation of the physical environment depends directly on the socio-cultural group of behaviors, adaptation strategies, communal choices, and the ecological milieu of the habitat. As we have seen in this study, there is definite information about the differences of culture with regard ecological adaptation strategies such as land uses, survival strategies, choices for occupation, social welfare, educational choices, family composition, life styles, privacy needs, communal bonds, attachment to traditions, and identity.

The parameters mentioned above play a major role in the formation of the social and physical structures of the settlements and also the identity of communities. These parameters also determine the choice of natural resource use and ecologic adaptation in a way that forms the life styles of future generations, even in a communication-based environment. It should finally be noted that, even though the economic-ecologic necessities may alter the social compositions and house forms, adaptation is the key process that organizes the complicated nature of these relations. Furthermore, although our world is moving towards global unity, differences in cultural structures are still the fundamental essences of human existence and should preserve the importance they deserve.

References

Berry, W. J. (1980). *Human Behaviour and Environment, Volume 4: Environment and Culture*. New York : Plenum Press.

Cooper, M., & Rodman, M. C. (1994). Accessibility and Quality of Life in Housing Cooperatives. *Environment and Behavior*, 26(1), 49-70.

Drakakis-Smith, D. (1990, 1991). *Economic Growth and Urbanization in Developing Areas, for the IGU Commission on Third World Development*. London: Routledge.

Hummon, D. M. (1988). House, Home and Identity in Contemporary American Culture. In S. M. Low & E. Chambers, *Housing, Culture and Design: A Comparative Perspective* (pp 207-228). Philadelphia University of Pennsylvania Press.

Lawton, M.P. & Nahemov, L. (1973). *Environment Design Research, Volume 1, Selected Papers* (Fourth International EDRA Conference). Stroudsburg, PA: Dowden, Hutchinson & Ross, Inc.

Rapoport, A. (1980). *Human Behaviour and Environment, Volume 4, Environment and Culture*. New York: Plenum Press.

The dialectics of urban play

Quentin Stevens
University of Queensland, Brisbane, Australia

Abstract. Not all aspects of urban social life are predictable and rational. Urban public space is characterised by tensions between efforts to regulate behaviour and stabilise meanings, and the diversity of everyday social practices. This paper examines such tensions, by focussing on how public spaces in Melbourne, Australia frame possibilities for play. Play includes unplanned, non-instrumental interactions between strangers, and explorations of the physical and symbolic texture of the urban landscape. The paper focuses on three dimensions of urban social life where spatial design has a critical influence: performance, representation, and control. These dimensions highlight how the meanings, desires, behaviours, and even the built forms of urban public spaces are shaped by a constant dialectical interplay between instrumentality, normativity, and play.

Keywords: public space, urban, behaviour, play, leisure

The dialectics of urban play

This paper focuses on the ways that urban public spaces frame tensions between the instrumental demands of social life and a wide variety of spontaneous, transgressive forms of playful behaviour. My study begins from the premise that the scope of urban life is not completely subordinated to the achievement of pre-defined, rational objectives. Indeed, urban conditions of density and diversity heighten tensions and contradictions between rational social organisation and other social desires, other outcomes. I draw from the theoretical approaches of Lefebvre and Benjamin the argument that non-instrumental playfulness thrives on a continuing dialectical negotiation with the various forms of discipline, exploitation, and spectacle which constitute the contemporary city. This paper uses play as a theme to explore the question of how urban, public conditions stimulate new experiences and enable the production of new identities, new meanings, and new social relations.

I present play as a series of definitions which frame it within that part of the social oeuvre which exists outside instrumentality, compulsion, convention, safety, and predictability. I follow a typology laid out by Caillois (1961), who defines four fundamental forms of play: competition, chance, simulation, and vertigo. Competitive play involves tests of strength and skill. It contrasts with instrumental work, by committing mental or bodily effort to non-productive purposes. Simulation involves disguising or forgetting one's usual self and one's place in the world by fabricating other identities and situations. New meanings are constituted through interpretive action. Play as chance means abandonment to uncontrollable and unpredictable circumstances. Many social activities in public are spontaneous and novel because they are derived from dynamic conditions of place, occasion, and individuals present. They provide opportunities for escape from predetermined and ritualised courses of action. Vertigo includes a wide variety of behaviours through which people escape normal bodily experience and self-control. Through disorientation, disorder, and destruction, people "lose themselves" in a purely physical mode of being, free of social meaning and purpose.

This definitional framework describes a variety of ways in which everyday actions elude subordination to rational order. Play concentrates our attention on practices which have a dialectical relation to the order, fixity, and functional and semiotic determinism of built form. Urban design often pursues such instrumental goals as comfort, stability, and legibility. My findings show how urban space also has a role in framing disorder, spontaneity, and change. My analysis examines where and how this thing called "play" actually happens; it involves a detailed examination of the structure of perception, bodily action, and social interaction in urban space, and the way actions both respond to and produce meanings.

Focussing on these various kinds of play activities thus tells us many new things about the role of physical space in our bodily experience and in our systems of discourse under urban conditions of density and diversity. This focused analysis of play aims to develop our understanding of what the diversity of possibilities in the city might actually mean at the level of individual experience. People's playful behaviour in urban public spaces is evidence that social identity continues to be produced with the body in a material environment. Through spatial practice, people retain a certain capacity to constitute new social meanings and produce their own reality. The study articulates many of the possibilities for mobilising identity which are latent in public space.

My analysis focuses on the ways that urban space reflects and produces the contradictions of the social superstructure, by framing social practices which stand in dialectical tension with production and social reproduction. To illustrate this dialectical tension, I will examine three dimensions of social life in public space: performance, representation, and control. These dimensions correspond to the three main components of my analysis of urban public space: how it frames relations between people, social meanings, and the actions of the body.

Performance

The city has long been understood as a theatre (Rudofsky, 1969, pp. 123-51). In urban public space people are always on display to strangers. The city brings together potential actors and audiences. However, playful performances in public do not often generate their own audience from scratch. As optional activities, such performances depend upon necessary activities (Gehl & City of Melbourne, 1994, pp. 13-38; Gehl, 1987, pp. 11-16) They appropriate and manipulate crowds which have already been gathered by the instrumental operations of the city. Physical space contributes to the framing of roles in public drama, defining where and who is "on stage." Yet public drama always develops dialectically, emerging from an interplay between spatial opportunities, the desires of various participants, and behaviours which mobilise these potentials.

On a Summer Sunday, loud pop music spills out of the "Sanity" CD store on Bourke Street Mall. The projected music similarly distracts, in an attempt to awaken desires in the body, so as to stimulate consumption (Crawford, 1992, pp. 13-18; Sennett, 1974, pp. 143-5). The music saturates public space. But commercial interests can only stimulate desire, they cannot dictate what actions arise from it. Some passers-by ignore the sound entirely, others receive it quite passively. One man in his fifties dances outside the front of the store. He's quite fit and quite uninhibited. His vertigo is clearly inspired by the atmosphere that spills across this threshold and into public space.

This man's act expresses the freedom people have in how they respond to the sensory stimuli which are compressed together in urban space. The man relies on the seepage of an instrumental musical performance across the threshold to lend atmosphere to his own display. Whilst most spectacular images in the city are framed for distant, passive perception, this music is engaged intensely through the body, in the presence of others. The man escapes everyday, practical behaviour by attuning his body to the music's rhythms. He is within the audible threshold of this store, but by remaining in the public space he retains the freedom to respond how he wishes, freedom to move, to jump and spin. He generally has his back to the store. Rather than yielding his attention to the store's merchandise, his use of the sound draws attention to him.

Although it is the music emanating from the threshold of this store which arouses the man's play, he draws upon stimuli both inside and outside the threshold. His enjoyment is also enhanced and given shape by the presence of many strangers in the public space of the Mall, who witness his act, comment upon it, and take part in it. He invites passers-by to join him, and several do. His enthusiasm rubs off, and this relies on the man being poised on the threshold of a public space which many others are using for their own purposes.

This example demonstrates that thresholds do not neatly separate the realms of inside and outside. People can orient themselves at the threshold so as to receive and mobilise various kinds of sensory and social stimuli in their playful acts.

In a diametrical situation, students from the College of the Arts put on a drama performance in one of the display windows of the Myer department store in the Bourke Street Mall (see Figure 1). This window theatre also confronts and transforms a general expectation about the edges of public space. This window, like many private street-level facades in the central city, is commonly used for artful displays of merchandise, which aim to distract the attention of passers-by, arouse their desires and stimulate impulse purchases (Crawford, 1992, pp. 12-16). The students' performance draws upon the same power to distract, but, again, subverts its instrumentality. They turn the window into something active, something with an unclear meaning and function.

Being behind this boundary of the windowpane means accepting limitations. The performers cannot easily be heard, and cannot touch or be touched. But they turn this to advantage. They emphasize visual expression through bodily exaggeration. The solidity of the boundary makes the proximity and unpredictability of the audience less threaten-

Figure 1. Drama performance in Myer department store window, Bourke Street Mall, Melbourne

ing. Their act includes dressing and undressing. They can press very close to the audience against the glass. They can perform without fear of interruption. This play act engages opportunities this boundary provides for heightened sensation of other people, while also controlling the risks of exposure. Parts of the performance involved the actors suddenly rushing forward and pushing themselves against the glass and thumping it. They also bellow loud enough that the crowd can hear. The boundary is a point of tension, and the performers play with the expectations this particular boundary condition establishes about the separation between them and their audience.

In both these instances, playful performance is framed within a setting which has actually been designed to serve instrumental commercial objectives. The blurring of boundary definitions frames possibilities of sensory perception which are then mobilised to shape particular kinds of non-instrumental social engagements.

Some forms of public performance arise through interactions between people within public spaces. Rather than separation, these performances require stages where people are brought together in public. Urban intersections are conspicuous public locations where strangers have close, unplanned bodily encounters in the presence of onlookers.

At the north-west corner of Collins Street and Swanston Walk, a man often stands holding a large signboard. His positioning indicates that he is keen to engage people. Attracting the attention of passers-by depends on visibility, both from a distance and up close. The man frames himself so as to capture that attention. At close quarters, he blocks direct passage, but even from a distance, the man's sign interposes between the pedestrians and their view of the path ahead. When he stands exposed just beyond the edge of the building's awning, his sign is quite luminous against its dark, enclosed understorey. The expansiveness and relative brightness of the skylit volume of the intersection serves to highlight whatever occurs there. The contrast is sharpest at this particular corner, where the waiting pedestrians often gaze idly out over the open space of the City Square diagonally opposite. The challenging, enigmatic images on the sign are framed between the immediate bodily discomfort of the crowd and the possibility of staring blankly into space. By standing out in the middle of the footpath, the man maximises the sign's visibility to people approaching from the other corners of the intersection and further along the streets.

This man stands directly in the flow of traffic, on one of the city's busiest intersections. The passing crowds are an essential element of this opportunity to express his opinion publicly. It is at such intersections that he can make contact with the greatest diversity of people, to generate friction, to stimulate debate. The compression of many people into the small space of a street corner brings about the transgression of personal boundaries. Interruption of the rational function of the footpath is essential to his performance. Every once in a while, someone stops and steps forward to engage him. There's no way of knowing who they might be or guessing their point of view. This element of chance, an escape into an unpredictable, relatively unrestricted social involvement, is part of the thrill of standing out there. Any performer's control over so public a stage is restricted by the behavioural freedoms of others, and relies on their assent. Audiences are not always willingly passive.

In summary, there are certain features of the urban landscape which clearly frame potentials for public performance. However, these are not the only sites where actors gain the attention of audiences and put on their show. Various acts of play illustrate constant tensions between displays which are instrumental and those that are not, between playful performances and more instrumental uses of public space, and between the actions of various parties who each enjoy the freedom of public space.

Representation

The public realm is a representational space layered with meanings and memories. As Lefebvre notes, (1991, p. 39) "[t]his is the space which the imagination seeks to change and appropriate." While built space has symbolic content, urban design is different to other arts because it is not an end in itself, it gets inhabited and put to both functional and expressive uses. Simulative play illustrates Lefebvre's argument that public space carries the tensions and contradictions of framing both action and meaning. In theorising how language carries meaning, Wittgenstein (1958, p. 150e) notes that in the game of chess, "the meaning of a piece is its role in the game." Similarly for public space, meanings arise and are engaged through use: that is, through interpretation and communication by members of the public. Whilst the figural and contextual contributions of an urban designer clearly cue certain possibilities for representation, ultimately the capacities which a built form has to represent are determined when these possibilities are brought together with the desires, ideas, and skills of actors, physical opportunities for communicating, and reception by audiences. Meanings are produced dialectically out of the interrelations of these factors.

Public parades are one example of behaviour creating meanings within space. Parades show that streets aren't just for efficient circulation; they have representational functions. Ritual parades along the axis of Swanston Walk pass by many of Melbourne's major institutions, drawing them together into narratives which bind social identity to place. But celebrations can themselves harbour tensions. In the 1950s the Moomba parade displaced the Labour Day parade, which had celebrated the proclamation of the eight-hour day. Moomba had subsequently grown to be seen as the "people's festival." Brown-May (1998, pp. 173-205) frames Swanston Walk at the centre of an unfolding dialectical struggle between formal civic processions, popular marches, the transgressions of carnival and everyday street activity. He highlights the constant efforts of local authorities to channel the surplus time and energy of the urban population away from excessive behaviour and disruptive, destructive, or transformative possibilities.

In the 2000 Moomba festival parade, a procession of decorated trams containing professional performers moves slowly along Swanston Walk. The themes on the trams are playful re-interpretations of aspects of local urban culture, and evocations of the transgressions and inversions of carnival. Traditionally, the Moomba parade had centred on the active participation of a large number of community groups. Hundreds of costumed people marched or danced along, accompanying thematic, musical floats which had been decorated by the groups themselves. The parade displayed and invigorated the reality of the city's ethnic and social diversity; a marked contrast to 2000's symbolisation of it. The procession of decorated trams, as a simulation of freedom, masks the production of behavioural controls. This year thus marked a tightening of the regulation of public leisure time, as a licentious, participatory public celebration was replaced by a spectacular simulation which people were supposed to passively watch. Public leisure is carefully choreographed, on the very day which is meant to sanctify the idea of public "free time": the Labour Day holiday. This transformation clearly reflects Debord's (1994, p. 12) analysis of contemporary public life, where "what was once intensely lived becomes mere representation."

A playful counterpoint to the 2000 Moomba parade was a small protest march which moved down the east footpath of Swanston Walk at the same time as the trams moved along the middle. This march featured a person dressed as a giant budgerigar and a small support group chanting "Bring Back the Bird!" (Figure 2). Later the same day a person wearing this costume jumps from the Princes Bridge where the parade route crosses the Yarra River. The protesters seek to draw attention to the cancellation of another event which had for many years formed a significant part of the Moomba festivities: the Birdman Rally. In this competition, people launched themselves in homemade, unpowered aircraft off the side of a city bridge and attempted to pilot them over a set horizontal distance. Many participants took off in nothing more than a funny costume, wildly flapping their arms in a ludicrous imitation of flying. The ungainly bird maquette ably represents the whimsical spirit of the contest itself. The Birdman Rally was a grand example of public play. It was a participatory event which brought together competitive display, intense, risky experience of the body in space, and sudden, dramatic wasting of energy.

As a context of meanings, the city is complexly layered, and while messages can be imposed through practices, they are not always readily received by diverse audiences, and can be contested and rewritten. In this example, playful behaviour appropriates, critiques and expands the social meanings which are written into urban space by ritual procession along a path.

The Birdman protesters struggle against the curtailing of behavioural excess at Moomba. They do so by turning the logic of the formalised procession against itself. The key here is bodily appropriation of the space of representation. This group attempts to look like a part of the main parade, through the use of a giant, fun, colourful figure. They harness the social concentration and excitement of the main parade and the symbolic power of this axis for their own purposes, by running in parallel to it. This protest reframes the social possibilities of carnival by inserting itself into a space and time which claims to represent carnival. The Birdman Rally undermines the Moomba organisers' attempts to inscribe a new social convention on the street. Through play, people continuously write new stories onto the urban landscape; they change the way we see ourselves reflected in our public spaces.

Figure 2. Birdman Protest Rally during Moomba Festival Parade, Swanston Walk, Melbourne

Spatial representations and practices also interrelate dialectically in cases where fixed symbols, such as public artworks, prompt people to playful action. Public art overturns the notion that the design of public space is instrumental to everyday practice, because it lacks "function" in the strict sense; it doesn't help achieve any preconceived instrumental task. It prompts playful action that explores new meanings.

The sculpture "The Three Businessmen Who Brought Their Own Lunch" stands on Melbourne's busiest pedestrian intersection. It consists of three very thin, life-size bronze statues in business suits with briefcases. People play with these statues in many different ways. They stand arm-in-arm with the figures, hug them, imitate their stiffness and their comical facial expressions. They shake their hands, pick their noses and pat them on the belly. Many of these playful engagements are transgressive of behavioural norms. People's playful contributions also imagine new roles and identities for the statues. A balloon is left attached to a hand. In winter one figure is given a woollen hat. All three of the figures have been designed with mouths pursed into deep circular holes. A woman puts her lit cigarette in the mouth of the rear statue, and she and a friend have a laugh, recognising that passing strangers are confronted by her contribution. Public statues usually expresses society's higher ideals. The cigarette transforms this sculpture into a promotion of something profane.

Many of the tensions of urban life are written into the Businessmen. The figures appear harried, expectant. Their formal dress and posture contrasts with the humour of their exaggerated features. The looks of surprise and apprehension on their faces, their frail bodies and unsteady, tilted stance suggest an inadequacy. They are figures of fun to be approached and interacted with; their meanings are not intellectually threatening or distancing. Play behaviour involving these statues illustrates that rather than just acknowledging existing meanings, people are also actually producing meaning within the built environment.

The examples of play I have considered in this section illustrate urban space as a representational medium through which social life is lived, where its values are both read and written. There are always contradictions and tensions between the wide variety of meanings found and created. Because social behaviour is not purely instrumental, it can be an active, interpretive and expressive response to meaning.

Control

The complex physical landscape of the city frames a great diversity of potentials for bodily action. Lefebvre (1991, p. 227) notes that space tends to localise and "punctualise" activities in ways that reproduce the social relations of capitalist production. The design of urban precincts and circulation spaces generally promotes instrumental goals such as production and commercial exchange. Discrete urban areas are also provided to facilitate the reproduction of social structure through relaxation and the harmless release of tensions. Urban planning and design generally limits the risk of incursions on such practical forms of behaviour, by organising and circumscribing human action in space, and by making alternative acts more difficult. Designing urban space around limited notions of function can be seen as a pervasive kind of social control.

However, as Lefebvre (1996, p. 129) argues, "[t]he satisfaction of basic needs is unable to kill the disaffection of fundamental desires." Functional, orderly urban space doesn't mean that play disappears. Indeed, design which satisfies function in a narrowly defined sense generally has unintended, non-instrumental consequences. Functional space even prompts play dialectically, producing desires for other kinds of actions. Playful behaviours do not arise through a tidy, rational chain linking intentions, actions, functions, and outcomes, and hence they are also difficult to regulate through rational strategies.

Dialectical relation between design, instrumental conceptions of function and playful practices in the city can be seen in the actions of in-line skaters and skateboarders. These playful modes of transport offer a critique of instrumental movement through urban space.

Early one Saturday afternoon, a family with two young children are touring through the city, all wearing in-line skates. The father pushes a third child along in a stroller with bicycle wheels. They pass through Block Arcade, a nineteenth century retail development. Rolling is vertigo, it's a different way of moving through and experiencing the city. This family, like the pedestrians in the arcade, are taking advantage of the quietness, the safety from vehicular traffic, from weather and from crowds on the streets. They're also taking full advantage of the smoothness of the arcade floor, which makes possible the sensory experience of going past its fine boutiques both at speed and up close. The family discovers and enjoys acting out new experiential opportunities that lie latent in this built environment.

Two in-line skaters race each other in circuits of the narrow plaza around the base of the Melbourne Central office tower, at exactly 5 p.m. on a Friday afternoon. They move fast across the smooth paving, jumping off flights of steps, swerving between trees and pedestrians. In this example, the risk and the thrill of moving fast is enhanced by the obstacles the urban space presents, including the tightness of the space, changes in level, the many workers leaving the building for the weekend, and by the limited reaction time when coming around the building's corners.

Such skating can be considered as a form of tourism which transgresses conventional notions of tourist behaviour. Most tourists perceive a city quite passively, but for skaters the urban landscape is not consumed as pre-packaged (Percy, 1975, pp. 46-56). Their leisure practice escapes the codification of leisure itself in the acts of promenading and sightseeing (Rojek, 1995). Skaters re-read a path which the average pedestrian takes for granted. The skater's perception of the city is reshaped by the wheels they travel on: it is more intense, immediate, subtle, and ephemeral than the walking sightseer's.

In public spaces in Melbourne, as in many other cities, local government attempt to control skating by retro-fitting the smooth edges of steps, seats, and planter beds with projecting metal angles. Behavioural restrictions are being written into space. Yet such interventions tend to modify practices, rather than displacing or preventing them. Skaters usually don't move on to easier, safer places.

One reason I have already considered is the desire for performance. Seeing and being seen in public space *is* a function; one through which both pleasure and identity are produced. Secondly, the identity of teenage skaters is constructed oppositionally, and this is enabled by spatial confrontations between their practices and those of people who disapprove, and also by direct confrontation of the prescriptive functionality of the built environment itself. In this context, damaging the lugs can in itself be seen as a function (Lefebvre, 1991, p. 177). Scratching and paint traces are other forms of resistance which give witness to different functions for steps. Thirdly, skaters don't skate because it's easy. For them, physical challenges are intrinsically enjoyable, and that's part of why they like the city so much. They develop new kinds of games, in dialectical response to the built properties of the space. The more the environment changes, the more new terrain there is to explore and contest. When design attempts to curtail or exclude behaviour, contradictions don't disappear: play follows the lines of tension. Human desires in the city are not just for openness, freedom, and safety; nor is exploratory behaviour only inspired by such conditions. Constraint and tension also stimulate action. Attempts to regulate playful behaviour challenges skaters to rethink how their bodily needs can be met through space. The prevention of simple skating moves often encourages them to move on to much riskier edges such as handrails and to jump out off ledges or over flights of steps. They seldom land safely, but that's scarcely important. Not all people use steps and handrails for safety. Risk is what makes it fun.

These observations show that space designed for a particular range of functions often turns out to be conducive to a broad range of other, more playful "functions." Somewhat paradoxically, the most practical and protective design features of urban spaces frame some of the most extreme experiences of escape through vertigo. In their simple utility, they provide the greatest challenges for playful bodily engagement, the greatest risks of speed and elevation.

Urban intersections are an example of how even the most instrumental allocation of space in itself engenders particular forms of spatial conflict, which can be acted out through play. Where the trajectories of people using different modes of travel and moving in different directions cross, there must be an intermittent discontinuity of action in time and space. The intersection where people have to stop is a gap in the functional time and space rhythm of the city. Thus the generally functional geography of the footpath and street become distorted. Crossing paths frame unplanned encounters, where actions are spontaneous and not entirely premeditated. In such ways, the fixed structure of the city contributes to the disorder of city life.

At rush hour on the last Friday evening of each month, hundreds of cyclists gather and ride through the centre of the city in an event called Critical Mass. They occupy the whole street, suddenly and temporarily blocking traffic at a series of major intersections. Critical Mass inverts the priority usually given to cars. Tension is maximised by holding the event when drivers are most keen to hurry home for the weekend and relax. By contrast, the cyclists are relaxed and enjoying themselves right there in the middle of the city, monopolising the city's most functionalised space.

A collective "bike lift" at a busy city intersection (Figure 3) serves an instrumental purpose, as a metaphorical show of collective strength. This behaviour is playful in a number of respects. As a collective practice, it pursues socialisation for its own sake. It is an excessive, frivolous, impractical, fun waste of cyclists' energy. It is a physical challenge, stimulated by others who participate and by the expectations of onlookers in cars and on the footpath. The act is exploratory, a playful response to the intensity of urban life, and the tensions it generates. It is a spontaneous, unexpected behaviour which tests a new set of relations between rider, bicycle, space, and the public.

Figure 3. Bike lift during "Critical Mass", Collins Street at Swanston Walk, Melbourne

Another significant feature of the bike lift is that it is transgressive, indeed confrontational. This play has a political dimension. Vehicles are supposed to stop before an intersection and systematically yield to other's needs, not occupy the space on a whim for their own pleasure. The unproductive, unserious, non-transportational act of the bike lift heightens Critical Mass's general contestation of the rules of the road. The political symbolism is amplified by the choice of location, directly in front of the Town Hall. Both participants and onlookers on the far kerb smile broadly. In acknowledging their enjoyment, they also lend legitimacy to the practice. The bike lift reveals that the accepted spatio-temporal structuring of the street is in fact only provisional. It relies upon an obedient flow of traffic for its constant reinforcement. It can be manipulated by moments which invert or suspend its rules and boundaries. A new form of social-spatial organisation is layered into those which preceded it. The bike lift can be compared to de Certeau's (1993) account of walking in the city: it is a practice which is "rhetorical." De Certeau (1993, p. 155) suggests that walking has the same relation to the rules of space that rhetoric has to the rules of language: it manipulates them to create its own logic. The bike lift demonstrates "the tactics of users who take advantage of opportunities."

These various relations between functional and playful uses of space show that the totality of behavioural possibility in urban space is a product of the *contradictions* which arise between limited conceptions of function, a dominant, instrumental mode of social production, and the diversity of interests comprising everyday urban life. Playful actions such as skating discover and perform new behavioural possibilities for commonplace functional spaces. Action defines the possibilities of function, and not just the other way around.

Conclusion

The varied incidents I have described in this paper illustrate play as a fundamental ingredient of the everyday life of a city. Although on one level they suggest spaces should be designed to serve the desire for play, they do not offer tidy, practical solutions for either curtailing or promoting play. My findings show that urban space isn't neatly staged, clearly meaningful or purely functional. But neither is social life itself teleological; new goals and means are constantly being revealed. Caillois' framework of play usefully maps a scope of human desires which cannot be explained by the concepts of production and reproduction. This study has illustrated the diversity of social practices which exist in a tension with rationality, with power, and indeed with idealised visions of unlimited freedom, such as those offered by retailers, and of social cohesion, as depicted by public celebrations.

Desire, needs, personal growth and freedom are not just abstract ideals confined to the realm of ideology, or instrumental goals over which actors strategise. Rather they are social experiences which arise and are constituted through practices, and which are lived, both in and by means of a material context. Social perceptions and interactions in the public spaces of the city inspire and give shape to desire, dialectically. Through a range of observations I have attempted to illustrate various ways that public spaces stimulate and frame desires. I have identified particular kinds of urban settings where such desires become manifest in practices. The three themes I have discussed here illuminate a range of ways in which urban settings stimulate play, ways in which people appropriate these settings to explore their playful desires.

In conclusion, there is no fixed correspondence between built form, human motivation, and social behaviour. Values, ideas, actions, and spaces define one another, are produced through each other. The design of urban public spaces enters into an extraordinarily complex matrix of perceptions, attitudes, meanings, and behaviours. The city is not fixed, not an ideal diagram, but a system in flux. Built form is the most stable part of this system, but this just makes its design and management the most problematic. Physical space can only mediate lived experience; it can't deliver satisfactions. Satisfactions come through the way life is lived in space, how the physical world, the social world, and the self are perceived, performed, embodied, felt. Urban design can't necessarily stimulate, can't make people free, make them imagine or make them take risks that expand their experience of the world and of self. The power which a setting provides to relate to other people, to signify or to act is complex and never guaranteed.

References

Brown-May, A. (1998). *Melbourne Street Life*. Melbourne: Australian Scholarly Press.
Caillios, R. (1961). *Man, Play and Games*. New York: Free Press of Glencoe.
Crawford, M. (1992). The World in a Shopping Mall. In M. Sorkin (Ed.), *Variations on a Theme Park*. New York: Hill and Wang.
de Certeau, M. (1993). Walking in the City. In S. During (Ed.), *The Cultural Studies Reader*. London: Routledge.
Debord, G. (1994). *The Society of the Spectacle*. New York: Zone Books.
Gehl, J. (1987). *Life Between Buildings*. New York: Van Nostrand Reinhold.
Gehl, J. & City of Melbourne. (1994). *Places for People*. Melbourne: City of Melbourne.
Lefebvre, H. (1991). *The Production of Space*. Oxford: Blackwell.
Lefebvre, H. (1996). *Writings on Cities*. Oxford: Blackwell.
Percy, W. (1975). The Loss of the Creature. In *The Message in the Bottle*. New York: Farrar, Straus and Giroux.
Rojek, C. (1995). *Decentering Leisure: Rethinking Leisure Theory*. London: Sage.
Rudofsky, B. (1969). *Streets for People: A Primer for Americans*. Garden City, NY: Doubleday.
Sennett, R. (1974). *The Fall of Public Man*. Cambridge: Cambridge University Press.
Wittgenstein, L. (1958). *Philosophical Investigations*. Oxford, Blackwell.

Gated communities and urban planning:
Globalisation or national policy

Sarah Blandy and David Parsons
Sheffield Hallam University, UK

Abstract. This paper investigates the tension between the implementation of urban policy in the UK and global trends in culture and lifestyle as represented by the development of gated communities. The huge growth of gated communities in America has been well-documented, as have the concerns that their development represents a retreat by the affluent into privatised security, at the expense of society as a whole (Blakely & Snyder, 1999). In the US this process has sidestepped conventional forms of governance, both in terms of planning control and in the provision of services once the development is completed (McKenzie, 1994). Gated communities are being built in England, although not on the same scale. It is clear that developers here are seriously interested in this type of housing scheme, given their phenomenal success in the States (Webster, 2001), and that gated communities are spreading throughout the world (Ritzer, 1998). This paper uses a case study of a gated community currently being built in a city in the north of England to investigate the extent to which decisions about the development and management of such schemes conform to government urban policy. We conclude that this form of housing is at odds with several key themes of stated policy. The paper goes on to look at why such schemes are built considering the impact of globalisation and by exploring the driving forces from the supply side through interviews with the developer and from the demand side through interviews with the residents. Finally we suggest that urban policy should consciously address the issues raised by this form of development, otherwise the market forces driving the trend will lead to a more divided, privatised and exclusive way of life without the repercussions being considered.

Keywords: gated communities, urban policy, global market forces, home ownership

Urban policy

At the broadest level the stated aim of urban policy can be summarised as a good quality of life for all, reducing inequality by combating social exclusion. The DTLR web-site lists the objectives of the organisation, one of which is "a high quality of life for all in our towns and cities" with the focus on helping the disadvantaged expressed as "the renewal of our most deprived communities" (DTLR, 2002). There is a very strong emphasis on reducing social exclusion which is defined as "What can happen when people or areas suffer from a combination of linked problems such as unemployment, poor skills, low incomes, poor housing, high crime environments, bad health, poverty and family breakdown." (Social Exclusion Unit website, 1998.) Searching for *social exclusion* on the DTLR web site generates six hundred and eighty three matches. The concept of social exclusion is now firmly embedded in UK policy to the extent that it is unlikely to be questioned. In analysing the media coverage of the New Deal for Communities, Hall (1999) points out that the phrase social exclusion was "accepted and deployed in all of the media reports examined." Given the concern that gated communities symbolise a more divided society as exclusive enclaves of privilege, it could be expected that there would be some consideration of their impact from the points of view of social exclusion in city wide planning policy and in consideration of individual schemes. The case study which follows investigates the extent to which such considerations are taken into account in practice. We now go on to investigate how the government envisage social exclusion being tackled.

Government statements recognise that social exclusion can only be reduced by taking a strategic approach at a city wide scale recognising the links between problems and the existence of systems and cycles in urban dynamics. The Social Exclusion Unit was set up explicitly by the Prime Minister to help improve government action to reduce social exclusion by producing "joined up solutions to joined up problems." (SEU, 1998) The National Strategy for Neighbourhood Renewal states that "only a joined up response will be effective in tackling the problems of deprived neighbourhoods. The need for this is particularly strong at the local level (i.e., the local authority level). It is at this level that many core services do their operational planning and at which many decisions about allocation of resources are made." (DETR, 2000) Understanding the effect that one area may have on another is seen as a key to solving these problems: "Neighbourhoods don't exist in isolation. The dynamics of place or city-wide solutions matter."

The concept of partnership has become fundamental in government policy for urban regeneration. Analysis of the Government's summaries of its *Area-Based and Other Regeneration Initiatives* (DETR, 1999) reveals that every one of the nine initiatives outlined has the term *partnership* as a feature of the delivery mechanism. The Urban White Paper, *Our Towns and Cities: The Future – Delivery on Urban Renaissance* reiterates the aim of a high quality of life for all, not just for the few. The Implementation Plan published in March 2001 saw this as being best achieved through "many people and organisations working together from local representatives at regional and district through to individuals in their communities. It suggests that the way forward is through "Community Strategies developed by Local Strategic Partnerships as the best framework for developing and implementing the changes needed in each area." (Urban Policy Unit, 2001)

The development of gated communities can be seen as the antithesis of partnership. The huge growth of gated communities in America has been well-documented, as have the concerns that their development represents a retreat by the affluent into privatised security, at the expense of society as a whole (Blakely & Snyder, 1999). Glasze notes the spread of gated communities to a number of other countries, including Latin America, East Asia, sub-Saharan Africa, the Middle East and Western Europe (Glasze, 2000). In the USA the development of gated communities has sidestepped conventional forms of governance, both in terms of planning control and in the provision of services once the development is completed (McKenzie, 1994). Given the way in which gated communities seem to be directly contrary to the aims of policy on partnership and a joined up approach at city wide level, it would seem reasonable for there to be some consideration of these aspects by public authorities as decisions about the development of such schemes are made. Again, whether or not this occurs in practice is investigated in the case study.

The concept of social balance can be traced through urban policy since the 1960s. Goodchild and Cole (1999) provide an historical chronology of the concept. It was a key aim of the Garden City movement which led to its adoption in the new town programme of the 1945 labour government. However there has been debate about the value of the concept. Goodchild and Cole (1999) point out that studies of the results of attempts to generate social balance are dubious about its value (Kuper, 1933; Orlans, 1953; Heraud, 1968). The point is made by Goodchild and Cole (1999) that a sense of community may be better fostered by social homogeneity.

Urban policy is clearly concerned with the concept of community. *The Implementation Plan for the Urban White Paper* (DETR, 2001) wishes to see "people shaping the future of their community." The questions are at what scale is some form of social integration to be achieved and through what means? *The National Strategy for Neighbourhood Renewal* (DETR, 2000) suggests that this should be addressed through a joined up strategy at city-wide level and the *Implementation Plan for the Urban White Paper* suggests that an aim is "town and cities which create and share prosperity." This relates back to the broad aim of equality. To quote Tony Blair, "over the last two decades the gap between the worst estates and the rest of the country has grown." (Social Exclusion Unit, 1998).

One of the means suggested for achieving social balance is mixed tenure (Page, 1993; 1994). Page promoted a joined up approach with mixed tenure as one of the main strands and this is reflected in the ethos of policy for urban regeneration. *The Guide to Good Practice in Estate Regeneration* suggests that there is "some evidence that tenure diversification which introduces owner occupation to estates can enhance long-term stability". This reflects the partnership approach which would see the three main types of housing developer, local authorities, private developers and registered social landlords working together. This is the case in several examples of regeneration schemes in the UK city in which the gated community we studied is situated. On two redeveloped sites formerly occupied by deck access housing, regeneration was implemented by provision of the land by the local authority following demolition of the multi storey housing. The new housing constructed was a mix of private developers building for owner occupation and housing associations building for let. The regeneration strategy for one of the most deprived council estates in the city includes selective demolition of local authority properties with a private developer building for sale on the land provided. Mixed tenure is generally seen as the means of achieving social balance in regeneration practice.

However, again gated communities seem not to be in line with this ethos. They are built solely for owner occupation and their exclusivity is a common feature of their market appeal. Clearly it would be possible to have a balanced community on a city wide level made up of homogenous single class enclaves but our analysis suggests that there is some contradiction between the aspiration for balanced communities achieved by mixed tenure and socially exclusive owner occupied gated communities. The extent to which this is taken into account in practice is explored in the following case study.

The case study

The empirical data on which this paper is based comes from a small-scale ongoing study funded by the British Academy, of a gated community in a northern English city. The development, the first of its kind in the city, is of an old hospital site, surrounded by an original stone wall, in a wealthy area not far from the city centre. Some of the hospital buildings were listed, including one originally used as a workhouse. When the Health Trust decided to sell the whole site, a detailed planning brief was drawn up by the local authority to ensure that the listed buildings were sensitively converted to new use. The property was bought by a large company which has invested £30m in its purchase and redevelopment as an exclusive residential development. The existing buildings have been converted to apartments and town houses, and new properties comprising both houses and flats, have been built, making 180 dwellings in all. There is fairly high-density occupancy with some nicely laid out garden areas in the centre. The development also includes leisure facilities and a swimming pool. The surrounding wall has been retained and has become *the* distinguishing feature of the development. There are now three access roads, all of which have electronically controlled gates. There are extensive CCTV cameras installed around the development, set up so that, for example, if anyone enters through a pedestrian access gate their movements through the site will be automatically tracked by a camera. A security guard will be on duty during working hours, and overnight a control centre takes over responsibility for security. This is the first gated community to be built in the city, and not surprisingly it has attracted a great deal of attention.

The study, which will be completed in the autumn of 2002, aims to discover why residents chose a gated community, paying over the odds in an area where house prices already exceed the national average; how residents accommodate the use and control restrictions integral to living in this development; and their expectations of community and of collective management and participation. The study also investigates the process of obtaining planning permission for such a development. So far the sales of 32 properties have been completed. Residents have been sent a questionnaire on moving in, and to date eleven responses have been received. We have interviewed two officers in the local planning authority, and two highways department officers who played key roles in the development. We have also interviewed the developer and two residents.

We have had access to the legal documents, which are in the form of a three-way 200-year lease between the developers, the leaseholders and the management company. Residents automatically become shareholders in the management committee through acquiring a share in the company when they purchase their lease. When the last plot is sold, the freehold of the whole site will be offered to the management company. The developers will retain a "golden share" for a further six months, but then bow out. Individual enfranchisement (whereby the owner purchases the freehold of the property) would be discouraged, but could not be prevented. The rights of the owners are restricted through the use of covenants in the lease; enforcement of the covenants will be the responsibility of the individual resident or the management company on their behalf.

Tensions within local planning policy

The planning authorities who are ultimately responsible for the spread of such developments appear powerless to resist their growth despite the apparent conflict with planning guidance and government policy.

The interviews with the planning officers showed that the aim of social equality identified above in urban policy was not taken into account at all. The exclusivity of the scheme by the wall and the gates was not a consideration. Both planning officers pointed out that the houses were so highly priced that the normal operation of the housing market was sufficient to ensure that the housing would be occupied by one class of resident.

Any control of the exclusiveness or otherwise of the scheme by the planning system would need to be based on a legally acceptable "material consideration." For this there would need to be an approved policy in the Unitary Development Plan for the City. There is no such policy in the current plan and the Issues Document (City Council, 2001) on which the new plan will be based does not include anything along these lines.

The alternative would be to rely on government guidance but the relevant Planning Policy Guidance Note 3 (PPG3) does not provide the necessary basis. The first planner interviewed stated, "PPG3 is very explicit about encouraging high density schemes, etc but it contains no tools to consider social exclusion." The second planner from the development control section of the City Council said that "the scheme broadly conforms to the policy outlined in PPG3. It is a brownfield site and the density is high." He also felt that the developer would successfully challenge any decision to control this aspect "the developer would have appealed and the department would have had insufficient grounds to defend its decision at Public Inquiry."

Thus the rhetoric about equality, inclusively and reducing social division in UK urban policy cannot now or in the future be taken into account in this city in general decisions as to whether the development of gated communities go ahead. The government's planning guidance does not provide sufficient legal basis and there is not and will not be an appropriate policy in the city wide UDP.

The gated communities scheme can be seen as a move away from the existing status of partnership in terms of provision of services. Highways in private housing estates of more than five houses are normally adopted by the City Council and become public highways maintained by the City Council. An officer from the Highways Department explained that this was traditionally seen as being in the interests of both the developer and the residents, "traditionally developers have wanted roads adopted so there are no maintenance charges to the residents." In the gated community, the developer did not want the roads adopted but to have their status as private roads maintained by a private management company. Although this runs contrary to the ethos of partnership the Highways authority found this to be acceptable. The department cannot legally insist on adoption in any case and it was in their financial interest. As the highways officer said, "we have limited resources and this is OK, it cuts down on maintenance costs."

Once the decision was made that the estate roads would not be adopted the gates were welcomed by both the planning and highway departments. From a highway point of view the officer said, "the highway authority wanted gates on so that it was clear to the public that they are private roads." The planning department was concerned for the management of the wider area, "new gates were seen as a way of handling parking – to differentiate the site from the wider area and to keep other local residents from parking on the site." Other physical aspects were also taken into account. The second planner said, "The main question about gates is whether they are physically safe and whether they intrude onto public space rather than any issue of whether the estate itself is part of the public realm." It was also pointed out by the officers that the gates and walls are in accordance with the guidelines in *Secured by Design*. This is a government approved police scheme to design security into new homes whose recommendations include "creating real or symbolic barriers for an estate" and "restricting access for the public through the estate to as few routes as possible."

Clearly the privatisation of formerly public services can be seen as a move away from partnership. This was discussed in general terms with both representatives of the planning department and that they would be concerned only about the physical aspects. "Questions such as the number, usability and location of bin stores would be taken into account but not who would operate the service. There is no concern about whether the estate is to be serviced by private security firms."

On the other hand the planning department did have the power to impose a degree of mixed tenure. As a result of a policy in the Unitary Development Plan the Council is able to insist on a proportion, usually 25%, of the dwellings in a scheme being affordable housing. This can be done by housing being built by the private developer but sold to a housing association which then rents to tenants. In the case of this gated community an agreement was made for the developer to pay a commuted sum which is a capital payment to the Planning authority in lieu of this obligation.

To sum up, although "gated communities are creating new forms of exclusion and residential segregation, exacerbating social cleavages that already exist" (Low, 2001). The case study demonstrates that these social dynamics are not taken into account in public policy decisions about their development and management. We suggest that this contradicts the thrust of the rhetoric about exclusion, community balance, partnership and mixed tenure that runs through governmental statements about a wide range of urban policy and legislation. The next sections explore the why they are built by investigating the motivations of the developers from the supply side and residents from the demand side.

Home ownership and housing types in England

Ideal housing in England has traditionally been represented by a detached house, owned on freehold tenure, and situated in a village or leafy suburb. This conforms with the old adage "an Englishman's home is his castle": a place of security and seclusion. However, many features of the gated community we have been studying seem to directly contradict these assumptions. A large proportion of the properties are apartments; others are described as townhouses, which are actually tall, narrow, vertical slices of an old building, joined on both sides to their neighbours. The residents of the gated community in this study have considerable economic power and, therefore, choice in the housing market. Yet they have, in the majority of cases, moved from freehold ownership to buy on leasehold (the developer retains freehold ownership of the property). Further, they readily accept the terms of their lease by signing up to a regime of restrictive covenants, surveillance and control which in many respects *minimises* their rights as owner occupiers.

Literature on the meaning of home suggests a variety of motives driving house purchases, from the search for ontological security (see Gurney, 1999; Morley, 2001) to the desire for an investment (Blandy & Parsons, 1999) which will maintain and increase its value in an increasingly commodified world (Ball, 1983). These ideas have been tested out in interviews with residents of the gated community.

Ontological security is a term first used in the field of psychoanalysis. The concept was developed by Giddens to mean a sense of confidence and trust in the world as it appears to be – "an emotional, rather than a cognitive, phenomenon ... rooted in the unconsciousness" (Giddens, 1991, p. 92). In Giddens' view, the need for ontological security is a deep psychological need which has become harder to fulfil in the rootless, urban modern world. During the 1980s Saunders (1990) investigated links between housing satisfaction and tenure. For Saunders, the private realm of the home is where people find ontological security, something increasingly sought in these days of alienation and the breakdown of traditional ties of kinship. He identified three aspects of satisfaction, namely home as haven, a place of safety and security; home as a source of personal autonomy and part of the roots of individual identity; and home as a source of status and pride. Saunders found that homeowners were more likely to associate positive images in his three areas of satisfaction, than renters of housing. He therefore made the case that home ownership confers more ontological security than renting. Saunders' ideas and research have subsequently been much criticised and debated (see overview of this literature in Hiscock et al., 2001).

Lawyers have also engaged with the psychological underpinning of the "need" to own your own home. Radin (1993) puts forward the view that the ideological role of property is to define an area of privacy, of personal autonomy and personal sovereignty. The conventional legal view is that "... the freeholder is an 'absolute owner' in a sense in which even a long leaseholder is not" (Harris, 1996, p. 72). This, in part, is because leasehold tenure is associated with flats rather than houses. In England, there are fewer blocks of flats used as dwellings than in mainland Europe, and fewer than in Scotland. Hiscock et al. (2001, 57) found, in their survey of over 6,500 people, that "Perceived autonomy appears to depend on the type of home"; in other words their respondents derived more ontological security from living in houses rather than flats.

Standard clauses in long leases for flats include covenants against causing noise, holding meetings, putting up posters, hanging up washing, and keeping pets. (Aldridge, 1994). Strangely, long leases are more restrictive of the residents' use of the property than weekly tenancy agreements for council flats. The lease in the gated community, for both flats *and* houses, restricts residents in how they may use the grounds and facilities. For example, it forbids children to play in any communal areas except the designated play area, and restricts the use of leisure facilities to those who permanently reside in the development. It also restricts residents' use of their own property, in clauses such as:

"Not to permit any laundry or other article to be hung or spread anywhere outside the Premises
To clean all the interior and exterior surfaces of the windows of the Premises at least once in every four weeks."

Living in leasehold flats and houses, with common responsibility for enforcement of intrusive and restrictive covenants, emphasises the interrelationship of residents. (For a discussion of the associated difficulties, see Clarke, 1998; Blandy & Robinson 2001). How is this compatible with Saunders' (1990) identification of the rights of disposal, control and the right of exclusive use and benefit, as essential characteristics of owner-occupation?

The right of disposal is still present for gated community residents, but in restricted form. The lease provides that properties may only be sold if the purchaser is prepared to take on the original covenants and restrictions, and on condition that the seller pays a proportion of the sale price to the management company. The right of control is considerably compromised by the restrictive covenants in the lease. The right of exclusive use and benefit is compromised by this, and also by the freehold ownership of the site, rather than plots being owned by individual residents. Of course, the real rights, status and expectations associated with a particular tenure change over time. The next section of this paper considers a possible reason for this change as related to gated communities.

Globalisation

Globalisation provides a possible explanation for the consumer choice of a gated community, which seems so far removed from the traditional ideal English home. This has produced what Beck describes as "a *single commodity-world* where local cultures and identities are uprooted and replaced with symbols from the publicity and image departments of multinational corporations." (Beck, 2000, p. 43, emphasis in the original).

The globalised media is feeding housing consumers with images of an aspirant lifestyle which gated communities represent, featuring modern design, security, and access to private leisure facilities (see Thorns, 2001). Robinson (2002, p. 23) describes this phenomenon as a "holistic corporate strategy of shaping lifestyle, in as many of its dimensions, and for the greatest proportion of its total duration, as possible." The Internet allows you to "log onto real estate sites and take a virtual tour through the properties not only in your own hometown but increasingly around the world." (Thorns, 2001, p. 4). Ideas about taste, not just for interior design, but the type of housing to aspire to, thus spread around the world.

The "urban aesthetic" dates back to New York of the 1980s when loft living became fashionable. Apartments created from existing, sometimes historic, buildings are strangely compatible with "new urbanist" traditional architecture. Some commentators have suggested this trend indicates a desire to return to a mythical time of neighbourliness and community, to create a sense of belonging. (see Morley, 2001). As a Chicago-based designer puts it: "The children of the sexual revolution are looking to put romance back in their lives and return to traditional lifestyles.... People seem to be looking for touches they would have remembered from visiting their grandparents." (quoted in Garber, 2000, p. 20).

In terms of design, we are inundated with magazines (*Ideal Home, Elle Décor, Better Homes and Gardens*, etc.) and television programmes (*Changing Rooms, Other People's Houses*, etc.) which tell us how we want to live. From around the world, we see identical images in publicity material for trendy housing developments, for which the key themes are luxury, leisure, sex (represented by idealised models of desirable young people), and security. In the presentation of this paper we will show a selection of these international images.

So far as luxury is concerned, the publicity materials often specifically use this term and "Stylish show home décor and smart fixtures and fittings offer the prospect of an obtainable dream, a touch of designer chic and celebrity living." (Halifax, 2001, p.11). Security is another key feature, and frequently the details are included in brochures; for example, CCTV, 24 hour guard system, walls and electronic gates, despite the fact that crime figures in the UK have been falling for the past ten years, as reported in successive British Crime Surveys (Home Office, 1995-2000). The images used and the security features described mean that "Potential buyers are sold, in short, the sense of community promised by a homogenous neighbourhood, responsibly managed, and protected from outsiders." (Robinson, 2002, p. 5). A Florida developer who specialises in gated communities explains how security is sold: "You go after the emotions. We don't go out and show a gate in the ad. But we try to imply and do it subtly. In our ad, we don't even show houses. We show a yacht. We show an emotion." (quoted in Garber, 2000, pp. 15-20).

Further security is provided, although rarely detailed in publicity material, by restrictions on residents' behaviour. We can see this as a direct import to this country from America. In the States, properties within the walls of gated communities remain freehold. However, this does not mean that there are no restrictions. The ownership deeds of 80% of American residents associations require membership of the association. Many American community residential associations reserve the right to vet potential purchasers. Purchasers who are accepted by the association must sign up to a number of enforceable covenants, conditions and restrictions, including: no basketball hoops to be erected on the front of buildings; no rooms to be let to tenants; pets to be limited to a maximum of three dogs or four cats over the age of six weeks, of maximum 30 pounds in weight, per house; only 35% of each front yard can be cemented over; and the familiar prohibition against hanging of washing out to dry. These covenants have led to numerous court cases between residents (see Kennedy, 1995).

Leisure is often a selling point for gated communities: exclusive access to swimming pools and fitness suites. Finally, hi-tech features within the dwellings as well as for the security devices are a common theme. The sales manager for the gated community we have been studying emphasises that "Each home will be equipped with digital TV and phone lines to complement today's lifestyles." (quoted in the local Property Guide, New Homes Supplement, 21.9.01).

The effects of this globalised selling is to change consumer attitudes. For example, in England there has been a traditional preference for older properties. Yet in 2000, 36% of people in the home buying market would consider buying a new property, an 11% increase over six years. They identified a number of attractive features, most of which had to do with design and convenience. (Halifax, 2001). Housing choice is part of lifestyle choice, and the word *lifestyle* appears frequently in publicity material. People around the world now want low-maintenance hi-tech homes, rehabilitated conversions of older properties (lofts from former factories, for example) are fashionable, and no one has time for gardening. In the higher reaches of the UK housing market, apartments are very popular. Urban living has become cool, and this means higher density occupation, very much in line with government policy (PPG3). A large survey of homebuyers in the UK showed that higher density homes, converted from existing buildings, were associated with the very desirable properties of elegance and individuality. (Halifax, 2001).

Views of residents and the developers

The developer of this gated community has described in interview how "we didn't start it from 'there's an emerging trend from America of gated communities, and therefore we ought to be ground-breaking'," but rather that a marketing angle was identified: "if it's enclosed, make it secure." As the walls already existed "adding gates and security wasn't a big deal." When asked about marketing research and the impact of globalisation, the developer responded, "did we look at the States, or south, and say this is the way to go? No...(It was) more: "what do we need to make ourselves distinctly different?... to enhance it above the competition."

The marketing material for this gated community emphasises, in line with globalised trends that the development is designed for a "Life of leisure, life of security, life of luxury, featuring hi-tech, loft style apartments...(which) offer lifestyle-inspired contemporary luxury." The developer's personal view was that "from a marketing point of view it's definitely the security, peace of mind." The sales negotiators have been specially trained to conduct potential purchasers on a "guided tour of show house, pointing out the hi-tech and convenience features."

This marketing has clearly paid off, as the case study shows that the housing in the gated community is being sold for prices around 15% higher than equivalent property in the surrounding area, which is already well above the national average. Analysis of respondents' questionnaires reveals that purchasers' reasons for moving to the development, in order of importance, are: security features, maintenance of property values, proximity to job equal with leisure facilities, and finally, moving into a community. None of the interviewees felt that the area was particularly dangerous or prone to crime, but security for cars and children were overwhelmingly the reasons they chose to live behind the gates.

In terms of globalised design and type of house, most interviews included comments such as:

"we wanted an older house, but having run (one) we didn't want the cost that incurs with that and the maintenance of things."

"the house itself I absolutely love, the way that it's been planned, the architecture of it, the fact that it is old but with all of the modern features as well."

One interviewee with definite opinions about taste in interior design commented rather sarcastically: "I think most people went through and chose exactly what was in the show home even though that's completely different to these houses."

All respondents in the gated community said that their legal adviser on the purchase had made clear the difference between freehold and leasehold tenure. In reply to a question asking for their understanding of the difference, most of those who replied wrote something along the lines of "we don't own the land on which the property is built." One interviewee gave high importance to the fact that leasehold tenure ensured long-term maintenance both of individual properties and of the communal grounds. "At least it's all going to look fairly uniform as well, rather than like say one person is doing one thing and somebody doing something completely different." The residents consider it far less important that their use of their own property is restricted, than that other residents are kept to their covenants. This may be because, as one resident said in interview, her "solicitor had explained, they've got very limited things they can actually do should you go against what's in the lease, anyway." For the residents of this particular gated community, leasehold ownership does not represent a compromised or second-rate form of owner occupation. Moreover, it seems that the search for ontological security has been superseded by the need for a secure environment and for more tangible and visible forms of security.

Conclusion

The central theme of this conference is the tension between the forces of globalisation and efforts to ensure local and regional identity and distinctiveness through planning controls. This case study suggests that UK urban policy is at odds with the international trend for the development of gated communities and that planning policy at local level is powerless to resist these globalised forces. Far from reducing social exclusion and creating balanced communities this form of development represents increasing social division in single class, single tenure enclaves. The local authority was not concerned about the reduction in the role of the public sector and in the case of highway maintenance welcomed it as a means of reducing expenditure. We suggest that rather than allowing this trend to gather momentum through market forces urban policy should consciously engage with the processes and decide on its approach to the issues. The issues to be addressed would include:
- the privatisation of the public realm including roads and communal areas
- the takeover of public functions and the decreasing role of local political life in what would otherwise be the public domain
- the increasing social fragmentation and division into single class estates
- the scale at which social inclusion or balance is to be achieved – city wide or estate by estate.

References

Aldridge, T. M. (1994). *The Law of Flats* (3rd edition). London: Longman.
Ball, M. (1983). *Housing policy and economic power*. London: Methuen.
Beck, U. (2000). *What is Globalisation?* Cambridge: Polity Press.
Blakely, E. J. & Snyder, M. G. (1999*). Fortress America: Gated Communities in the United States*. Washington DC: Brookings Institution Press.
Blandy, S. & Parsons, D. (1999). *Power, choice and the commodification of housing*. XXVll Congress Proceedings, International Association for Housing Science.
Blandy, S. and Robinson, D. (2001). Reforming Leasehold: Discursive Events and Outcomes, 1984-2000. *Journal of Law and Society*, 28(3), 384-408.
City Council. (2001). *Review of the Unitary Development Plan, Full Listing of Issues*.
Clarke, D. (1998). Occupying "Cheek by Jowl": Property issues arising from communal living. In S. Bright & J. Dewar (Eds.), *Land law: Themes and perspectives*. Oxford: Oxford University Press.
DTLR. (2002). Available online at http://www.dtlr.gov.uk.
DETR. (2001). The Implementation Plan for the Urban White Paper. London: HMSO.
DETR. (2000). *National Strategy for Neighbourhood Renewal*. London: HMSO.
DETR. (1999a). Available online at *www.regeneration.detr.gov.policies/area/summaries.html*.
DETR. (1999b). *Urban Renaissance: Sharing the Vision*. London: DETR.
Garber, M. (2000). *Sex and Real Estate: Why We Love Houses*. New York: Anchor.
Giddens, A. (1991). *Modernity and Self Identity: Self and society in the late modern age*. Cambridge: Polity Press.
Glasze, G. (2000). Des Societes Fragmentees. *Urbanisme,* 312, 70-72.

Goodchild, B. & Cole, I. (1999). *Social balance and mixed neighbourhoods: a review of discourse and practice in British social housing.* Sheffield: Centre for Regional Social and Economic Research, Sheffield Hallam University.

Gurney, C. M. (1999). Pride and Prejudice: Discourses of Normalisation in Public and Private Accounts of Home Ownership. *Housing Studies*, 14(2), 163-183

Halifax New Homes Marketing Board. (2001). *New Homes Today.* Halifax: Halifax plc

Hall, T. (1999). *Mapping the Spaces of Government Intervention: The Media and the New Deal for Communities.* Cheltenham: Geography and Environmental Research Unit.

Harris, J.W. (1979). *Property and Justice.* Oxford: Oxford University Press.

Heraud, B. J. (1968). Social Class and the New Towns. *Urban Studies*, 5, 33-53.

Hiscock, R., Kearns, A., Macintyre, S. & Ellaway, A. (2001). Ontological security and psycho-social benefits from the home. *Housing, Theory and Society*, 18(1-2), 50-66.

Home Office. (1995-2000). *Annual British Crime Survey.* London: The Stationery Office.

Kennedy, D.J. (1995). Residential Associations as State Actors: Regulating the Impact of Gated Communities on Nonmembers. *The Yale Law Journal,* 105(3), 761-793.

Kuper, L. (1953). Blueprint for living together. In L. Kuper (Ed.), *Living in Town.* London: Cresset Press.

Low, S. (2001). The Edge and the Center: Gated Communities and the Discourse of Urban Fear. *American Anthropologist,* 103(1), 45-58.

McKenzie, E. (1994). *Privatopia: Homeowner Associations and the Rise of Residential Private Government.* New Haven: Yale University Press.

Morley, D. (2001). Belonging – Place, space and identity in a mediated world. *European Journal of Cultural Studies*, 4, 425-448.

Orlans, H. (1952). *Stevenage, a sociological study of a new town.* Westport Connecticut: Greenwood Press.

Page, D. (1993). *Building for Communities.* York: Joseph Rowntree Trust.

Radin, M. J. (1993). *Reinterpreting Property.* Chicago: University of Chicago Press

Ritzer, G. (1998). *The McDonaldisation Thesis: Explorations and Extensions.* London: Sage.

Robinson, P. S. (2002). *Local Expressions of Globalising Capitalist Society: The Privatisation and Fortification of Public Space.* Paper presented to the Gated Communities Conference, Johannes Gutenburg University, Mainz, 5-8 June, 2002.

Saunders, P. (1990). *A Nation of Home Owners.* London: Unwin & Hyman.

SEU (Social Exclusion Unit). (1998). *Bringing Britain together: A national strategy for neighbourhood renewal.* London: DETR.

Thorns, D. C. (2001). *The Making of Home in a Global World.* Paper presented to the Managing Housing and Social Change conference, City University of Hong Kong, 16-18 April, 2001.

Urban Policy, DTLR (2002). Available online at http.www.dtlr.gov.uk/about ktlr/

Urban Policy Unit. (2001). *Our Towns and Cities: The Future. An implementation Plan.* London: HMSO.

Watt, P. & Jacobs, K. (1999). *Discourses of Social Exclusion: An Analysis of "Bringing Britain together: A national strategy for neighbourhood renewal."* Paper presented to the Discourse and Policy Change conference, University of Glasgow, 3-4 February, 1999.

Webster, C. (2001). Gated Cities of Tomorrow. *Town Planning Review*, 72, 2.

Performance and appropriation of residential streets and public open spaces

Maria Cristina Dias Lay and Jussara Basso
Federal University of Rio Grande do Sul, Porto Alegre, RS, Brasil

Abstract. The study aims to identify the compositional and contextual factors that affect appropriation of urban open spaces, in order to assess which of these factors influence more intensively type and intensity of use in the urban spaces evaluated. Performance evaluations were carried out in streets and public open spaces located in three residential areas characterised by differences in cultural and socio-economic aspects, and aspects related to physical characteristics, located in the city Campo Grande, Brazil. Methodological procedures included mental maps, interviews, physical measurements, observations of behaviour, and questionnaires, which were responded to by residents in the sampling areas and users of the three public open spaces investigated. The analysis of relationships between individuals' characteristics, environmental attributes and level of appropriation of streets and public open spaces suggest that type and intensity of use of residential streets are more strongly affected by compositional factors, while contextual factors affect the level of appropriation of public open spaces more intensively. The results contribute to the existing knowledge and outline recommendations that might support the social life dynamics of urban spaces.

Keywords: residential streets, public open spaces, compositional factors, contextual factors, appropriation, performance evaluation

Introduction

Urban open spaces – such as streets, plazas and parks – are spaces intentionally provided to encourage a specific group of behaviours, giving a place in which particular activities can take place. According to Francis (1989), these spaces are the common ground where public culture is expressed and community life developed, reflecting the users, their private beliefs and public values. The importance of open spaces in residential environments seems to be paramount not only for the development of community life, but also to fulfil residents' needs for recreational and functional activities. Additionally, the provision of green threads and neighbourhood parks can bring leisurely appreciation of nature, and are highly valued parts of cities (Francis, 1987b; Halprin, 1963; Ulrich, 1979). Therefore, the network of urban open spaces, besides providing the channels for movement, has the potential to promote social life, contact with nature, and fulfilment of sports activities and healthy life. The use of urban spaces for performing this spectrum of activities, as Gehl (1987) remarks, is a critical ingredient in making spaces meaningful and attractive, and subsequently, *lively* environments: *use* or *liveability* (e.g., Appleyard, 1981) is a criterion often employed to measure success of open space, and it is also one of the prerequisites for a successful urban open space. According to Whyte (1980), when urban spaces are not used, they are unsuccessful spaces.

Literature shows that several factors might affect the ways appropriation of urban open spaces occurs. Francis (1987) argues that open-space research originated from public awareness of the social failure of many urban open spaces, largely failing to serve their intended uses and users (e.g., Hester, 1989; Jackson, 1981). Indeed, research over the past decades, focusing on previously neglected aspects of open-space quality, has provided recognition of the social and psychological benefits of urban open spaces (e.g., Gold, 1972; Rutledge, 1986). It is also noticeable that appropriation of urban open spaces occurs differently among cities, as well as between areas in the same city. For example, one of the first things that motivated Whytes' research on urban spaces was to understand why some city spaces work for people and some do not, even when spaces are located in neighbourhoods that ranked very high in density of people. Hence, when examining the origins and current realities of public life and public space in the literature, the discussion of the apparent decline in public life emerges. One could list numerous reasons mentioned in the literature for the decline in the proportion of public life in streets or any other type of public open space, related to the changing technological and sociopolitical configuration of our society (i.e., the automobile, television, economic imperatives which make small businesses unprofitable, increased crime and violence on the street, and so on). The supposition put forward by Carr et al. (1992) is that public life, like any other aspect of the culture, reflects the process of adaptation of culture to environment. Therefore, instead of decline, public life might be taking new forms. But how strongly are new forms of public life being affected by socio-economic and cultural factors or by the fit or misfit between the physical characteristics of the setting and users' needs?

Assumptions in the literature have suggested a bias towards cultural or physical factors in affecting appropriation of urban open spaces. Regardless of the argument of Carr et al. (1992) that public spaces are different in terms of appropriation due to *cultural forces* (involving basic and functional needs, social community life and symbolic sense of public life), and the *physical structure* of the place (which affects the conditions offered to perform the demanded activities – involving environmental qualities, such as appearance, security, accessibility and environmental fit), researchers tend to bias their arguments in favour of physical and management factors as a means of achieving successful public spaces. Similarly, other authors (e.g., Francis, 1987; Gehl, 1987; Whyte, 1980) emphasise that type and intensity of use of public open spaces depend on those environmental qualities that favour the taking place of static or dynamic activities. A central question posed by many researchers is whether people are free to achieve the types of experiences they desire in public spaces. Following Lynch's (1981) argument that people should have access to a public space, freedom to use, change, and even claim the space, as well as to transfer their rights of use and modification to other individuals, Francis (1987a) defends the need for democratic streets, defined as the ones that are well used and that invite direct participation, provide opportunities for discovery and adventure, and that are locally controlled and broadly accessible. As residential streets represent informal spaces with a great potential vitality, increasing interest on how streets are used has emerged among researchers. For example, the liveable street movement, pioneered by Whyte (1980) and Appleyard (1981) recognises the importance of the street environment for the social life of cities, while Levine (1984) discusses the broad theme of street "liveability" or "sociability."

Commenting on the spatial structure of streets, Ellis (1991) mentions the numerous entries contained in dictionaries, suggesting the particular, continuing, multiple nature of the street: at once a road and a place, inseparable from the buildings that flank it. Similarly, Gutman (1991) calls attention to the elements streets have in common, which constitute features common to definitions of the street in both the architectural and the urban cultures. That is, apart from being a social fact, the street is three dimensional, and includes not just the road or sidewalk surfaces but also the buildings located along it, the street furniture, and other structures that mark its length or define its beginning and end. The interdependency of all the physical elements and functions underlies most of the essential qualities of the street, both good and bad. As the many dictionary variations suggest, they might still have the potential for enriching the individual activities and the collective conception of the urban surroundings shared by the residents. This provokes the examination of all facets of these basic urban elements, from their functions to their formal properties.

Looking more specifically to cultural aspects, Rapoport (1987), for example, claims that walking and other street activities are primarily culturally based in that they result from unwritten rules, customs, traditions, habits, and the prevailing lifestyle and definition of activities appropriate to that setting. According to Rapoport, it is culture that structures behaviour and helps explain the use or non-use of streets and other urban spaces. However, he notes that those settings that are appropriate will further influence the activities that occur in a street, and acknowledges that physical variables might play a part in the appropriation of streets. Accordingly, Gans (1962) points to the important role that attitude, expectation, and social-class homogeneity play in the decision to interact with surrounding environment and engage in neighbouring relationships, as in most working class communities reported upon in the literature. Nonetheless, Gans' emphasis on the relevance of social-class homogeneity in interacting with surrounding environment or engaging in neighbouring relationships is contradicted by Rapoport's (1987) claim that in high-status areas the streets tend to be frequently empty. Therefore, although social-class homogeneity has been identified by Gans as critical to friendship formation, it does not follow that the presence of homogeneity will automatically allow design features to influence behaviour, or that all homogeneous social groups desire neighbouring relationships. The confrontation of arguments further suggest that there are other factors affecting appropriation of residential streets that could be related to life style of the different social groups, or to physical characteristics of each setting.

Therefore, the set of assumptions discussed so far suggest that there is a relationship between appropriation of urban open spaces, cultural and socio-economic aspects of potential users (compositional factors), and aspects related to physical characteristics of the setting (contextual factors). Nonetheless, whether appropriation of urban open spaces is more strongly affected by compositional factors than contextual factors, or vice-versa, understanding how urban design could better fit users' cultural and socio-economic characteristics and environmental needs remains vague. These are topics not yet studied in enough detail; therefore, this study intends to contribute to existing knowledge. This investigation explores these relationships and investigates which compositional and contextual factors are more relevant in affecting types and intensity of use in residential streets and public open spaces.

Methods

The survey was designed with the purpose of investigating how compositional factors, such as users' socio-economic characteristics, life-style related variables and contextual factors, related to physical characteristics of the settings that affect evaluation of quality of urban spaces, such as appearance, security, accessibility and environmental fit, may influence appropriation of urban open spaces. The means and methods of measuring the variables were investigated and tested through a comparative study of residential streets and public spaces in a sample of three neighbourhood areas in the city of Campo Grande, capital city of the state of Mato Grosso do Sul, Brazil. With a population of approx. 600,000 inhabitants, Campo Grande is a city characterised by its recent expansion, with a considerable migrant population originating from other towns in the state of Mato Grosso do Sul, other Brazilian states, and neighbouring Latin American countries. The three neighbourhood areas selected are characterised by differences in residents/users origins, household income and density of the areas.

Data collection

A combination of multi-method techniques was used for data collection. The investigation started with preliminary observations of how people use urban open spaces in three previously selected neighbourhood areas. Mental maps and interviews with users in residential streets and parks were undertaken in order to find out what were the perceived limits of their neighbourhood, and to identify which public open spaces were included within the perceived limits. Moreover, it was identified where users came from and how frequently they used it. This information, summed to the established sampling selection criteria, allowed the selection of streets and public open spaces in the three neighbourhood areas. In order to understand the role that those places play in people's lives, and why spaces are used or ignored, further methodological procedures were adopted for data gathering, which included physical measurements, observations of behaviour, and questionnaires completed by residents in the selected streets and users of public open spaces. A total of 240 questionnaires were completed (80 respondents in each area). The questions were formulated to discover residents' and users' perceptions and attitudes towards the physical environment, cultural and socio-economic characteristics and life-style. Systematic observations were undertaken in order to generate data about people's activities and the relationships needed to sustain them; about regularities of behaviour; about expected uses, new uses and misuses of a place; about behavioural opportunities and constraints that environments provide. Spatial behaviour was observed and recorded in each of the settings, using a behavioural map procedure. Behavioural data were collected in each setting during weekdays and weekends, on a day and time-sampling basis, during the period of two weeks. Twenty-eight behavioural maps of each setting were drawn with AutoCAD software; this consisted of a record of the number of individuals engaged in each of a number of predetermined behaviour types. The overlaying of behavioural maps by days, time-sampling (mornings and afternoons) and the total behavioural data, allowed a detailed behavioural portrait of each setting. Focused interviews were further administered with the objective of understanding issues related to people's attitudes toward their residential environment and public open spaces which could not be fully understood through the other methods.

Data analysis

The data from behaviour mapping, survey of the settings and interviews were analysed on the basis of frequencies. According to the nonparametric nature of the data, for the statistical analysis of the questionnaires, cross tabulations, Kruskal-Wallis tests, and Spearman rank correlation coefficients were applied, providing measures to assess how strongly appropriation of urban open spaces were related to contextual and compositional factors.

Sampling selection

The criteria adopted for sampling selection were intended to select three neighbourhood areas with marked differences on the following characteristics: a) residents/users socio-economic and cultural background, b) density, c) urban configuration, and d) intensity of use of urban open spaces. Moreover, each neighbourhood area should have a public open space, located up to 400 metres walking distance from any point within the delimited investigation areas. According to Halprin's (1963) definition of urban spaces, the spaces investigated in each area are characterised as residential streets, recreation park (small park, mainly devoted to children's play and recreation), neighbourhood park (medium park, provided for recreation, refreshment and appreciation of nature), and central park (the major park in the city centre, which performs weekend vacation function for city dwellers). Streets with combined characteristics, such as dwelling type, appearance and comfort of sidewalks, relationship between the dwelling and the street, were selected for data collection in each area.

Characteristics of the selected neighbourhood areas

Neighbourhood Area 1 – Itanhanga

This area is provided with a neighbourhood park of medium size, with the purpose of recreation and appreciation of nature (***POS1**- Public Open Space 1*), and presents the highest household income among the three areas (Figure 1). Streets with two distinct combined characteristics form the area; therefore two sub-areas were selected. Sub-area *Buzios*, with approx. 20 inhabitants per hectare, is formed by detached houses in large lots enclosed by walls that totally obstruct visual contact between the dwelling and the street; the streets are long and curve, with few intersections. Sub-area *Santa Tereza* has approx. 90 inhabitants per hectare, with houses located in smaller lots than in Buzios, and a Condominium with three storey high blocks of apartments, built during the 70s. A variation of walls and up to three metre high fences densely covered with vegetation encloses the lots. Partial visual contact with the street exists at ground level in the condominium. The two sub-areas are provided with local shops at walking distance from the dwellings.

Figure 1. Neighbourhood Area 1 – Itanhanga

(a) Street in Santa Tereza   (b) Example of footpath   (c) Inside neighbourhood park

Formerly barrier free, fences up to two metres high surround the neighbourhood park (POS1) of 18.000m^2 since 1997. It consists of a large landscaped area with many large trees, crossed by two narrow streams. With the exception of the playground area, the park is very well maintained, and is further provided with a walking track, a stage for multiple activities and a few sitting spaces. Access for animals and the use of bikes, skates and formal ball playing are prohibited in the park. There are two gates to demarcate the entrances, which are controlled by gatekeepers.

Neighbourhood Area 2 – Copatrabalho

Characterised as a low-income, working class neighbourhood, the area is provided with a recreation park (***POS2** – Public Open Space 2*), the smallest among the three public open spaces investigated with 9.100 m^2 (Figure 2). The recreation park is centrally located in the neighbourhood, mainly devoted to sports activities, and characterised by large paved areas with three playing courts, playground area, several sitting spaces with benches and tables under large trees, and a boule court. It has two buildings used as community centre and police station, and artificial lighting is adequately provided.

Formed by a grid of narrow streets, a central avenue divides the residential area of approx. 106 inhabitants per hectare, where local shops, amenities and public transport are provided. The detached houses are located in small lots, enclosed by low and high fences with a front yard. Footpaths are narrow (1.50m wide), and are frequently obstructed, adversely affecting pedestrian circulation.

Figure 2. Neighbourhood Area 2 – Copatrabalho

(a) Example of street (b) Sitting spaces (c) Playing court

Neighbourhood Area 3 – Horto Florestal

Similarly to Itanhanga, streets with two distinct combined characteristics are predominant in the neighbourhood; therefore two sub-areas were selected (Figure 3). Sub-area *Vila Carvalho*, built during the 20s, is one of the oldest allotments in the city. Located in the city centre, the residential streets have small shops and amenities along them. The lots are large, and density in the area is approx. 43 inhabitants/hectare. Mature trees in the streets are abundant, providing adequate thermal comfort in the area. Sidewalks are wider and more accessible than all the other residential areas investigated.

Figure 3. Neighbourhood Area 3 – Horto Florestal

(a) View in the park (b) Sargento Amaral (c) Vila Carvalho

In sub-area *Sargento Amaral*, the row houses were built in the 70s. The streets are narrower and with less vegetation than streets in Vila Carvalho, and are exclusively residential, with local shops at walking distance. In the two sub-areas, low and high fences limit lots, and at least 80% of visual contact between the dwelling and the street is preserved. The central park in the city (***POS3*** – *Public Open Space 3*), with 48.000m^2, provided to perform weekend vacation function for all city dwellers, is located in this middle-class neighbourhood near the city centre. Located in the former Botanical Garden, vegetation is abundant. It has a forest with mature trees, greenhouse, walking lane, playground area, boule courts, places for refreshment, a large paved space for large public events, a stage, covered arena, bike track, several sitting spaces and a skate rink. There are two major buildings inside it: the Municipal Public Library and an elderly centre, which offers several special activities. This public open space is very well maintained and managed, with skills in activity programming and event programming. The park is totally enclosed by fences, with limited access through four gates.

Results

The differences of influence of socio-economic characteristics and environmental attributes on appropriation of residential streets and public open spaces were identified. They were measured by the observed differences on types and intensity of appropriation occurring in each category of open space, according to users' and residents' characteristics, and the physical characteristics and performance evaluation of each space made by residents and users through questionnaires and interviews. Results indicate that compositional factors influence more strongly type and intensity of use of residential streets than contextual factors. Conversely, in relation to public open spaces, results show that type and intensity of use is more influenced by contextual factors than compositional ones. Based on the results obtained, some of the considerations formulated are discussed below.

Appropriation of residential streets

The residential streets investigated illustrate three ways in which dwellings can blend into public space. They are further discussed according to the type of interface existing between the dwelling and the street. The first type of street, located in the low-income neighbourhood (Neighbourhood Area 2 – Copatrabalho), is formed by houses with front yards mostly defined by railings (66%) and low fences (20%) between the house and the footpath (soft interface), with visual accessibility to what is going on and to be seen, which allows them to be informed and entertained by the street life. In this residential area, residents have the opportunity to engage in various social encounters, all in a very relaxed way, or occasionally bring a chair out on to the footpath. The streets are intensively used for pedestrian circulation, social and recreational activities, by all age groups. Children walk to the main road to get public transport to school, and residents tend to purchase goods in the local shops.

Houses with front yards mostly defined by low (17%) and high fences (68%) form the second type of street, located in the middle-income neighbourhood (Neighbourhood Area 3 – Horto Florestal). Nonetheless, despite the existing soft interface between the house and the footpath, the use of streets to perform social and recreational activities is moderate when compared with appropriation of streets in Copatrabalho. It is noted that the aged residents, who have lived in the same street for a long time, tend to maintain informal conversations on the footpath. Albeit the significant number of car users, pedestrian movement is frequent in the area, which is located near the city centre.

Houses without front yards connected to the street, mainly separated by high walls (66%) – hard interface – and high fences (29%) form the third type of street, located in the high-income neighbourhood (Neighbourhood Area 1 – Itanhanga). In this case, the houses, as well as the street, tend to look more forbidding and uninviting, and privacy and fear of crime is evidently a major concern, where the prevalent interest is in screening off family dwellings from the surroundings. Behavioural maps show that streets are rarely used for pedestrian circulation, as residents tend to leave their dwellings by car. Social and recreational activities are not performed in the streets, and it is noted that the predominant users, for circulation purpose only, are domestic employees working in the dwellings, when they arrive to work or eventually leave their working place to buy basic goods in the local shops. Most residents shopping activity is made in large shopping centres, far from the neighbourhood. In Itanhanga, the streets look dull and lifeless because of the absence of any activities in the public space, further increasing fear of crime. To avoid these adverse consequences, security guards are paid by dwellers, in order to guarantee security in the area.

Influence of contextual factors on appropriation of residential streets

In NA1 (Itanhanga), although the landscape is very attractive, the design of the streets is not functional, and the hard interface between the street and the dwellings completely isolate them. Ornamental plants are located in the middle of the predominantly narrow footpath (between 1.50 and 2.00 metres wide), adversely affecting passage. The physical attributes, such as pavement on the footpath, vary according to the taste of the dwellers. Lighting is adequate and benches are not provided. Despite the presence of security guards, fear of crime is high (security is perceived as inadequate by 51% of respondents). Residents do not use the streets to perform activities, but they evaluate them as very satisfactory (82% of respondents are very satisfied with visual appearance of streets), mainly because they enjoy to see them attractive and well maintained when they pass through by car. Results suggest that the positive environmental attributes evaluated by residents did not influence their behaviour, as appropriation of streets is very low. The differences in configuration existent between the two sub-areas did not affect the results.

In NA2 (Copatrabalho), the streets are narrow, with footpath 1.50 metre wide, with uneven pavement and the presence of physical barriers, which disrupt pedestrian circulation and other activities. Artificial lighting and thermal comfort due to the presence of trees are adequate, but there is no specific furniture, such as benches, to support activities, which are predominantly performed simultaneously with vehicle circulation. Steps and folding chairs are used for social activities. Nonetheless, although visual accessibility and surveillance from the dwellings is high, perception of security in the streets seems to have been adversely affected by the presence of non-residents in the area (mentioned by 44% of respondents), who come from neighbourhoods nearby to use the recreational park. Despite their inadequacy, residents positively evaluate the streets, and find them enjoyable (89% of respondents are very satisfied with living in the neighbourhood). This suggests that the intense appropriation of streets in Copatrabalho was not positively influenced by their environmental attributes nor affected by their inadequacy.

In NA3 (Horto Florestal), the streets are more adequate than in NA2 in terms of width, quality of pavement and presence of physical barriers. Artificial lighting and the presence of mature trees provide adequate comfort. As in NA2, benches are not provided to support social activities among the residents, who eventually use folding chairs to seat in front of their houses. Surveillance in the area is good and visual accessibility to the street is preserved to a certain extent. According to 77% of respondents, the area is perceived as a more secure neighbourhood, when compared to NA1 and NA2, despite its proximity to the busy city centre, the presence of prostitutes nearby and the higher number of police reports on burglary and related infractions in the area. Nonetheless, most houses have their windows and doors protected by iron bars. The residential streets are evaluated positively, but appropriation is moderate. This indicates that positive evaluation of environmental attributes might have not interfered with the level of appropriation of streets. Evaluation of the environmental attributes did not differ between residents of both sub-areas investigated.

Therefore, it is noted that adequacy of streets in terms of pleasantness, security, dimensions, pavement or comfort are not sufficient conditions to guarantee appropriation. On the other hand, the inadequacy of streets is also insufficient to abort appropriation in low-income neighbourhoods. Nonetheless, correlation tests indicate that *footpath width* ($r_s = 0.1433, p = 0.026$), *quality of pavement* ($r_s = 0.2648, p = 0.000$), *sitting spaces* ($r_s = 0.2401, p = 0.000$) and the *provision of mature trees* ($r_s = 0.1429, p = 0.027$) can affect *intensity of use of streets*. That is, if these attributes were better provided, favouring adequate spaces to perform social and recreational activities, streets might be more intensively used, increasing social contact among neighbours, whenever any cultural disposition exists for activity in that given place, such as in NA2 and NA3. A notable example is the Dutch Woonerf, or "play street," viewed as an effective way to reduce traffic speed and provide for social activities such as ball play, sitting, and communal use of neighbourhood space (e.g., Appleyard, 1981). Moreover, a correlation was found between *level of satisfaction with the neighbourhood* ($r_s = 0.2492, p = 0.000$) and *pleasantness* ($r_s = 0.2496, p = 0.000$) with *intensity of use of streets*, suggesting that the more residents use the streets, the more satisfied they are with the streets and with living in the neighbourhood.

Influence of compositional factors on appropriation of residential streets

When considering the influence of demographic, cultural and socio-economic factors such as household income, stage in lifecycle, education, occupation, place of origin, time of residence and life style (defined by habits of leisure, shopping, means of transport, and type of social interaction) on appropriation of residential streets, it was possible to unveil relationships between these factors and appropriation of streets in the three neighbourhood areas. Some authors tend to include many of these features in the concept of lifestyle. Actually, lifestyle represents a choice among the alternative ways of allocating the resources available, and influences activities directly, and is highly dependent of socio-economic factors. Out of the factors considered, it was noted that household income (high, middle or low), stage in lifecycle and time of residence were the factors mainly determining peoples' lifestyle, further affecting appropriation of streets, while no relationship was found between education, occupation and place of origin, and type and intensity of use of streets. Moreover, it seems that the major factor in this respect, which tends to be a common characteristic shared by the population of each neighbourhood area, is household income. The stage in lifecycle of the population in each neighbourhood is statically significant, indicating that in NA2 the number of children in the household is higher (57%) than in NA1 (with 40%) and NA3 (30%)

In the case of NA1, its high-income population is formed by people from Campo Grande and from the southern region of Brazil, with different periods of residence in the area. The cultural differences among them did not play a major role, as lifestyle is similar to all residents: the streets are only used for circulation, the means of transport is by private car, and activities such as shopping, leisure, education and work are performed outside the neighbourhood. In NA2, the low-income population is formed by people from other parts of the state of Mato Grosso, and people from Paraguay and Campo Grande, who moved to the area approximately in the same period. Again, despite their cultural differences, the number of children and youngsters in the neighbourhood is high, and lifestyle is similar: streets are intensively used for recreational and social activities; pedestrian circulation is intense, as most residents use public transport and local shops. Nonetheless, it must be emphasised that the intense use of streets might have been further affected by the poor quality of the dwellings, with reduced size and high occupation rate, and lack of adequate ventilation: people also stay outside in order to have more space to perform activities, especially children, and better environmental conditions. In NA3, where the majority of the middle-income residents were born in Campo Grande, the number of children per household is smaller than in NA2, and the neighbourhood presents the largest aged population among the three areas investigated. In fact, this is the most homogeneous group of residents, and lifestyle is very similar too; use of streets for recreation, socialising and pedestrian circulation is moderate, the predominant means of transport is by car, and shopping and leisure activities are performed mostly outside the neighbourhood.

The question of the identification of residents with their neighbourhood was further raised. It was assumed that the awareness of a positive attachment contributes to place identity and thus, in a positive case, influences self-esteem. That is, if place identity is lacking, then it is to be expected that public places and facilities are experienced as strange and aversive (e.g., Rapoport, 1986), in which there is no longer any consensus as to the social rules of use and the contact between human beings.

As expected, it was noted that residents who, for whatever reason, do not identify very much with the area outside their home, and have the financial means and chances to leave the neighbourhood, tend to satisfy their needs outside of their residential area in special facilities. In NA1, the high-income neighbourhood, together with the avoidance of, or merely the brief functional use of the public areas in the residential areas, a retreat into the private sphere taking place was found, which is followed by a reduction of the interpersonal contacts in the public and semi-public area. Nonetheless, it had been incorrectly assumed that reduced social contact in the neighbourhood would prevent, in turn, chances for creating or maintaining social networks and thus a potentially effective social support system in the immediate and surrounding residential area. On the other hand, it was found that although those residents do not use the streets for social contact, they belong to the same social network developed outside the neighbourhood, sharing the same leisure facilities, clubs, shopping centres, schools or workplace, and common friends, which guarantees an effective social system, totally independent of their residential environment. The findings corroborate Webber's projections (in Levitas, 1991) derived largely from the network analysis of today's upper middle class. This group spends little time in immediate surroundings, does not equate proximity with neighbouring, and makes its friends among a far-flung society generally related by profession or business interests, which in this case, coincidentally live in the same neighbourhood. Fully stimulated by the amount of information they must process for their professions and to maintain their life-style, this group would appear to require very few of the amenities that have been suggested as appropriate to diversity on the street (e.g., Jacobs, 1961).

In the case of low-income neighbourhoods, where residents do not have other alternatives at their disposal, Levitas' (1991) observation that the street turns into a social space was confirmed. Residents in these specific areas tend to structure their lifestyle in such a characteristic manner that induces a more intense use of public spaces to perform social activities. It was noticed that the more time people spend on the street or in areas visible from the street, the higher the number of instances when neighbours actually meet and exchange a greeting or engage in a conversation. These interactions are important not only in enriching the lives of the occupants but also in helping to form a loose web of acquaintances and friends who are more inclined and able to co-operate in such tasks as child supervision and protection of property from vandalism and theft, and who are likely to feel a greater sense of pride towards their street, and of belonging in their neighbourhood. The importance of social interactions on sense of place is further corroborated by correlations found between *friendship in the neighbourhood* and *pleasant neighbourhood* ($r_s = 0.3286$, $p = 0.000$), and between *satisfaction with the neighbourhood* and *intensity of activities in residential streets* ($r_s = 0.2496$, $p = 0.000$).

Appropriation of public open spaces

POS1 – Itanhanga

The public open spaces investigated illustrate different types of urban parks, in terms of size, potential users, provision of activity settings and accessibility. Despite its highly potential use, the medium sized neighbourhood park (POS1) located in Itanhanga, provided for recreation and appreciation of nature, shows low intensity of use. According to information obtained through interviews, before this park was fenced, it was intensively used by people living in the neighbourhood and by people of diverse income groups living in other parts of the city, and was one of the favourite places for family day out. After the park was fenced, the location of the playground area became less accessible from the two entrance gates, with consequent minimal use. Moreover, the fact that bikes, skates and play ball are not allowed in the park since its enclosure adversely affected the presence of youngsters and children in the park, despite its proximity to a school. Behavioural maps show that adults mainly use it for walking. Activities in the stage for multiple activities rarely occur, but the few sitting spaces provided along the walking track are often occupied, especially when located under mature trees. Considering its size, this park is under occupied.

POS2 – Copatrabalho

Without hard physical demarcation, the recreational park (POS2) located in Copatrabalho is half the size of POS1, and is intensively used by residents of all age groups in the neighbourhood and people living nearby, to perform all predicted activities, during week days and weekends. The playing courts, playground area, sitting spaces and boule court are in continuous use, as the number of users is very high. Mainly youngsters also use the setting during the night. The park and the few refreshment places provided near the local shops are the main source of recreational and social activities in the neighbourhood, due to the lack of other leisure alternatives nearby, and the economic constraint faced by the population. Moreover, it must be emphasised that the provision of such successful and demanded spaces in low-income neighbourhoods is scarce in the city. Contrary to POS1, this park is over occupied.

POS3 – Horto Florestal

The central park (POS3) located in Horto Florestal is heavily used during weekends by city dwellers of all age groups, when programmed activities and major events occur. It was observed that apart from the bike-cross site, all activity settings provided for recreation and socialising are constantly occupied during weekends. With a specific entrance, the bike-cross track is confined by fences, and the access of bikes to other parts of the park is forbidden, limiting this type of activity. The diversity and quality of amenities provided turn the park into a very dynamic place during weekends. During weekdays, intensity of appropriation is lower, and residents in the neighbourhood area predominantly use the park, for activities such as walking, boule games, and a few children play activities and skating. The public library and the elderly centre attract different groups of users to the more segregated and quiet area in the park.

Influence of contextual factors on appropriation of public open spaces

In POS1 (Itanhanga), the landscape is carefully planned and nature is exuberant, positively affecting users' evaluation of visual appearance, which is perceived as very satisfactory by 98% of respondents. Security, in spite of its physical and visual segregation in relation to its surroundings, is also perceived as very satisfactory (by 84% of respondents), probably influenced by the presence of gatekeepers. However, visual and functional accessibility have been adversely affected by railings and the reduced number of gates. Complaints were made in relation to the concentration of mature trees in certain areas, and the mismatch between sun- protected spaces and sitting places. Apart from that, the need for more benches was often mentioned. But users were very satisfied with the park as a whole. As Rapoport (1986) argues, when choice is available, settings are preferred on the basis of environmental quality: people choose settings. Which setting is preferred and chosen, and which works well, depends on its being supportive of desired activities and behaviours of potential users. In the case of POS1, the fact that children and youngsters are not regular users might have attracted the current group of users, who seems to prefer more quiet and isolated settings. Nonetheless, the size and quality of this setting, as well as the lack of similar settings in the city suggest that it should be more overtly shared among the population. It can be assumed that people that choose not to use the park, and consequently did not participate in the investigation, are not as satisfied with its performance as the limited number of actual users is. So, it appears to be the case that, despite its excellent landscape qualities, the park is too restricted, and is not as successful as it was in the past, when railings did not exist and access was open. According to interviewees, by that time, a more diverse group of users frequented the park, more children, and especially more families, to have contact with nature. A correlation was found between *scarcity of activities* and *intensity of use* ($r_s = -0.3092, p = 0.02$), indicating that if more assorted activities were available, use of this public space would increase.

Despite its detailed and pleasant design, the intensity of use in POS2 (Copatrabalho) seems to have adversely affected the perceived visual appearance of the setting, due to the resulting unachievable demand for maintenance (66% of respondents are satisfied with visual appearance and only 24% of respondents are satisfied with overall maintenance). Correlations were found between *visual appearance* ($r_s = 0.3018$, $p = 0.011$), *supply of friendly spaces* ($r_s = 0.3592$, $p = 0.000$), *provision of adequate spaces* ($r_s 0.3409$, $p = 0.005$) and *intensity of use*. Nonetheless, the great diversity of activity settings provided in this relatively small space was insufficient to support the huge demand for recreational opportunities in the neighbourhood, which has by far the largest youngster population among the three areas investigated. The need for large paved areas resulted in scarce number of mature trees to provide shelter, which together with the need for more sitting spaces, were users' major complains. Visual and functional accessibility are very satisfactory, but the lack of control over it affected user perception of security in the park (49% of respondents are dissatisfied with security). As this park is one of the best spaces provided for recreation in low-income neighbourhoods, people living in areas nearby, with similar socio-economic conditions, tend to hang around too.

The major POS3 (Horto Florestal), inserted adjacent to the city centre, turned into a very accessible setting for the population, as access to the city centre by public transport is very satisfactory, as well as parking areas for private vehicles are adequately provided. Despite its physical delimitation by railings, the entrance gates to the park are better positioned than in POS1, leading straight to the most demanded activity settings, with partial visibility through the railings of most activities going on, however functional access is limited. Security in the park is perceived as more satisfactory during weekends than during weekdays, when the number of users significantly decreases. The good maintenance and management positively affected users' evaluation of visual appearance and overall satisfaction. Nonetheless, when large events occur, the sitting spaces available are insufficient and excessively exposed to the sun, which makes them totally inadequate due to the high temperatures reached during the day in this region of Brazil, as the exuberant vegetation with mature trees is located far from most activities arena. Even with the inadequacies identified, the park is positively evaluated, showing high levels of satisfaction. It seems that the quality and diversity of activities settings and programmed events provided partially overcome the negative impact of inadequate environmental attributes, but it is further confirmed that diversity of activities, sufficient sitting spaces and adequate furniture are necessary conditions for effective use of POSs. Similarly to POS1, a correlation was found between *scarcity of activities* and *intensity of use* ($r_s = -0.2684$, $p = 0.018$), further corroborating the argument mentioned above. In addition, when statistically analysing POSs in general, it was found that *lack of adequate sitting spaces* is correlated to *intensity of use* ($r_s = 0.1296$, $p = 0.045$).

Visual appearance and security are two of the environmental attributes affecting user satisfaction and users' decisions to choose to use a certain public open space, as is the case of POS1 and POS3. But regarding POS2, users' decisions to choose this specific setting seems to be more strongly based on what there was available; that is, what better alternatives existed to better fulfil their needs of contact with nature, perform recreational activities and achieve social contact, according to their economic constraint.

Moreover, it must be stressed that access control, claimed by Newman (1972) to be a means of increasing security, constrained accessibility in POS1 and POS3, confirming that the ability to enter spaces is basic and proportional to their use (e.g., Carr et al., 1992), as corroborated by the correlation found between *difficulty of access* and *intensity of use* ($r_s = -0.3537$, $p = 0.002$), indicating that the higher the difficulty of access, the more intensity of use decreases. Public spaces closed to the public by means of fences or guards may discourage entry. When limits to access exist in the form of gates or gatekeepers, or impediment to the physical access of particular groups, the use of a space is severely restricted, and people's rights are limited, as in POS1. In POS2, where fences do not exist, a correlation was found between the *possibility of enclosure of public space* and *intensity of use* ($r_s = -0.2597$, $p = 0.034$), indicating respondents' perception that if fences were provided, intensity of use might be adversely affected. The existence or lack of rules and regulations is important to the achievement of freedom for action, as asserted by Lynch (1981). It involves the ability to carry out the activities that one desires and to use a place as one wishes, but with the recognition that a public space is a shared space. In POS1, most desired children activities were forbidden, inhibiting the use of this public space by this group of users and by families in general. Nonetheless, although current users seem to enjoy isolation, the size of the park allows the co-existence of diverse activities, fulfilling the needs of the different age groups and potential users.

In addition to physical access, it was confirmed that visual access or visibility is important in order for people to feel free to enter a space. Visibility was particularly important in judgements of the safety of a space, indicating that the public's perception that a space is free of persons who threaten users is an important consideration for its use. For example, according to interviewees, the two metre fences and bushes that surround POS1 made it difficult for passers-by to see into the space and thereby discouraged entry.

Influence of compositional factors on appropriation of public open spaces

The study indicates that appropriation of POSs in Campo Grande tends to increase among people originated from the Southern and Central-west regions of Brazil. The socio-economic characteristics of users varied according to socio-economic characteristics of residents in the neighbourhood area where the public open space is located, as in the case of POS1 and POS2. In POS3, weekend users come from all parts of the city, attracted by the magnitude of events in the park, while most week-day users are residents in the middle-income neighbourhood area. As POSs tend to be largely located in middle and high-income neighbourhoods, it was found that the majority of users of public open spaces are middle and high income. But results clearly point to the massive demand for provision of public spaces in low-income residential areas, which are the most in need and are the more densely populated. No correlations were found between appropriation of POSs and compositional factors.

Conclusion

Results show that street life depends heavily on the quality of social relations in neighbourhoods, which is affected by residents' life style, which is mainly defined by their economic conditions. For low-income people, their evaluation of urban life seems to be affected more by their social relations than by the physical environment. Accessibility of places for activities was found to be an important aspect of the urban scenery. Everyday the movement between these places forms an activity pattern, demonstrating much of the physical conditions established for living in a neighbourhood by urban physical structure.

Despite this, it was also confirmed that any predisposition that exists for activity in residential streets or public open spaces could be greatly helped by appropriate design, further supporting social life dynamics of urban spaces. The physical environment, as part of the material setting people live in, is both a condition and a consequence of the pattern of social relations in an area. As a condition, it allows for social contacts, or makes them difficult, even impossible. As a consequence, it is shaped, partly, by patterns of social relations, their content, intensity and frequency. The physical environment reflects its usage by people – their activities, their wellbeing or their ill feelings. And the changes it undergoes are, to a certain extent, reflections of this use. Therefore, planning should preserve and even support social relations in residential areas, as they are a prerequisite for collective action in the neighbourhood and for its development. If the dynamics of streets for social activities is desirable, this can be achieved as long as physical conditions favour social gathering, such as traffic control, provision of sidewalks with adequate width, functional accessibility, levelled paving, and social connection between dwellings and adjacent public streets, which further supports surveillance.

Visual appearance and security are two of the environmental attributes affecting user satisfaction with public open spaces, but users' decisions to choose to use a specific setting is based on what diversity of activities exist to fulfil their needs, sufficient sitting spaces, adequate furniture, and freedom for action. Moreover, it was found that visibility and access control by means of fences or guards constrain accessibility, confirming that the ability to visualise and enter spaces is basic and proportional to their use.

References

Appleyard, D. (1981). *Liveable streets*. Berkeley: University of California Press.
Carr, S. et al. (1992). *Public space*. Cambridge: Cambridge University Press.
Ellis, W. C. (1991). The Spatial Structure of Streets. In S. Anderson (Ed.), *On Streets* (pp. 115-130). Cambridge: MIT Press.
Francis, M. (1987a). Urban Open Spaces. In I. Zube and G. Moore (Eds.), *Advances in Environment, Behaviour and Design*. New York: Plenum Press, Vol. I.
Francis, M. 1987b. Some different meanings attached to a public park and community gardens. *Landscape Journal*, 6, 100-12.
Francis, M. (1989). Changing Values for Public Spaces. *Landscape Architecture*, January/February, 54-59.
Gans, H. (1962). *The Urban Villagers*. New York: The Macmillan Company.
Gehl, J. (1987). *Life between buildings*. New York: Van Nostrand Reinhold.
Gold, S. D. (1972). Non-use of neighbourhood parks. *Journal of the American Institute of Planners*, 38, 369-378.
Gutman (1991). The Street Generation. In S. Anderson (Ed.), *On Streets,* pp. 249-264. Cambridge: MIT Press.
Halprin, L. (1963). *Cities*. New York: Reinhold Publish Corporation.
Hester Jr., R. T. (1989). Social values in open space design. *Places*. New York: Design History Foundation V. 6 (1).
Jackson, J.B. (1981). The public park needs reappraisal. In L. Taylor (Ed.), *Urban Open Spaces*. New York: Rizzoli.
Jacobs, J. (1961). *The Death and Life of Great American Cities*. New York: Vintage Books.

Levine, (1984). Making City Spaces Lovable Places. *Psychology Today*, June, 56-63.
Levitas, G. (1991). Anthropology and sociology of streets. In S. Anderson (Ed.), *On Streets*. Cambridge: MIT Press.
Lynch, K. (1981). *A Theory of Good City Form*. Cambridge: MIT Press.
Newman, O. (1972). Defensible Space: Crime Prevention Through Urban Design. New York, The Macmillan Company.
Rapoport, A. (1986). The use and Design of Opens Spaces in Urban Neighbourhoods. In D. Frick (Ed.), *The Quality of Urban Life*, pp. 159-175. New York: Walter de Gruyter.
Rapoport, A. (1987). Pedestrian streets use: culture and perception. In A. Vernez Moudon (Ed.), *Public streets for public use*. New York: Van Nostrand Reinhold.
Rutledge, A. (1986). *A visual approach to park design*. New York: McGraw-Hill.
Ulrich, R.S. (1979). Visual landscapes and psychological well being. *Landscape Research*, 4, 17-19.
Whyte, W. H. (1980). *The social life of the small urban spaces*. Washington: The Conservation Foundation.

Acknowledgement
To Brazilian National Research Council (CNPq) for post-doctoral scholarship.

V. Children and the environment

Evaluating links intensity in social networks in a school context through observational designs[1]

M. Teresa Anguera Argilaga[1], Carlos Santoyo Velasco[2] & M. Celia Espinosa Arámburu[2]

[1]Universidad de Barcelona, Spain and [2]Universidad Nacional Autónoma de México, Mexico

Abstract. Children spend several hours daily at school and a range of interactive processes take place in this setting. Systematic study of social interaction in situ is important for developmental psychology and for the scientific understanding of social processes. Assessment strategies need to be designed to identify behavioral mechanisms involved in the regulation of social interactions and social preferences.
The methodological contribution of this work is strategically directed to the empirical study of social networks based on a behavioral perspective through the observational design, and our design is diachronic/nomothetic/multidimensional (D/N/M). The use of polar coordinates analysis allows us to identify the intensity of social connections in a social network. This analysis allows the reduction of data from the value *Zsum*, introduced by Cochran (1954) and applied by Sackett (1980). Polar

---

[1] C. Santoyo and M.C. Espinosa thank CONACYT (40242-H) for financial support for the empirical research. M.T. Anguera thanks DGES for financial support for the evaluation research (BSO2001-3368).

coordinates analysis allows the combination of diachronic and synchronic perspectives through the representation of the values obtained as vector modules and angles (polar coordinates). The nature of the interactive relationship, depending on the quadrant in which it is located, is represented by the angle, while the degree of significance depends on the vector module.

Data are discussed from the point of view of the flexibility and precision deriving from an adequate observational design and its possibilities for behavioral sociometry.

Keywords: preschool social interactions, observational designs, polar coordinates analysis, behavioral sociometry

Behavioral sociometry and social interactions

The study of social interaction in situ requires methods that extend beyond the individual as the focus of analysis (Cairns, 1979; Kinderman, 1996). The school environment provides rich opportunities for the accommodation or transformation of children's social preferences and social adjustment, which are key components of the socialization process. The understanding of behavioral mechanisms involved in social interchanges is necessary to the advancement of scientific knowledge about social behavior. By studying social networks it is possible to identify links, cliques, and, in general, social preferences in a group for some "popular," "isolated," and "rejected" children. This information allows us to explain the stability or change in social preferences for specific children.

Traditional sociometric status procedures of nomination are poorly suited to identifying the behavioral mechanisms of mutual control that regulate bidirectional dyadic associations. New research strategies are required to clarify the contents, directionality, and effects of social interaction in children.

Functional analysis offers a fresh view of "affiliative organization" (Strayer & Santos, 1996; Santoyo, 1996), which differs radically from "classic" sociometric procedures. In order to design a field research of social interactions, we propose a strategy based on observational designs, specifically on a "sociobehavioral framework" (Santoyo, 1994).

In this work we assume an "ego-centric" approach (Scott, 1991), where the focus of analysis is the network of children anchored around a specific target or child.

Polar coordinates analysis and choice of methodology

Polar coordinates analysis, first described by Sackett (1980), is a double data reduction strategy which provides a vector representation of the complex network of interrelations between the different categories (or configurations of field formats) that make up an *ad hoc* system produced to code the behavioral flow deriving from any activity or situation.

The structure of the polar coordinates technique is based on the complementariness of two analytical perspectives: *prospective* and *retrospective perspectives* (Anguera & Jonsson, 2002; Anguera & Losada, 1999; Hernández Mendo & Anguera, 1998; Santoyo & Anguera, 1993):

Prospective perspective

A criterion behavior is established for each analysis. This behavior is proposed as the trigger of a series of connections with a whole set of categories, known as matching behaviors In this paper the criterion behavior was the participation in play of each child of a specific group in a preschool setting.

The first part of the process follows the same operations as the simple lag technique for sequential analysis (Bakeman & Gottman, 1989; Bakeman & Quera, 1995; Sackett, 1978, 1980). Starting from a number r of lags specified in each case, the tables of matching frequencies and matching probabilities (expected and observed) are produced and the corresponding Z statistics are calculated. This provides us with a matrix, $k \times r$, where k is the number of categories in the system (or the number of field format configurations), and r the number of lags taken into account (Santoyo & Anguera, 1993).

Thus far only the **prospective perspective** has been considered, that is to say the result of considering the forward evolution of the stream of social exchanges of the individual with their *social* environment.

Retrospective perspective

This perspective aims to establish the extent to which previous events in the behavioral flow demonstrate a significant intensity in the connection of each of the k categories in the system to the focal behavior, which in this retrospective perspective takes the role of the matching behavior.

That is, we want to know whether the association between each of the categories of the system or field format configurations (now regarded as the criterion behavior) is either excitatory or inhibitory (that is, statistically significant or non-significant) with respect to the focal behavior, which is now in the role of the matching behavior. Anguera (1997) proposed that there may be a genuine retrospectivity through the criterion behavior and "backwards", considering negative lags, and detecting the consistency of actions of order n previous to the criterion behavior. If E is the criterion behavior, then A, B, C, and D would be matching behaviors, but only considering negative lags.

In agreement with this suggestion, behavioral patterns obtained through retrospective sequential analysis provide us with a mirror image of how the last, last but one etc., behaviors previous to the criterion behavior maintain a stable relationship with the others, and consequently the actions that are "preparatory" to the event of the criterion behavior, bearing in mind that each behavioral pattern is composed exclusively by the codes (category codes or field format configuration codes) that were shown to be excitatory or significant in each of the negative lags considered

Construction of vectors

With the z values of relative indexes of sequential dependence (Bakeman, 1978), we can apply an extremely powerful technique for the reduction of data by calculating the *Zsum* statistic described by Cochran (1954), consisting of $\sum z / \sqrt{r}$. That is, for each of the columns corresponding to each of the z value matrices – which are independent of one another –

the sum of those values divided by the square root of the number of lags considered throughout the process.

The *Zsum* statistic is based on the principle that the sum of a number, r (since there are as many as there are lags), of independent z scores follows a normal distribution, with $\mu = 0$ and $\sigma = 1$. Consequently, we obtain as many *Zsum* as there are categories in the perspective, the prospective, and the retrospective; the ones corresponding to the former are termed criterion *Zsum* and the latter matching *Zsum*.

Each *Zsum* can be either positive or negative, and so the set of signs will determine in which of the four possible quadrants the categories will be located in relation to the focal behavior adopted. Each quadrant is characterized as follows:

Quadrant I (+ +). Mutually excitatory criterion and matching behaviors.
Quadrant II (– +). Inhibitory criterion behavior and excitatory matching behavior.
Quadrant III (– –). Mutually inhibitory criterion and matching behaviors.
Quadrant IV (+ –). Excitatory criterion behavior and inhibitory matching behavior.

Figure 1. Polar coordinates analysis: Decision on the quadrant for the *Zsum*.

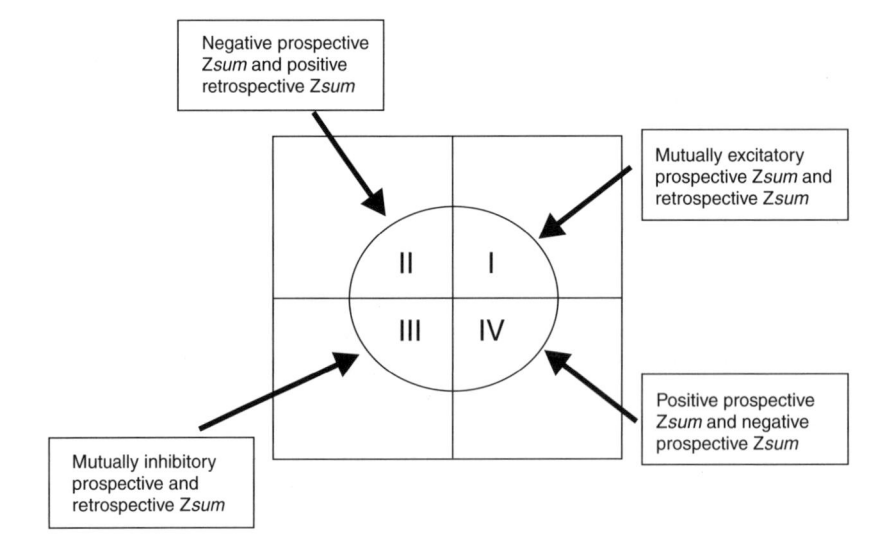

The most genuine and useful representation provided by this data reduction technique is the one known as polar coordinates – after which the analysis technique is named. It is produced in vector form (with a module or length, and with a specified angle). Because categories are in different quadrants depending on the type of relation between the criterion behavior and the matching behavior, it is possible to determine the distance from the origin (0,0) of the *Zsum* coordinates and the point of intersection (determined by the criterion *Zsum* value on the horizontal axis and the matching *Zsum* value on the vertical axis), which is the value of the **radius** or **module** (which, when > 1.96, is statistically significant at a level of 0.05, and, when > 2.58, is statistically significant at a level of 0.01), and which is calculated by adding together the square root of the prospective

Zsum and the square of the retrospective *Zsum*; the **angle** is calculated as the retrospective *Zsum* arc sine divided by the radius.

So we can trace all the vectors corresponding to each of the codes depending on the focal behavior considered, and every category or field format configuration can be considered as a focal behavior when necessary for the evaluation. The technique is thus a powerful tool for overcoming the numerous methodological obstacles that face the attempt to identify all the interrelations between all the category codes or field format configurations.

Empirical study: Objectives and planning

This study focuses on a new methodological aspect in the assessment of the relationships of social interactions inside a group: the intensity of links inside social networks. We want to know the differential degree of intensity with which these connections take place in a group of preschoolers in different contexts: the classroom and the playground.

Observational methodology has repeatedly been shown to be the most appropriate for the analysis of the social interactions and is the approach that we will use in this study. First, it is necessary to locate the study in a certain observational design, considered as a conceptual structure formed by the crossing of three basic approaches: Temporal recording criterion, observed units, and dimensionality of observed behavior (Anguera, Blanco & Losada, 2001).

In our case, the recording criterion approach involves follow-up of the development of social exchanges; the observed units approach is considered as nomothetic because there are a group of observed children, and the dimensionality of observed behavior approach allows us to consider a multidimensional position, since in the analysis of social interactions our instrument permits the study of individual and social behavior in natural settings, making it possible to identify the events and situations which contribute to and maintain social interactions and preferences.

Accordingly, our study can be considered as a *follow-up/nomothetic/multidimensional design (F/N/M)* (Anguera, Blanco & Losada, 2001), characterized as the most complete design for the information included. We planned follow-up of a certain number of scheduled sessions over time, we observed N units, and we contemplated diverse levels of response (dimensions). Inside this F/N/M design, certain decisions have to be made concerning the type of recording, kind of parameters, use of observation instrument, observational sampling, control of quality data, and analysis of data.

From among the various data analysis techniques that are possible in this design (Anguera, Blanco, & Losada, 2001), for this study we chose polar coordinates analysis, because we wanted to know *the intensity of* relationships between a focal behavior of a *target* child, and the other children in two school contexts.

Each child is a potential emitter and receiver in the network, in which unidirectional, bidirectional, or multidirectional exchanges are possible and are always assigned a positive or negative polarity depending on the definition of each category in the observational instrument (Santoyo & Espinosa, 1987; Santoyo, Espinosa, & Bachá, 1995). The values of Z_{sum} in the analysis of polar coordinates will allow us to draw the corresponding vectors, and will show the quality and degree of intensity of the relationships established in the social exchanges in the group in each context (classroom and playground).

Method

Subjects
A target girl was selected and the whole set of her social interactions with 29 possible peers was observed over 3.5 consecutive years.

Setting
All the observations were obtained from the classroom and playground of a private school in Mexico City. The first wave of observations was obtained when the target was in a preschool setting, and the last wave was obtained when the target was in second grade of an elementary level in the same school.

Sampling
The target was observed in the classroom and playground every six months during a period of 3.5 years. During each wave of observations six 15-minute sessions were held in the classroom on different consecutive days and an average of four 15-minute sessions in the playground.

Reliability
Throughout sessions, interobserver agreement was always higher than 80% (Cohen Kappa > .74).

Observational System
The Observational and Behavioral System of Social Interaction, OBBSI (Santoyo, Espinosa & Bachá, 1994) is an instrument that permits the systematic study of individual and social behavior in natural settings. OBBSI permits the identification of the events and situations that contributes and configures social interaction. This instrument also permits the analysis of the organization of social interchanges, where the acts of some person are supported and mutually coordinated with the ongoing activity of the other. OBBSI is also a useful system for the development of structured work aiming to understand and explain behavior as a functional model based on the social interaction analysis in situ.

Summary of Characteristics of OBBSI
OBBSI is an exclusive, exhaustive behavioral categories system, with intervals of five seconds; it is a flexible system of categories; it is constituted by representative behavioral categories of the actions that subjects exhibit in classroom and playground (i.e., group play behavior); it is a sequential "event-based" record in which observers write the order of occurrence of the events in the interval. More importantly for this specific study, it detects the direction of social interaction in order to identify who initiates an interchange (peers involved in social interaction and the direction of the interchanges, specifically in group play).

The data were analyzed using *SDIS-GSEQ* (timed data) (Bakeman & Quera, 1995) and *Excel* software.

Results and discussion

The results show the length and angle of vectors reflecting the interactions between the categories proposed as focal category and the rest for the whole set of children.

Polar coordinates analysis allows us to distinguish and interpret the vectors in function of their length (significant relation at $p < 0.05$ and $p < 0.01$ level) and angle (kind of relation, according the quadrant of vector). For this reason, we analyzed these relations, and with the same focal behaviors, in classroom and playground.

Tables 1 and 2 show us the prospective and retrospective values of Z_{sum}, with their signs, and the length and angle of vector, for both contexts.

Table 1. Module and angle of the vectors corresponding to the interrelations between the focal category (Positive social episode from Ele) and the rest of the players in classroom. Significant vectors ($p < 0.05$) are lightly shaded, and significant vectors ($p < 0.01$) are heavily shaded.

Child	Prospective (Criterion $Zsum$)	Retrospective (Matching $Zsum$)	Quadrant	Radius	Angle
Dig	1.92	1.89	I	2.69	44.5
Mlo	0	0			
Sat	0	0.56	I	0.56	90
Oro	−5.53	−4.42	III	7.07	218.6
Ir	0	0			
JL	0.54	0	IV	0.54	0
Ela	0	0			
Isa	3.38	3.5	I	4.86	45.99
Pma	2.02	2	I	2.64	44.71
Cec	0.53	0.55	I	0.76	46.06
Fda	8.71	8.98	I	12.51	45.87
Ro1	0.53	0.55	I	0.76	46.06
Ro2	1.24	1.41	I	1.87	48.67
Mir	0.96	0.99	I	1.37	45.88
Alja	0	0			
Olv	0	0			
Kar	−14.03	−16.78	III	21.87	230.10
And1	1.47	1.51	I	2.10	45.76
And2	1.76	1.51	I	2.31	40.62
Jla	0	0			
Leo	11.89	11.01	I	16.20	42.79
Gllo	0	0			
Cra	5.88	5.98	I	8.38	45.48
Ele	3.63	2.47	I	4.39	34.23
Ko	−18.9	−18.71	III	26.59	224.71
My	2.7	4.62	I	5.35	59.69
Mg	4.46	4.53	I	6.35	45.44
Mu	0	0			

Table 2. Module and angle of the vectors corresponding to the interrelations between the focal category (Positive social episode from Ele) and the rest of the players in recess zone. Significant vectors ($p < 0.05$) are lightly shaded, and significant vectors ($p < 0.01$) are heavily shaded.

Child	Prospective (Criterion *Zsum*)	Retrospective (Matching *Zsum*)	Quadrant	Radius	Angle
Dig	3.87	–1.23	IV	4.08	342.36
Mlo	1.69	1.79	I	2.46	46.64
Sat	–22.11	–20.09	III	29.87	222.25
Oro	7.99	7.28	I	10.80	42.33
Ir	–42.23	–39.62	III	57.90	223.17
J L	2.97	2.92	I	4.16	44.51
Ela	0	0			
Isa	–4.6	–4.5	III	6.43	224.37
Pma	0	0			
Cec	0	0			
Fda	3.62	3.73	I	5.19	45.8
Ro1	1.31	1.38	I	1.90	46.49
Ro2	6.77	8.56	I	10.91	51.65
Mir	0	0			
Alja	1.31	1.38	I	1.90	46.49
Olv	0	0			
Kar	0	0			
And1	0	0			
And2	8.35	8.79	I	12.12	46.47
Jla	0	0			
Leo	0	0			
Gllo	2.17	2.24	I	3.11	45.90
Cra	0	0			
Ele	0	0			
Ko	–10.27	–8.03	III	13.03	218.02
My	6.23	5.24	I	8.14	
Mg	0	0			
Mu	0	0			

Figures 2 and 3 represent the significant vectors that indicate the interrelations between these focal categories and the rest of children.

The methodological strategy demonstrates its utility because it permits the detection of the intensity of the target's social links with each of the possible peers in their social ecology in two settings: the classroom and the playground. This strategy allows provides efficient quantitative criteria to decide which are the strong links for the target in the group, instead of subjective or random criteria.

Empirical data show that the target interacts with different peers in different ways depending on the school context. For example, target (ela) shows mutually excitatory

Evaluating links intensity in social networks in a school context through observational designs 295

Figure 2. Significant vector representation of the interrelations between the focal category player (Ele in positive social episode [EPS]) and the rest of children in the classroom.

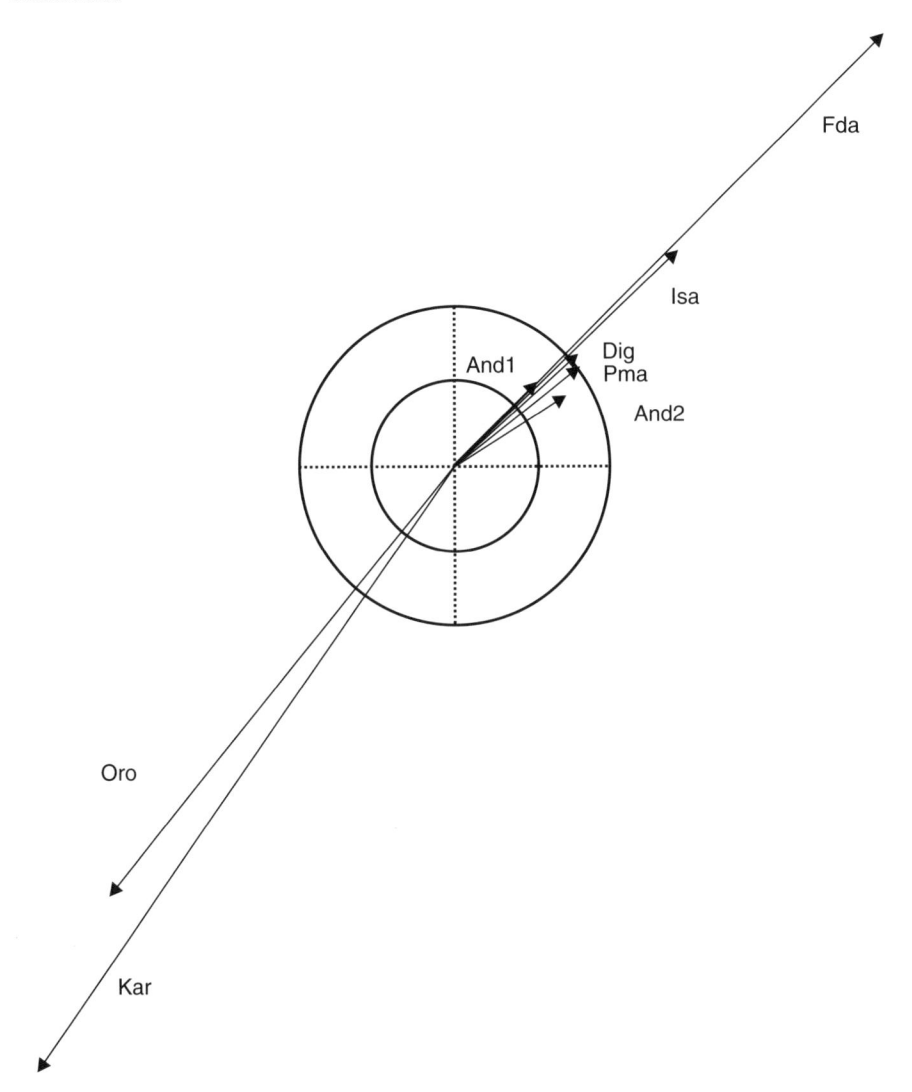

social interactions with "and2" & "fda" in both settings (classroom and playground). However, the intensity of the relationship is stronger in the playground than the classroom for "and2", while the reverse is true for "fda".

Other links shows an interesting phenomenon: the asymmetry of the relationship. For example, the target's relationship with "isa" is mutually excitatory in the classroom but mutually inhibitory in the playground, while the relationship with "oro" is mutually in-

Figure 3. Significant vector representation of the interrelations between the focal category player (Ele in positive social episode [EPS]) and the rest of children in the playground.

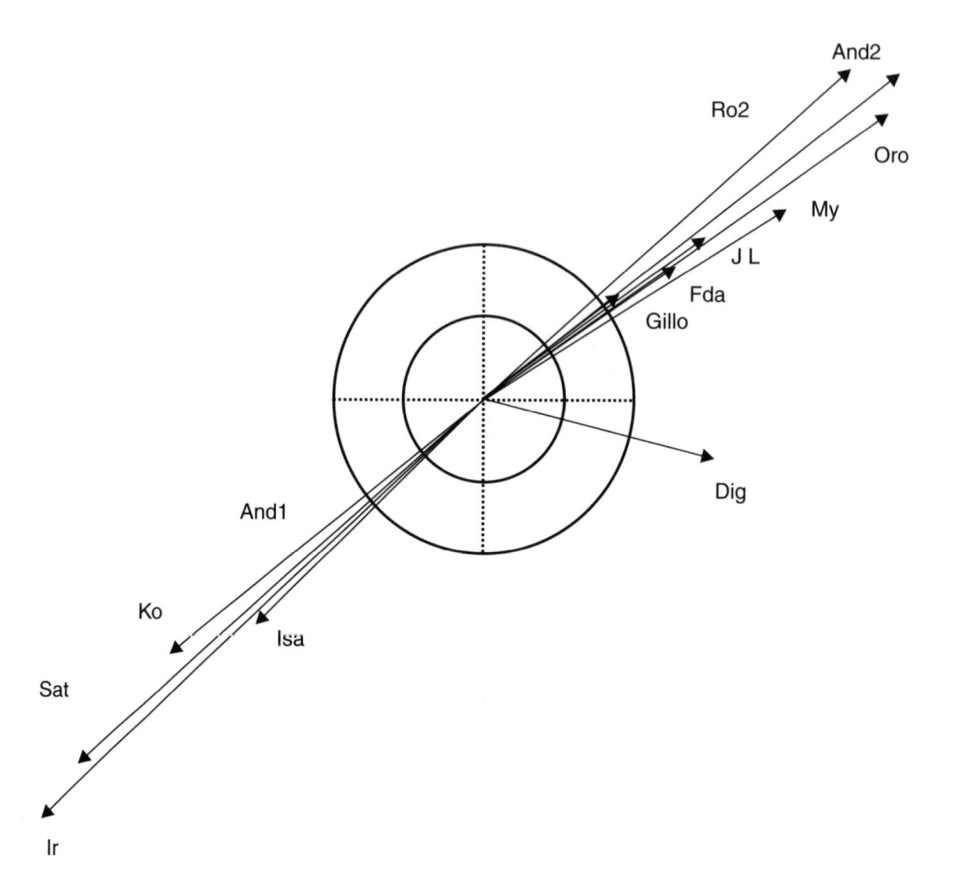

hibitory in the classroom but mutually excitatory in the playground. In a third example, the relationship with "dig" is mutually excitatory in classroom but asymmetrical in playground. Finally, with the other peers the target only maintains relationship in a single context (playground **or** classroom), either mutually excitatory or mutually inhibitory.

Information of this kind is very useful for developmental and social competence analyses. For example, the main links that the target maintains with her peers seem to depend on the fact that such relationships are symmetrical, mainly excitatory.

Some general developmental implications can be derived: Reciprocal interactions are among the main mechanisms that explain the links that subjects maintain in their social networks. However, the study of asymmetrical relationships offers interesting information about the nature of the bonds that children maintain with specific peers. Future studies should analyze social effectiveness, social reciprocity, and social responsiveness (Santoyo, 1996) as indexes which could help to explain social adjustment in school settings.

In addition, longitudinal studies should be conducted to evaluate pathways in social networks. These studies should evaluate the stability or changes in social preferences in the social ecology of children.

References

Anguera, M.T. (1997, April). *From prospective patterns in behavior to joint analysis with a retrospective perspective*. Colloque sur invitation *"Methodologie d'analyse des interactions sociales"*. Paris: Université de la Sorbonne.

Anguera, M.T., Blanco, A., & Losada, J.L. (2001). Diseños observacionales, cuestión clave en el proceso de la metodología observacional [Observational designs: Essential question in the process of observational methodology]. *Metodología de las Ciencias del Comportamiento,* 3(2), 135-161.

Anguera, M.T., & Jonsson, G.K. (2002, June). *Detection of real-time patterns in sports: Interactions in football*. Third Meeting of the European Research Group on "Methodology for the analysis of social interaction". Milan: Catholic University of Milan.

Anguera, M.T., & Losada, J.L. (1999). Reducción de datos en marcos de conducta mediante la técnica de coordenadas polares [Data reduction in behavioral settings through polar coordinates technique]. In M.T. Anguera (Ed.), *Observación de la conducta interactiva en situaciones naturales: Aplicaciones [Observation of interactive behavior in natural situations: Applications],* pp. 163-188. Barcelona: E.U.B.

Bakeman, R. (1978). Untangling streams of behavior: Sequential analysis of observation data. In G.P. Sackett (Ed.), *Observing Behavior, Vol. 2: Data collection and analysis methods,* pp. 63-78. Baltimore: University of Park Press.

Bakeman, R., & Gottman, J.M. (1986). *Observing interaction. An introduction to sequential analysis*. Cambridge: Cambridge University Press.

Bakeman, R., & Quera, V. (1995). *Analyzing interaction: Sequential analysis using SDIS and GSEQ*. New York: Cambridge University Press.

Cairns, R.B. (1979). *Social Development: The origins and plasticity of interchanges*. San Francisco: Freeman.

Cochran, W.G. (1954). Some methods for strengthening the common χ^2 test. *Biometrics,* 10, 417-451.

Hernández Mendo, A., & Anguera, M.T. (1998). Análisis de coordenadas polares en el estudio de las diferencias individuales de la acción de juego [Analysis of polar coordinates in the study of individual differences of play action]. In M.P. Sánchez & M.A. Quiroga (Coords.), *Perspectivas actuales en la investigación psicológica de las diferencias individuales [New perspectives in psychological research about individual differences],* pp. 84-88. Madrid: Centro de Estudios Ramón Areces.

Kindermann, T.A. (1996). Strategies for the study of individual development within naturally-existing peer groups. *Social Development,* 5(2), 158-174.

Paredes, T.V., & Santoyo, V.C. (1998). Mecanismos conductuales en la regulación de mapas socioconductuales [Behavioral mechanisms on the socio-behavioral maps regulation]. *La psicología social en México [Social psychology in Mexico],* 7, 489-494.

Pulido, R.M., Fabián, T.A.L., & Santoyo, V.C. (1998). El estudio diferencial de las redes sociales: factores microrregulatorios y de desarrollo [The differential study of social networks: Microregulatory and developmental factors]. *La psicología social en México*

[Social psychology in Mexico], 7, 461-466.
Sackett, G.P. (Ed.) (1978). *Observing Behavior (Vol. 2): Data collection and analysis methods*. Baltimore: University of Park Press.
Sackett, G.P. (1980). Lag sequential analysis as a data reduction technique in social interaction research. In D.B. Sawin, R.C. Hawkins, L.O. Walker & J.H. Penticuff (Eds.). *Exceptional infant. Psychosocial risks in infant-environment transactions,* pp. 300-340. New York: Brunner/Mazel.
Santoyo, V.C. (1994). Sociometría conductual: El diseño de mapas socioconductuales [Behavioral sociometry: The design of sociobehavioral networks]. *Revista Mexicana de Análisis de la Conducta*, 20, 183-205.
Santoyo, V.C. (1996). Behavioral assessment of social interactions in natural settings. *European Journal of Psychological Assessment*, 12(2), 124-131.
Santoyo, C., & Anguera, M.T. (1993). Evaluación ambiental: Integración de estrategias flexibles en situaciones naturales [Environmental assessment: Integration of flexible strategies in natural situations]. In M. Forns & M.T. Anguera (Coords.), *Aportaciones recientes a la evaluación psicológica [New contributions to psychological assessment]*, pp. 121-135. Barcelona: P.P.U.
Santoyo, V.C., & Espinosa, A.M.C. (1987). El sistema de observación conductual de interacciones sociales [An observational system of social interactions]. *Revista Mexicana de Análisis de la Conducta*, 13, 235-253.
Santoyo, V.C., & Espinosa, A.M.C. (1988). El análisis conductual de las preferencias sociales [A behavioral analysis of social preferences]. *Revista Mexicana de Análisis de la Conducta*, 14, 29-39.
Santoyo, V.C., & Espinosa, A.M.C. (1991). Decisiones metodológicas para el análisis contextual de la interacción social [Methological decisions on the contextual analysis of social interaction]. *Revista Mexicana de Análisis de la Conducta*, 17, 85-104.
Santoyo, V.C., Espinosa, A.M.C., & Bachá, M.G. (1995). Extensión del sistema de observación conductual de las interacciones sociales: Calidad, dirección, contenido y resolución [Extension of the behavioral analysis system of social interactions: Quality, direction, content, and resolution]. *Revista Mexicana de Análisis de la Conducta*, 11, 55-68.
Strayer, F., & Santos, A. (1996): Affiliative structures in preschool peers. *Social development*, 5, 2, 117-130.

Projects and policies for childhood in Italy

Antonella Rissotto
Institute of Cognitive Sciences and Technologies, National Research Council, Rome

> **Abstract.** This contribution has two aims. First to briefly describe the cultural and legislative context produced in Italy by the ratification of the United Nations Convention on the Rights of the Child. Second to present some of the outcomes of the Children's City project.
> This project was started in 1991 and today it involves 46 cities in Italy, several in Argentina and a few in Spain. It is aimed at improving the urban environment from the perspective of children. The study of actions carried out by the Italian cities in order to promote children's participation and children's autonomous mobility highlights the positive and negative aspects of the project.
> The main limits of the project are connected with the following factors: project's dependence on temporary tasks of politicians; the emphasis given to the educational dimension of interventions which shifts attention away from the city's transformation and from children's empowerment; the experimental character of the interventions which results in reduced impact on the urban environment and on the child population. The main positive aspects of the project are related to changes in the perception of childhood, which motivate administrators to make innovative choices and encourages the engagement of larger components of the community.
>
> **Keywords:** urban environment, children's empowerment, independent mobility, children's participation, network of cities, legislative actions, evaluation

Introduction

Over the past few decades, especially in the industrialized western countries, a profound change has taken place in the relationship between children and their living environment. Increased road traffic, pollution and crime, the fragmentation of the urban fabric, the loss of community feeling, have transformed the cities into places that are unsuitable for children to play in. In many cities, the reduction in the range of urban spaces in which they can play (Bozzo, 1995; Gaster, 1991), has been paralleled by the construction of rigorously horizontal spaces, fenced in, and equipped and furnished in a stereotyped fashion, in which only the games for which they were designed can be played (Marillaud, 1991). These spaces seem to cater more to the adults' demand for safety than to the children's play needs (Ader & Jouve, 1991; Alexander, 1977; Tonucci & Rissotto, 1998), as they penalize play expressiveness and encourage adult–child relations (Danacher, 1991).

The modern city cannot be explored freely by children. This has resulted in part from parent's fears about the safety and security of their children (Blakely, 1994; Harden, 2000; Scott et al. 1998), in part from environmental pollution and in part from the "adultization" of childhood because children's times is filled with organised activities such as sports, music and scheduled activities (Francis and Lorenzo, 2002). The spread and extent of the restrictions placed on children's autonomous movement (Hillman, 1993), led Gaster (1995) to assert that the concept of home range can no longer be applied to the interaction between the child and his or her living environment. The reduction in children's autonomy of movement affects their parents' habits (Gershuny, 1993), reduces the children's opportunities for getting regular physical exercise (Armstrong, 1993) and is detrimental to the acquisition of environmental knowledge (Hart, 1979; Spencer, 1992; Torell & Biel, 1985; Torell, 1990).

Projects and intervention on behalf of children

Growing awareness of the consequences for children of changes occurring in the cities has led to the development of projects and interventions aimed at improving the relationship between the child and the environment in which he or she lives.

Cities' Projects for Children

In the 1991, Italy ratified the International Convention on Children's Rights. Civil society associations first responded to this document by starting up a number of projects. The *Città educative* (Educational Cities) project saw the light with the drawing up of a charter indicating the fundamental principles on which the educational choices of cities ought to be based. The city administrations participating in the project are pledged to reappraise the city's management as a function of the cultural development needs of all its citizens, in the first instance, of the children. The Italian UNICEF Committee started up the *Sindaci difensori dei bambini* (Mayors as children's champions) project to promote greater awareness of childhood, starting from the institutional level closest to the citizens. In 1991, the *La città dei bambini* (Children's City) project of the CNR Institute of Psychology got off the ground with the aim of promoting a change in the urban environment starting from the needs of the younger citizens. The *Città sane* (Healthy Cities) project is included in the Health for all Project for the Year 2000 and Beyond project of the World Health Organization (WHO) to improve the mental, social and physical well-being of persons living and working in the cities. The Italian Healthy Cities – WHO Network, which today comprises over 130 cities, was set up in 1995. *Democrazia in erba* (Budding Democracy) is an association set up for the purpose of promoting and coordinating the Children's Town Councils, an experience of participation by young people in the cities' decision-making processes.

Starting from different standpoints, these projects have stimulated a collective reflection on the characteristics of the urban environment as well as on the children's conditions and have given rise to a set of initiatives to enhance the relationship between the children and the environment in which they live. Such projects have also contributed to the spread of a new conception of childhood since, in agreement with the UN Convention, the child is considered as a citizen having rights, and capable of playing an active role in the community to which he or she belongs. This new view of childhood has had a strong impact on the Italian legislation.

Legislative actions

In 1997 the first Government Action Plan for Childhood and Adolescence was launched. This plan, of three years' duration, was aimed at promoting the realization and integration of actions in favour of childhood carried out in all national governmental sectors. The following interventions are among the results produced by this Plan.

(1) *Law 285/97 "Provisions for the promotion of rights and opportunities of childhood and adolescence"*, promoted in 1997 by the Ministry of Social Solidarity, made substantial funds available to the municipalities for positive action to promote the rights of childhood and adolescence, for the enhancement of enjoyment of the urban and natural environment and for the development of well-being and the quality of life of minors (Moro, 1998; Ricci, 2000).

(2) *Neighbourhood Contracts.* This is an agreement between the Ministry of Labour and Social Solidarity signed in 1998 which provides the municipalities with resources for mission-oriented urban and social rehabilitation action in peripheral slum areas through citizens' participation, starting from children and elderly persons.

(3) *"Sustainable cities for little girls and boys".* This project, promoted in 1997 by the Ministry of the Environment, is aimed at promoting the sustainable development of the environment through the participation of children in the city's life. This is the scope of law 344/97 which provides for the establishment of a form of recognition for cities that show the greatest commitment to promoting urban sustainability by improving the living conditions of children insofar as they are indicators of urban quality (Ministero dell'ambiente, 1998; 2000).

The Children's City project

The Italian cultural and legislative horizon seems to be characterized by a fresh and significant awareness of both the conditions of the cities and the needs of childhood. However, this awareness has not actually transformed Italian cities into places that are particularly suitable for children or for their autonomy of movement (Giuliani et al., 1997; Prezza, et al., 2001), for their acquisition of environmental knowledge (Rissotto & Tonucci, 2002) or for their participation (Alparone & Rissotto, 2001). A preliminary assessment of the application of law 285/97 confirms this situation and indicates that, although innovative, this law has had only a partial effect on the child–city relationship (Baraldi, 2001).

In order to appreciate the extent to which children can actually influence the environment in which they live and to illustrate the factors standing in the way of transforming innovative policies in favour of childhood into concrete initiatives, one specific project has been selected, Children's City, and an assessment has been made of the initiatives promoted by it in support of children's autonomy and participation.

One methodological choice of the Children's City is to address the project to the mayors by considering them the main references. Mayors are requested to include in their political agenda urban changes functional to children's needs (Tonucci, 1996).

Local development of the project is entrusted to a laboratory that follows two main action lines:
(1) to involve the children in the decision-making and planning processes by setting up a Children's Council and Participation in Urban Planning experiences; and
(2) to initiate and implement the initiative Let's Go to School on our Own. The purpose is both to allow children to regain their autonomy of movement and the city local government officials to reappraise urban mobility in connection with sustainable development.

These types of initiatives have been considered strategic because of the widespread consequences that they possibly introduce in the urban system.

In 1996, a working group was set up to promote and coordinate the Cities Network activity, to run training workshops on the project topics, to provide for the publication of a newsletter, to run a centre of documentation on the action carried out in the cities and carry on research activity in children's autonomy and participation (; Rissotto, 2001Tonucci & Rissotto, 2001).

The Cities Network

Forty-six Italian cities[1] are participating in the project: 61% have less than 50,000 inhabitants; 26% have between 50,000 and 100,000; 13% have more than 100,000. Altino, with 2,560 inhabitants, is the smallest city in the Network, while Rome, with 2,662,500 inhabitants, is the largest.

The project is under way in 15 of the 20 Italian regions. The northern regions account for 35% of the cities, the centre for 48% and the south for 17%.

Law 285/97 fostered the cities participation in the project: 63% joined in the last four years.

The oldest city in the Network is Fano, where the project began 11 years ago; the most recent is Rome, which joined in 2001. Three municipalities, which suspended their activities for several years, have rejoined the project after a change of administration. Six cities (13%) have lost contact with the Network.

Three cities of the Network have obtained recognition as a sustainable city for little girls and boys from the Ministry of the Environment. The action taken by 9 cities of the Network has been entered into the good practices register of the Ministry of the Environment.

---

[1] Also several Argentine and Spanish cities are participating in the project, but these have not been taken into account in the present study.

Sixty percent of the cities have opened a laboratory dedicated to the local development of the project.

The activities of the Children's City network

The evaluation of the project, which regards the initiatives carried out in the Network of Italian cities, was based on an analysis of the materials produced by cities and collected in the *Children's City* Documentation Centre.

Overall, we have examined materials which documented 65 initiatives: 29 of them referred to Participation in Planning, 14 to Children's Councils and 22 to Let's Go to School on Our Own. The materials documenting these initiatives were analysed with regard to: objectives, duration of the activities; age of children involved; types of adults involved; sources of funding; results or ongoing projects (table 1).

Table 1. Analysis of cities' initiatives regarding the Children's City project

	Participation in Planning $N = 29$	Children's Councils $N = 14$	Let's Go to School on Our Own $N = 22$
Duration of activity	24% under way for 4–5 years	7% under way for 11 years	5% under way for 7 years
	38% under way for 2–3 years	45% under way for 4–5 years	14% under way for 4–5 years
	10% under way for 1 year	21% under way for 2–3 years	53% under way for 2–3 years
	28% stopped after 16.5 months	20% under way for 1 year	14% suspended and re-started
		7% stopped after 12 months	14% stopped after 12 months
Children's age	Average age 8.5 years (range 3–14 years)	Average age 11.5 years (range 9–14)	Average age 9.5 years (range 6–13)
	53% 6–11 years	72% 9–11 years	95% 6–11 years
	25% 11–14 years	28% 9–14 years	5% 11–13 years
	15% 6–14 years		
	7% 3–11 years		
Adults involved	*Adults with institutional role:*	*Adults with institutional role:*	*Adults with institutional role:* 49% technical experts and local governmental officials
	44% teachers	47% teachers	
	40% technical experts and local governmental officials	33% technical experts and local governmental officials	41% teachers

Table 1. continued

	Participation in Planning $N = 29$	Children's Councils $N = 14$	Let's Go to School on Our Own $N = 22$
Adults involved	16% representatives of public bodies (e.g., juvenile justice centres)	20% representatives of public bodies (e.g., juvenile justice centres)	10% representatives of public bodies (e.g., city police)
	Adults with no institutional role:	*Adults with no institutional role:*	*Adults with no institutional role:*
	61% self employed	75% self employed	31% tradesmen
	27% representatives of associations	25% representatives of associations	27% self employed
	12% volunteers		27% volunteers
			8% parents
			6% representatives of associations
Sources of funding	65% city budget	67% city budget	63% city budget
	31% law 285/97	29% law 285/97	32% law 285/97
	4 % sponsor	4 % sponsor	5 % sponsor
Results or on going initiatives	*Types of projects:*		*State of the play:*
	56% green areas		45% running without specific support
	18% squares, streets, road signs		18% at start up
	16% school courtyards		14% under experimentation
	10% other (e.g. Children's City Laboratory furnishings)		14% suspended
			9% generalized to other schools
	203 projects produced	670 children involved	42 schools involved
	64% of the children's projects realised	86% of the cities have produced documents indicating attention paid to children's contribution	68% of the cities has carried out action in support of the initiative

The objectives

Not all the cities of the Network have developed all the three types of activities. The implementation of the project's objectives has varied a great deal among the cities. In some cases cities have proceeded with a minimal implementation of the objectives of the project, implementing symbolic initiatives (e.g., awareness-raising campaigns concerning children's rights which have not produced other initiatives on behalf of childhood). In other cases, innovative strategies were devised to take into account the children's point of view (for example, several cities adopted new administrative procedures rapidly to implement the projects conceived of by children).

The duration of the activities

In general, the local government officers believe that the participation of children or the possibility of taking into account their needs for autonomy are exceptions rather than the rule. For this reason the duration of the initiatives is an indicator of the change of attitude of the administrators in the face of the importance of the children's point of view on the city policies.

The duration of Participation in Planning, Children's Councils and Let's Go to School on Our Own corresponds to the number of years since the cities began these types of activities. In Participation in Planning, every year a different theme is given to the children for them to work upon. Work on this lasts for the length of a school year. The length of time elapsing between the selection of the theme to be to worked upon and its realisation can vary a great deal because of notable differences in the time required by local governments for the realisation of these projects. In the Children's Councils each child takes part in the initiative for two years and 50% of the children's councillors are elected each year.

If we compare the three types of activities, it can be seen that Children's Council lasts the longest, while Participation in Planning is the activity which suffers most from continuity problems. In three cities included among those which interrupted the activities of Participation in Planning the local government officials carried out the children's projects and since then have not started other projects of this type. In the other cities the interruption of the initiatives produced more negative consequences because they were stopped after the planning phase without implementing what the children had planned.

In most cases, but not always, the interruption of the Children's Council, Participation in Planning and Let's Go to School on Our Own was linked to the change in the local government officials. The fact that the Children's City project depends on the transitory nature of political power is one of its most vulnerable areas. The interruption of an activity does not always mean that it is halted completely. In some cities Let's Go to School on Our Own was re-started by the new local government officials and was helped in this by the availability of funds provided by law 285/97.

Adults involved in the three types of initiatives point out as critical factor of the planning the evaluation of the amount of time which is required to the children engaged in the activities and to organise them. In fact it must be taken into account not only the time necessary to develop the activities, but also the time which children need to perform the required tasks, that are: planning; being a 'spokesman' for other children; creating an environment which fosters their autonomy. However setting time limits is meant to be an innovative element in strategies demanding children's collaboration. The respect of time limits, both those necessary for children to plan and those necessary for the local government officials to realise the children's projects, to accept children's contribution or to create conditions which foster their autonomy, becomes important when the children's work it is not considered just a didactical exercise but a real undertaking.

The age of the children involved

Citizens under the age of eighteen have little or no possibility of influencing the planning and running of the city. This is however generally true for children. Ten european countries have organisational structures to promote the participation of young people, but only four (France, Hungary, Italy and Romania) have structures which promote also the involvement of children (Matthews, 2001). The presence of children as young as three years old in the participatory projects realised by the Children's City is thus indeed innovative.

The cities of the Network involved age groups in a different way in the three types of initiatives. Participation in Planning is the activity with the widest range of age groups. This is the result of the link between the activity and the school world. Teachers have viewed participatory planning as providing an opportunity for continuous initiatives throughout different school years. In Let's Go to School on Our Own and in Children's Councils mainly children aged from 6 to 11 were involved. The age of children who have been involved in the mobility initiative lies in the age group most heavily penalised by the drastic decrease in autonomy of movement that has been documented by several researchers. The data concerning boys and girls aged 11-13 years are worrying because they show that the age at which boys and girls are allowed to go to school on their own is continually rising.

Types of adults involved

All three activities require the intervention of different adult figures: teachers, who play an important role in the development of all the activities; technical experts (architects, engineers, urban planners, psychologists, etc.); local government officers and different members of society. This aspect is particularly evident for Let's Go to School on Our Own, which requires a change in the physical characteristics of the city and the re-acquisition of the sense of community, but it is present also in the two other initiatives.

Three interesting results which emerge from the analysis of the types of adults involved in the activities are: the diversification of the professional roles in a city which takes childhood into account, the integration of the activities undertaken by these professionals and the involvement of the community.

The plural nature of adult figures involved may be considered one of the successes of the Children's City. The project promotes the idea that the relationship between a child and his or her environment and the change of this environment should not just be the responsibility of local government officials who normally look after children from the educational, health and social viewpoint. This relationship should also be the responsibility of the city as a whole. The initiative Let's Go to School on Our Own, in fact, requires the involvement of different sectors of local government including technical experts and local government officials in charge of security and mobility within the city. In the same way, children's participation promotes the dialogue between the children, the local government officials and the technical officers who have responsibility for the city's planning, ecology and public works.

Often cities rely on self-employed professionals to carry out the project. This decision has been justified by the local government officials as being due to their lack of competence in the specific themes of the initiatives.

Sources of financing

The data held at the Documentation Centre of the Children's City project are unfortunately incomplete regarding the financing of the activities. The analysis of this factor is limited to the source of funding.

The most interesting data is that approximately 60% of the finance comes from the city's budget and only a small part from sponsorship. The use of the city's budget funds to realise the proposals put forward by children or to take on board their needs as an important priority in the programming activities within the cities reveals that an important change has occurred in the administrators' attitude towards childhood. In one city, for example, some projects have been implemented by using a part of the funds set aside for street maintenance. In another city a pedestrian cycle path had been planned for some time but was given priority only after the launch of the initiative Let's Go to School on Our Own.

Although sponsors only play a small part, their contribution should be highlighted because it does provide evidence of a new awareness of children's rights and needs.

Results of on-going projects

The type of spaces given over to children's planning influences the social visibility of their participation and offers information on the type of skills which adults believe children possess. From this point of view, the less interesting initiatives are those regarding the planning of school playgrounds (16% of the projects), places which are important above all or exclusively for children. Seventy-two per cent of the projects are for green or open spaces. This shows that children are considered capable mainly of planning areas which are dedicated to play, meeting or relaxation activities. Street planning or the planning of street signs (18% of the projects) highlights the link between the promotion of children's participation and their mobility. The remaining 10% of projects cover different themes such as the furnishing of the laboratory for the Children's City or the realisation of products to promote communication between adults and children.

Before analysing data regarding the state of play of the initiatives it should be emphasised that it is easy to give a picture of Participation in Planning but not of the Children's councils or of the Let's Go to School on Our Own initiative because only in the first activity does the children's work lead to a specific project being realized. In Participation in Planning, children drew up 203 projects and 64% of these were realized. For the

Children's Council, we examined materials which show the promotion and upgrading of the children's contributions, the acceptance of the children's proposals, the organisation of meetings involving children councillors and local government officials or the formal recognition of the existence of the Children's Council. Eighty-six per cent of the cities produced documents of this type. For the initiative Let's Go to School on Our Own we examined materials which document the realisation of activities supporting the autonomous mobility of children such as the increase in street signs, the creation of pavements, the introduction of traffic lights, campaigns for motorists, etc. Sixty-eight per cent of the cities realised activities of this type.

A large percentage of the children's proposals were accepted by the local authorities, which often also realised activities favouring the mobility and the involvement of children in the cities' decision-making processes. But this did not always happen. A problematic aspect of the Children's City is the attitude of adults, in particular local government officials, who assign an exclusive or mainly educational significance to the project. This attitude is based on widespread social and cultural stereotypes linked to the idea of the child as an individual who is not fully formed, someone to be educated and to be protected. This point of view sees children as such as lacking in skills and thus the promotion of their participation would reach its main objective in training and not in sharing with them techniques and instruments used in planning and decision-making. In the same way children are recognised as needing protection but not autonomy. The emphasis put by local government officials and by the other adults involved in the educational dimension leads to a reduction of the importance attributed to changing the environment or to the empowerment of the children. In some cases children's projects are not realised simply because this is not necessary for their education. In other cases, the initiative Let's Go to School on Our Own may remain at its initial stage for a long time because this allows the children to acquire the knowledge of their environment and postpone the moment in which the children can move independently and the local government has to provide answers to their demand for security and environmental change.

Another problematic aspect of Children's City is the experimental nature of the activities. This is particularly evident for Let's Go to School on Our Own but it is also relevant to the participation activity. Only 45% of the experiences promoting the autonomy of children are now routine, that is to say an activity which runs without the support of the local government. An even smaller percentage (9% of the initiatives) has spread to other schools in the city. Thirty-two per cent of the experiences are at the initial or experimental stage, with the support of the local government, and have been there between one and three years.

The experimental dimension of the activities allows the local government officials to begin the new activities in a controlled manner and to link the realisation of these activities to the needs that the children express and to the characteristics of the environment. This approach however limits the impact of the activities themselves on the city and on the child population because they seldom spread throughout the city after the experimental stage. Participation in Planning has seen the realisation of 130 projects, the Children's Councils have involved a total of 670 children and no city has set up local district Children's Councils to promote an increase in the number of children involved. The Let's Go to School on Our Own experience has involved a total of 42 schools.

From this data, we can see that the activities realised by the Children's City Network have interesting and innovative characteristics but also a limited impact on the relationship between the child and his or her environment because they are often restricted to a specific area of the city (one school for Let's Go to School on Our Own, one urban space for the participatory planning) and involve a limited number of children. The fact that the activities do not become more widespread is indeed the principal weak point of the project. For this reason the organisers of the Children's City should pay more attention to the specific factors standing in the way of a permanent and more widespread inclusion of the project among the city's policies.

References

Ader, J., & Jouve, H. (1991). Jeu et contexte urbain [Play in Its Urban Context]. *Architecture & Comportement, 7,* 115-119.

Alexander, C. (1977). *A pattern language.* New York: Oxford Press.

Alparone F. R., & Rissotto A. (1998). *Il consiglio dei bambini nella V Circoscrizione di Roma [The children's council in the 5th District of Rome].* Paper presented at the Second National Congress of AIP (Social Psychology), Florence, 30 September-2 October.

Alparone, F. R., & Rissotto, A. (2001). Children's citizenship and participation models: Participation in planning and Children's Councils. *Journal of Community and Applied Social Psychology,* 11, 421-4234.

Armstrong, N. 1993. Independent mobility and children's physical development. In M. Hillman (Ed.), *Children transport, and quality of life.* London: Policy Studies Institute.

Baraldi, C. (2001). I diritti dei bambini e degli adolescenti: una ricerca sui progetti legati alla legge 285 [Children and adolescents' right: a research regarding the projects connected with bill 285]. Roma: Donzelli.

Blakely, K. S. (1994) Parents' conceptions of social dangers in the urban environment. *Children's Environments Quarterly,* 11, 16-25.

Bozzo, L. (1995). Il gioco e la città [The play and the city]. *Paesaggio urbano [Urban landscape],* 2, 30-33.

Danacher, A. (1991). Contraintes de l'espace ludique aménagé [Constraints of the space for play]. *Architecture & Comportement,* 7, 153-165.

Francis, M., & Lorenzo, R. (2002). Realms of Children's Participation. *Journal of Environmental Psychology,* 22, 157-169.

Gaster, S. (1995). Rethinking the children's home range concept. *Architecture & Comportement,* 11, 35-42.

Gaster, S. (1991). Urban children's access to neighbourhood. *Environment & Behavior,* 23, 70-85.

Gershuny, J. (1993). Escorting children: Impact on parental lifestyle. In M. Hillman, (Ed.), *Children, transport and the quality of life,* pp 62-76 London: Policy Studies Institute.

Giuliani, M. V., Alparone, F. R., & Mayer, S. (1997). Children's appropriation of urban spaces. *Urban Childhood International Conference,* Trondheim, 9-12 June, 1997.

Hart, R. (1979). *Children's experience of place.* New York: Irvington.

Harden, J. (2000). There's no place like home: The public/private distinction in children's theorising of risk and safety. *Childhood,* 7, 34-45.

Hillman, M. (Ed.). (1993).*Children transport and quality of life*. London: Policy Studies Institute.

Marillaud, J. (1991). Jeu et securité dans l'espace pubblic [Play and security in public space]. *Architecture & Comportement*, 7, 137-145.

Matthews, H. (2001). *Children and Community Regeneration: Creating better neighbourhoods*. London: Save the Children.

Ministero dell'ambiente [Ministry of the Environment] (Ed.). (1998). *Guida alle città sostenibili delle bambine e dei bambini [Guidebook to sustainable cities for little girls and boys]*. Rome: Ministero dell'ambiente.

Ministero dell'Ambiente [Ministry of the Environment] (Ed.). (2000). *Le bambine e i bambini trasformano le città: progetti e buone pratiche per la sostenibilità nei comuni italiani [Children transform the cities: projects and best practices for the sustainability of Italian Municiplaties]*. Florence: Litografica I.P.

Moro, A. C. (Ed). (1998). *Infanzia e adolescenza diritti e opportunità Orientamenti alla progettazione degli interventi previsti nella legge n. 285/97 [Childhood and adolescence: Guidelines to plan interventions for bill 285]*. Florence: Centro nazionale di documentazione e analisi per l'infanzia e l'adolescenza.

Prezza, M., Pilloni, S., Morabito, C., Sersante, C., Alparone, F. R., & Giuliani, M. V. (2001). The influence of psychological, social and urban factors on children's independent mobility and relationship to peer frequentation. *Journal of Community and Applied Social Psychology*, 11, 435-450.

Ricci, S. (Ed.). (2000). *Il calamaio e l'arcobaleno: orientamenti per progettare e costruire il Piano territoriale della L. 285/97 [The ink pot and the rainbow: guidelines to plan and to build the Territorial Plan of bill 285/97]*. Florence: Centro nazionale di documentazione e analisi per l'infanzia e l'adolescenza.

Rissotto, A., & Tonucci, F. 2002. Freedom of movement and environmental knowledge in elementary school children. *Journal of Environmental Psychology*, 22, 65-77.

Rissotto, A. (2001). Da bambino farò un parco: quando i bambini progettano [From child I will make a park: when children plan]. In L. Amodio, C. Majorano, & C. Riccio (2002), *I bambini trasformano la città: Metodologia e buone prassi della progettazione partecipata con i bambini [The children transform the city: Methodology and best practices of participatory planning with children]*. Rome: Ministero dell'Ambiente.

Scott, S., Jackson, S., & Backett-Milburn, K. (1998). Swings and roundabouts: Risk, anxiety and the everyday world of children. *Sociology*, 32, 689-705.

Spencer, C. (1992). Life span changes in activities, and consequent changes in the cognition assessment of the environment. In T. Garling & G. W. Evans (Eds.), *Environment, cognition and action: An integrated approach*, pp 295-306. New York: Oxford University Press.

Tonucci, F. (1996). *La città dei bambini [The children's city]*. Bari: Laterza.

Tonucci, F., & Rissotto, A. (2001). Why Do We Need Children's Participation? The Importance of Children's Participation in Changing The City. *Journal of Community and Applied Social Psychology*, 11, 407-419

Torell, G. (1990). *Children' Conception of Large-Scale Environments*, Gothenburg: Gothenburg University Press.

Torell, G., & Biel, A. (1985). Parental restrictions and children's acquisition of neighborhood knowledge. In T. Garling & J. Valsiner (Eds.), *Children within environments: toward a psychology of accident prevention*, 107-118. New York: Plenum Press.

Web sites

Centro nazionale di documentazione e analisi per l'infanzia e l'adolescenza
http://www.minori.it
Carta delle Città Educative
http://www.comune.torino.it/citedu/intro.htm
Statuto dell'Associazione Internazionale delle Città Educative
http://www.comune.torino.it/citedu/statuto.htm
Le città sostenibili delle bambine e dei bambini Ministero dell'ambiente
http://www.cittasostenibili.minori.it/
Associazione Democrazia in Erba
http://www.cittasostenibili.minori.it/guida/pag95.htm
Contratti di quartiere
http://www.minori.it/archivi/pubbli/altre/appe4.html
UNICEF Italia
http://www.unicef.it/
Sindaci difensori dell'infanzia
http://www.unicef.it/comita.htm

The Body Goes to the City project:
Research on safe routes to school and playgrounds in Ferrara

Marcello Balzani[1] & Antonio Borgogni[2]

[1]University of Ferrara, Italy; [2]UISP Ferrara, Italy; University of Cassino, Italy

Abstract. The research activity presented here is the fruit of a co-operation between an association (UISP Ferrara, The Body Goes to the City project), a faculty (the Faculty of Architecture of the University of Ferrara) and local authorities (La Città bambina [Children's City] project). The common objective – to regain the possibility of children gaining autonomy with regards their personal mobility – has stimulated common strategies of intervention. These projects started from sociological research concerning liveability as well as safety, and have involved both school children and the whole neighbourhood, and were further developed in workshops in schools. Finally, the data collected in the studies and the workshops was adopted by town planners for the elaboration of the final project.

The particularities of this research action are the common long-term intervention methodology, which – also thanks to the common training of operators with different professional backgrounds – allows the conservation of similar educational approaches and critical viewpoints; the adoption of the body as an analyser of the quality of life; the particular attention that the planners pay to the perceptive and qualitative survey of the neighbourhood.

The results of the sociological research, which has clearly revealed the children's lack of autonomy in their everyday movements, are in some ways similar to those obtained in other studies. However, they appear even more worrying considering that they have been gathered in Ferrara, a city considered one of the most liveable cities in Italy and a "city for cyclists" on the European level.

The word that characterises the results of the study is "limit". We reckon that as the situation persists, it shall not only modify the autonomy of the children, who increasingly adopt their parents' models of behaviour and do not perceive the city as a place where they can live and play, but, in the long run, even their very desires are limited.

Keywords: body, shared strategy, perception, co-operation, compli-city

Where did we start from?

The idea of a shared town-planning project around the topics of mobility and safety arose from an experience of applied architectural survey and techniques of representation in a part of the city of Ferrara. An initiative of the project The Body Goes to the City" (Il corpo va in citta) of UISP Ferrara, realised in co-operation with the project Children's Cities (La città bambina) of the Municipality of Ferrara and the Faculty of Architecture of the University of Ferrara, as part of a research project entitled *Safe School Routes in the Giardino Neighbourhood*, aimed at finding different town planning interventions for the promotion of safety and accessibility of pedestrian paths and cycle lanes.

The Body Goes to the City is a plan for shared urban development, involving citizens in the decision-making process concerning re-designing urban spaces by means of needs analysis, common planning, information and activities.

A key characteristic of the project is the proposal to use the body as a means of analysing the quality of life, the idea that letting the body express itself is an indicator of high levels of user-friendliness.

The body discussed in this context is the one that makes everyday movements, and which co-ordinates its activities to climb steps, to play, to ride a bicycle, to skate, to sit down, to shake hands, make friends and to do sports.

Safe School Routes in the Giardino Neighbourhood

The Giardino neighbourhood is an urban area delimited by the old city walls, very near to the railway station and closely connected with the old nucleus of the town. There, as in an alchemical melting pot, several experiments of interpretative readings and community participation projects are being made involving the residents and the schools of the area, and considerably promoting the integration of the environment. (One of the many examples of such projects was our sociological research, realised through questionnaires filled in by all the children of the elementary schools of Govoni and Poledrelli, interviews with selected representatives of the area, laboratories with the parents and meetings with activists and associations.) An important urban area, with all the structural and qualitative requisites to be adapted to new needs, so as to gradually lose its role as a disorientating and anonymous "container-park", as it is presently characterised.

To develop the kind of project we had in mind, we needed a versatile working group: our group is composed of educationists, architects, sociologists and psychologists.

Our research was actually composed of *two* different studies, one of which is mainly sociological, while the other concerns mainly town planning. These two studies were then joined in order to realise an initial re-planning of the areas. With this first plan, workshops were then organised with school children, residents and stakeholders living in the area. Finally, after this phase of elaboration, we obtained a definite plan.

One part of the sociological research was realised with questionnaires concerning school runs, compiled by the children of two elementary schools of the area. We also had questionnaires for the parents, made several in-depth interviews and organised workshops and focus groups.

The sociological study

"Pulled here, pushed there"

The context of the results of the questionnaires consists of a city considered one of the most liveable in Italy, the "city of bicycles"; a project that allowed synergies between the local authorities and associations and an elevated level of attention and sensitivity towards children.

These were the ingredients behind the results of the questionnaires that, accompanied by interviews with the selected representatives of the schools and the neighbourhood, provided us with some useful information for a complete revision project of the neighbourhood. Its first step was to make the pavements and cycle lanes safer, as well as coming up with ideas for a project to be realised with the children of the two elementary schools.

The T-shirts

According to the results of the questionnaire, children appear to be pulled by the T-shirt in their wish for the comfort of the car and by the shorts or the skirt in their wish to move autonomously.

Some answers are particularly significant in expressing this ambivalence of desires: while uncertain between in their seeing "the comfort of the car" and "going by bike" as "enjoyable things in the school run," the children indicated the number and the invasion of cars as the main problems in "going to and from school". At the same time, however,

Table 1. Nice things about the school run (three answers were allowed)

Comfort of the car	108	20%
Riding the bike	110	20%
Calm streets	83	16%
Playing in front of the school after school	184	16%
Going through green areas	65	12%
Walking	38	7%
Taking the school bus	17	3%
Nothing	9	2%
Does not know	21	4%

Table 2. Problems around the school when entering and leaving the school (three answers were allowed)

Cars that invade pedestrian areas and cycle lanes	132	26%
Too many cars stopped or moving	130	25%
Lack of parking spaces	104	20%
Lack of playing areas	46	9%
Lack of bus stops near the school	41	8%
Lack of school bus	28	5%
Lack of green spaces with trees	33	6%

they complain of the lack of parking space, which three children out of four would like to have "just next to the school."

The equal percentage of those finding the comfort of the car and riding the bike (both 20%) as nice things about the school run is significant.

In total, 51% mention cars as a problem in one way or another; the lack of parking space is, however, also mentioned.

In this ambivalent relationship the children have with themselves, we can distinguish marked conditioning from grown-ups (is it really the children that see the lack of parking space as a problem?), but also – and here the analysis goes beyond the statistics – a negative tendency to absorb the problems of the parents and a desire for a sedentary way of life, which doesn't seem to us to be a classical patrimony of childhood.

The parents

The relation between parents and children is closely connected with everything expressed above. When looking into the opinion of the children as to the "reasons that lead to coming to school escorted", there is a considerable difference between the perception that the children have about the parents' fear of possible "nasty encounters" or of traffic (in total 57% of the answers) and the fear that the children themselves have regarding the same factors (18%).

The parents' fear of traffic and possible nasty encounters ("stranger danger") emerge as the main reasons (30% and 27% respectively).

Table 3. Reasons that lead to escorting children to school (according to children)

The parents' fear of "stranger danger"	116	30%
The parents' fear of traffic	106	27%
The child's fear of "stranger danger"	43	11%
* Other reasons	38	10%
The child's fear of traffic	29	7%
Does not know	28	7%
Lack of suitable buses	10	3%
Lack of a school bus	11	3%
Not interested	7	2%

The Body Goes to the City project   317

These are fears that children recognise in their parents when it comes to their autonomy. It is interesting to notice that the same fears are reduced to one third if we ask what the children themselves feel (11% and 7% respectively), and that girls appear – surprisingly – to be less fearful than boys.

The percentage of escorted children is very high (88 and 90%). About 10%, however – a small but qualitatively important number – travel to school at least sometimes alone.

The considerable difference in the percentages of children going to and from school escorted by an adult between the ages ten and eleven (from 85% to 52%) makes one wonder about why this change cannot take place earlier.

Table 4. How does the child go to and from school?

The child goes to school		
Escorted	243	88%
Sometimes alone, sometimes escorted	23	8%
Alone	8	3%
No answer	1	1%
The child comes home from school		
Escorted	248	90%
Sometimes alone, sometimes escorted	16	6%
Alone	8	3%
No answer	3	1%

Table 5. Means of transport used for getting to school

Car	150	55%
Bike	75	27%
On foot	37	13%
Bus	2	1%
School bus	6	2%
Other	3	1%
No answer	2	1%

Car is by far the most common means of transport (55%), while bike (27%) and walking (13%) account for 40% collectively. Almost no one uses the buses, either school buses or public buses.

Note that the number of children coming to school by bike includes the children that are transported to school by bike, those who are escorted while driving their own bikes, and those who go to school by bike alone.

One risk emerges clearly: the perpetuation of a sequential logic, according to which "I parent do not give you autonomy; you do not try and do not do; consequently, you do not feel capable of doing, but express your difficulties in reading the city; I tend to give you even less autonomy, you absorb my fear and wish less autonomy".

Boys and girls

There are no considerable differences between boys and girls in the percentages of being escorted to and from school, and neither in their desire to go to school alone. Also the reasons why they think they are being escorted to school are the same. This causes one to reflect on the perceptions children and parents have regarding the importance of the child's sex when it comes to moving autonomously in the city. The sensation is that reaching sexual equality in this field is closely connected with the refusal of the right to "cope alone" in the city.

In the city

Nine children out of ten go to or from school escorted by an adult. The number is high, but it includes, however, differences that shed some light of hope: six out of ten would like to go to school alone, four out of ten go on foot or by bike (among these, one alone and three under adult supervision), while four out of ten, if they could choose, would go by bike. Alternatively, other elements induce anxiety: one child out of four does not wish to go to school alone, and more than a half go there by car.

Table 6. The children's desire to go to school alone

Always	89	32%
When the weather allows	80	29%
Never	70	25%
Does not know	31	11%
No answer	5	3%

In total, 61% of the children answer that they would like to go to school alone; among these 32% would like to do so always, and 29% only when the weather is nice. On the other hand, 25% do not want to go to school alone.

The information is revealing: six children out of ten would go to school alone, if it were made possible. One out of ten, however, do not even wish to do so, that is, they are not interested, or are indeed afraid of a moment of autonomy.

Compared to other studies carried out in cities, the attractiveness of the bicycle appears even more notable here; this is especially true for girls and when it comes to using

Table 7. Desired means for the school run

Bike	109	40%
The usual one	98	36%
Car	20	7%
School bus	17	6%
Walking	14	5%
Does not know	8	3%
Bus	6	2%
No answer	3	1%

a bike as a substitute for the usual means of transport. In fact, among those who normally go to school by car the "bike fever" is at 41%, while among those who normally walk it is 43%, and among the children who already go by bike 84% indicate it as their preferred means of transport.

It is worth noticing, also comparing to other research done in cities, that 40% say they would like to go to school by bike. Note that only 7% would like to travel by car, while only 5% would like to go on foot.

The attraction of bicycle becomes, however, even more considerable if we analyse the crosses between the answers that reveal, for example:
- the attraction the bike has to girls;
- the fact that 4 children out of ten among those who currently go to school by car would like to use the bike and only 3 wish to continue travelling by car.

Children are well aware of problems related with traffic and capable of recognising the risks and dangers, the enjoyable elements of the city and the actions that could better it.

In the parks

When it comes to green spaces and the furniture of public spaces, once again a fact that often tends to be forgotten emerges: children especially want spaces with natural elements, such as water, trees and plants. In any case, useful spaces, such as sports fields and cycle lanes, are more requested than swings and slides and games. The most frequent complaint is about the untidiness of these spaces.

Despite their getting little attention, the indications coming from children are very concrete, as is often the case. Their cry has been losing its strength for some years:

Table 8. Things that children would like to have in green areas (out of 1082; five choices were allowed)

Trees	139	13%
Fountain	111	10%
Sports ground	108	10%
Sand, water games	83	8%
Benches	80	7%
Cycle lane	76	7%
Small tunnels and caves	72	7%
Slide, play structures	70	6%
Nets where to jump		6%
Bushes, lawn	59	5%
Parties	54	5%
Covered playing spaces	37	4%
Bar, ice cream shop	39	4%
Walls that you can paint on	31	3%
Toilet	28	3%
Lighting	24	2%

I want spaces and not furnished adventure playgrounds, I like swings and slides and all the structures but you adults think that with them you have resolved everything. I want a place that becomes meaningful only when I'm there also, that I share with others, be it more or less peaceful, that I can feel is mine, and where I can invent things, games, fantasies and stories. Up a tree or on a hill I can become Tarzan, hide, smell the trees and the grass, but in a jungle of plastic and wood I can only do what has been planned, in a pirate ship I can only become a pirate, my behaviour has already been planned.

From the statistical point of view, it is useful to note once again that the percentages are drawn from the total number of answers. Another reading, however, calculated on the number of respondents, tells us that 139 children out of 275 (51%) mentioned the wish for trees. The dispersion of the wishes regarding the furnishing elements is therefore understandable and depends on the possibility to choose five different elements. The average number of choices per respondent was 3.9.

Even if the answers have been distributed rather evenly between the different options, it is important to stress once again how the natural elements always appear as the very first choices of children: they prefer trees, plants, water in different forms or a small sports ground rather than ready-made playing structures.

Children and adults

An initial indication of the measures to be taken emerges from the study: it will not be sufficient to create safe school routes to notably increase the percentage of children going to school by themselves. As has been testified by other studies, too, the impression is that any town-planning solutions that might be adopted, would probably be insufficient to effectively change the behaviour.

The conclusions drawn from the interviews with the selected representatives further increasingly directs our attention towards a community involvement: on the one hand, in fact, there is a lot of pessimism about the possibility of the behaviour regarding escorted school runs being changed in the future, in spite of any town-planning interventions; on the other, the children's loss of autonomy doesn't awaken anxiety. Few people see the school run as an educational moment in itself, as a moment of exchange between equals, a moment of chat, of discoveries, of rituals, of making friends and learning social behaviour.

The aim of the "Percorsi sicuri" (Safe School Routes) project, of which this research is a first phase, is to change the behaviour of accompanying children and to increase the number of children going to school alone or that, at least, do parts of the school run on their own.

The limit

The conclusive idea that emerges from the results is that of there being a limit.

The changes in the liveability of the city have lead to a change in the mentality of the adults that modifies not only the behaviours, but also the desires of the children. The city that a constantly growing percentage of children wish for is not a city "in the pocket" to be explored any longer; the city awakens fear, and the children wish to see it through the car window, or in any case always under adult supervision.

We have to take action immediately.

In the section of the city of Bologna in the newspaper *La Repubblica* a father writes:

> "They've painted blue lines in Piazza Trento e Trieste [the sign of a parking meter area: translator's note] today. It was only two modest flower beds, a few trees, a small unused asphalt field (...) and lots of rubbish (...) as in so many other places in the city. But it was still a small space – the only one in the neighbourhood – where the children could go for a ride on their bikes, play some ball, skate, socialise (...). My daughter learned to ride a bike in Piazza Trento e Trieste; my son will probably have to learn in the corridor of our home."

The town-planning study

From a critical survey of a part of the town to a proposal for safety and the revival of pedestrianism

The critical architectural survey

Starting from the above-mentioned point of view, the urban context representation effort, realised in collaboration with the Faculty of Architecture, was oriented towards the recovery of elements in the built-up environment that would help orientation and "navigability" by offering points of identity, of identification and of belonging to a context. In fact, by means of a systematic survey, a great quantity of information could be acquired: morphological, on the one hand, to describe the built context, and critical and qualitative on the other, to examine the levels of accessibility (correspondence with the services required by the Italian law, plurisensory perceptual evaluations, identification of situations of low environmental safety, efficiency of the structure of paths in facilitating the orientation and the navigability of the public connective texture, urban furniture for pedestrians, etc.). The analysis was made with appropriate survey charts. Particular attention was also paid to the representation, which aimed to integrate typical GIS methods with public administration description supports by 2D and 3D modelling and representations of urban environment.

Metaproject proposals

The results and proposals of the study, which have been presented in *Paesaggio Urbano* (Balzani & Conficconi 2002), are preliminary results of the research still underway. They are solutions that correspond mostly to the needs that emerged concerning the main path that crosses the quarter, and that represent a different, creative approach. At the moment all the information regarding the environment, obtained in the survey using a single questionnaire, is being registered and integrated. This information will be useful firstly as a technical mediator in the choices that have been made in the preliminary "meta-project" planning, and secondly as a general operative support capable of stimulating and integrating different levels of planning actions (mobility, rehabilitation of the urban furniture, paving, integrated pedestrianism etc.) into the systemic network of interrelations between the blocks and the streets.

Conclusion

Only by structuring workgroups composed of various professions and by connecting them with the local (municipal) urbanistic offices it is possible to transpose in an urbanistic project the ideas emerging from the participation of the citizens and in particular from the researches carried out with children. The integrated work, consisting in the sociological research on one hand and the town planning survey on the other, the results of which are then interlaced in the workshops, allows the needs and the rights to emerge so as to be interpreted and transformed in a project. The urbanistic project, which is then presented to Local Authorities, thus becomes a real conclusive synthesis of the 'research itinerary', synthesis that includes the results of the sociological research, the interpretation of these results and the elaboration carried out in the workshops.

Acknowledgments

An integral part of this presentation was taken from the master's thesis of Barbara Conficconi, discussed at the Faculty of Architecture of Ferrara in July, 2000, with Marcello Balzani and Antonio Borgogni as thesis supervisors. The research effort was made possible also by the manifold environmental information obtained from the results of the Integrated Course in Architectural Survey II and Techniques of Representation in Architecture (academic year 1999–2000 and 2000–2001, Ferrara), taught by architect Marcello Balzani and engineer Gabriele Tonelli in collaboration with architect Maria Pia Sala, Antonio Borgogni (UISP Ferrara), Barbara Conficconi and, for the Municipality of Ferrara, Anna Rosa Fava (Città Bambina).

References

Accessibilità e ambiente costruito [Accessibility and the built environment]
. (1997). Monographic issue of the review *Paesaggio Urbano*, 2.
Balzani, M. (2002). "Il rilievo del brutto" [Surveying ugliness]. In R. Maestro (Ed.), *Il Bello ed il Brutto, strategie per la difesa della città [The beautiful and the ugly, strategies for defending the city]*, pp. 109-136. Florence: Edizioni Polistampa.
Balzani, Marcello, & Barbara Conficconi (2002). Il quartiere giardino a Ferrara. Dal rilievo critico di una parte di città una proposta per la sicurezza e il recupero della pedonalità urbana [The Giardino neighbourhood of Ferrara. From the critical survey of a part of a city a proposal for the safety and the recovery of urban pedestrianism]. *Paesaggio urbano, 1*, XX-XXVI.
Balzani, Marcello. (1998). Arredo e degrado urbano [Urban furniture and decay]. In A. Ubertazzi (Ed.), *Il dettaglio urbano. Progettare la qualità degli spazi pubblici [The urban detail. To project the quality of public spaces]*, pp. 259-268. Rimini: Maggioli editore.
Baraldi, C., & Guido M. (2000). *Una città con i bambini [A city with children]*. Rome: Donzelli.
Borgogni, Antonio. (1997). Storie di città e di zuccotti rubati [Stories of cities and of stolen zuccottos]. *Infanzia, 8*, 10-20.

Borgogni, A. (2001). Il corpo va in città [The body goes to the city]. *Paesaggio Urbano,* 1, 26-29.

Borgogni, A. (2001). Tirati di qua e di là [Pulled here, pushed there]. *Urbania,* 1, 3.

Borgogni, A. (2002). Poter giocare. Dagli spazi di gioco in città a città da giocare [Licence to play. From urban playgrounds to cities to play]. *Paesaggio urbano,* 1, 12-16.

Camy, J., Adamkiewics E., & Chantelat, P. (1993). Sporting Uses of the City: Urban Anthropology Applied to the Sports Practices in the Agglomeration of Lyon. *International Review for the Sociology of Sport,* 2-3, 175-186.

Consonni, G. (1989). *L'internità dell'esterno [The internality of the external].* Milan: Clup.

Galimberti, U. (1987). *Il corpo [The body].* Milan: Feltrinelli.

Gennari, M. (1995). *Semantica della città e educazione [Semantics of the city and education].* Venice: Marsilio.

La Cecla, F. (1995). *Bambini per strada [Children on the street].* Milan: F.Angeli.

Lauria, A. (1994). *La pedonalità urbana. Percezione extra-visiva, orientamento, mobilità [Urban pedestrianism. Extravisual perception, orientation, mobility].* Rimini: Maggioli.

Maciocco, G., & Tagliagambe, S. (1997). *La città possibile [The possible city].* Bari: Dedalo.

Merleau Ponty, M. (1945). *Phénoménolgie de la perception [Phenomenology of Perception].* Paris: Gallimard.

Moravia, Sergio. *L'enigma della mente [Enigma of the mind].* Bari: Laterza, 1988.

Environmental comfort and school buildings:
The case of Campinas, SP, Brazil

Doris C.C.K. Kowaltowski, Silvia A. Mikami G. Pina, Regina C. Ruschel, Lucila C. Labaki, Stelamaris R. Bertoli & Francisco Borges Filho

State University of Campinas, Brazil

> **Abstract.** This paper presents the results of an extensive post occupancy study[1] of 15 schools in the city of Campinas, SP, Brazil. The learning environments were analyzed as to environmental comfort conditions and possible simple solutions to improve the quality of the learning environment. Classrooms and recreation areas were observed and critical comfort conditions were measured with equipment. School directors, teachers, employees, and students were questioned as to their perception and evaluation of the comfort conditions, and given the opportunity to express their satisfaction and desires about their learning spaces.
>
> **Keywords:** school buildings, learning environment, post-occupancy evaluation, satisfaction and desires, intervention possibilities

Introduction

This paper presents the results of a research project on school environments, conducted in the city of Campinas, Brazil. Many international and local studies exist which have

---

[1] The study was supported by FAPESP, the Research Funding Agency of the State of São Paulo. (Projeto Fapesp Programa Ensino Público, Processo 97/02563-8, "Melhoria do Conforto Ambiental em Edificações Escolares Estaduais de Campinas")

evaluated the learning environment with different goals in mind. This study was undertaken to analyze comfort conditions of schools and develop ideas and simple solutions to common problems.

Many studies exist establishing a relationship between the physical comfort conditions in schools and the learning capacity of students (Gifford, 1997; Yannas, 1995). Comfort conditions can affect the attention span of students, speech comprehension, and the legibility of information presented. Spatial configurations and dimensions can inhibit the development of activities deemed positive for learning. All aspects of comfort therefore play a role in the learning environment and should be optimized as to the spatial quality they provide.

In countries where codes on environmental comfort and building regulations are rigorous, schools generally reach an adequate standard of comfort conditions. In these countries, research on the relation of environmental conditions and learning effectiveness concentrates on wider issues of environmental psychology. Specific details which contribute to the creation of a better total learning environment are tested, including the physical space, equipment, audio-visual material, and human relations. Studies proceed in evaluating architectural configurations and the spatial conditions of classrooms. School size and grouping are analyzed. Furniture layout is related to attention span of students. Special in-depth studies are conducted on the effect of natural light or its absence. Urban noise interference with behavior in the classroom is analyzed (Gifford, 1997).

Other studies concentrate on defining, with more precision, important comfort parameters, such as internal classroom temperature, that improve learning performance. Research on energy efficiency is also seen as more and more important in relation to environmental education and the sustainable building questions arising (Gifford, 1997; Yannas, 1995).

Some results of the relation between environmental comfort conditions and school performance, or increased learning by students, show architectural elements and detailing to be important. Evidence suggests that noise interferes with learning while it occurs, even after the noise is gone in the case of long-term exposure. Girls are more susceptible to noise than boys. When the task is more difficult, noise interferes more with performance. Interference with information processing, lowering of the perception of control, and increases in blood pressure can be registered in noisy classrooms. Thus, proper acoustic conditions are shown to be of extreme importance for adequate communication and to increase levels of learning by students. Behavior-modification techniques, as well, have shown, in many cases, to be important for control reinforcers in classrooms (Gifford, 1997).

The effect of illumination on student behavior and performance is less evident than in the case of acoustics. Since most research is conducted in adequately lit rooms, the influence of the type of lighting is not strong. But, as with noise, lighting may have important effects on specific subgroups of individuals. Light conditions show, in overall results, to affect basic cognitive and motor activities. Thus a mixture of natural and artificial light, representing a large spectrum of light, should be the goal for classroom illumination conditions (Gifford, 1997).

Space density and furniture arrangements are extremely important for classroom performance and related behavior. Choices of activities are curtailed in overcrowded classrooms, thus reducing the students learning experience. Increased social density has been shown to lead to higher aggression and withdrawal. Classroom performance is, however, always strongly related to teaching styles and methods and age grouping. These factors and architectural features, such as the open classroom, may counteract a high-density factor (Gifford, 1997).

In sum, school architecture has a variety of influences on students. Having an attractive school is associated with better classroom performance. The school size issue must always be carefully considered. Students in large schools have more opportunities, but often learn as spectators and not as participants, which is more prevalent in small schools.

Research on the school environment in Brazil is mainly concentrated on minimum comfort and maintenance conditions. Thus, we find that in the state school system in the State of São Paulo, few schools adopt innovative teaching arrangements. Traditional classrooms of 35 to 40 students prevail, with small individual desks in rows. Most studies show that comfort conditions are below standards for all comfort aspects (thermal, acoustics, lighting, and functional details). Thus classrooms are shown to be overcrowded, hot in summer, and without uniform illumination levels. Glare problems are common. Communication suffers from high reverberation levels and outside noise interference (Kowaltowski et al., 2001; Ornstein & Borelli, 1995). Since schools are mostly built for a smaller total student population than the actual students present, many adaptations are made in school buildings, often without technical advice.

On the national level, a recent study showed that 13% of the population considers school buildings to have serious problems and that the lack of school books and equipment contributes to a low standard of education in the country (INEP/MEC, 2000). The same study pointed out that school building defects are related to: leaking roofs, badly finished walls and floors, lacking or broken doors and windows, and especially to the precarious state of bathroom installations. These conditions get worse in locations in the northeast of the country.

When analyzing the history of school architecture in Brazil and particularly in the State of São Paulo, one can recognize periods of monumental buildings intended for the education of the local elite. Today school buildings are regulated by a state foundation (FDE, 1990), and school designs must follow standard rules and regulations. Although this situation would imply a minimum construction standard, this paper presents data from a recent POE (post-occupation evaluation) and user-satisfaction study showing that many schools still fall short of offering comfortable conditions and thus can be said to interfere with the classroom performance of its users.

Methodology

The methodology used to collect data on the comfort conditions of school buildings in the Campinas (SP, Brazil) region, was based on POE methods of observation, application of questionnaires to users, and technical tests and measurements of comfort parameters. To evaluate the school environment in the state school system of the city of Campinas, 15 school buildings of the 150 state schools in the region were evaluated.

The sample definition was based on random selection of a structured total of school buildings in the region. The sample structure took into account age grouping of pupils in each of the 150 schools. Forty-three spaces (classrooms and recreational spaces) were observed, measured, and tested at three distinct times (8:00, 12:00, and 16:00). Questionnaires were given to 15 directors, 48 staff, 56 teachers, 1,414 students with reading and writing abilities, and 358 first grade students. Comfort satisfaction and desires for the school environment were assessed. For first grade students, the questionnaire used drawings representing the various satisfaction levels for each question. Open questions were answered through drawings as well by the children.

Environmental comfort conditions were assessed through technical measurements. Functional aspects of the classroom were observed. User behavior, type and arrangement of furniture and equipment, occupation density, and dimensions of rooms were registered. This data led to the technical evaluation of the visibility of the blackboard in each room, the ergonomic adequacy of furniture and equipment in relation to user age groups, and the general level of organization of classrooms in relation to observed activities.

Buildings were observed to assess the quality of construction techniques. Building materials and construction types were observed. Maintenance conditions were evaluated, especially in relation to bathrooms, and storage and cleaning facilities. Fire safety and security details were tested. The neighborhood was characterized and urban infrastructure was evaluated.

Comfort parameters in relation to illumination and visual communication in the classroom were observed and measured. Light levels (lux) were taken in various points in the room. Glare was recorded in relation to desk tops and blackboard surface area. Observations were made in relation to the existence and position of curtains and external sun protectors.

Thermal comfort parameters were recorded through temperature (dry, wet bulb, and radiant) and air speed measurements. Shading devices were once more observed and their position. Ventilation conditions were analyzed. The presence of mildew was registered.

Verbal communication was assessed through acoustic parameters. The conditions of open or closed windows or doors were recorded. Equipment noise, such as ceiling fans, was registered. Noise levels were measured. The origin of outside audible disturbances was analyzed. Thus school activities or neighborhood problems were distinguished.

Data was collected and structured into a data bank. With this system, cross-referencing is possible and simple solutions to improve comfort conditions can be devised through data analysis. A school improvement and maintenance system was developed to allow staff direct access to information on specific spaces in individual schools or general data on the most common problems and their solution options (Faccin et al., 2001).

Results

School comfort conditions

All of the 15 school buildings were found to have comfort problems. Most buildings showed that schools lack sufficient financial support, which is most apparent in the maintenance conditions of buildings. Many schools have insufficient numbers of classrooms, no library or special rooms for laboratories or audiovisual equipment. There is a serious lack of storage space, apparent in improvised deposit solutions under stairs, at the back of corridors, and even classrooms.

A comparison between the school buildings and their architectural programs shows that the programs were insufficiently dimensioned and lacked detailing. Programs did not take into account growing demands on school buildings and the changing concepts of teaching in general. Therefore, all schools were found to be based on the repetition of standard sized (7m x 7m) classrooms.

To accompany the dynamics of today's educational systems, school buildings need complex restructuring and in most cases more functional space. These modifications and additions should have technical support at the design and execution stages. The results of this study point out some major deficiencies. The study thus has the intention of orienting school principals in the task of improving the teaching environment in their individual schools through incremental progression.

When analyzing the comfort conditions in detail, thermal comfort is shown to be a major problem. Most classrooms are facing east and have serious problems with direct sun penetration. This causes over-heating and thermal discomfort during most of the year. A lack of adequate ventilation for summer climate conditions was also observed and measured in most schools. Since winters are mild in the region, the only serious problem was that of lack of wind control in the covered recreational areas, which are used for the lunch break by the children. Although the thermal conditions are not ideal, some simple solutions can be suggested. First and foremost, windows should have sun protection in the form of external shading devices. Cross ventilation must be made possible through open doors or the introduction of additional openings in all classrooms, after acoustic interference is assessed. User participation in improving comfort conditions must be stimulated. Windows and curtains were found to be closed in many classrooms in overheated situations. The participation of students was stimulated by the study itself, which, to the research team, was a promising sign of the possible introduction of improvements. Students became aware of the environment and with increased perception were ready to intervene.

Illumination levels in most classrooms were under the recommended light levels of local standards. These light levels were often due to dark surface colors, closed window curtains, and high lot walls in close proximity to windows. In some cases, large vegetation was observed to interfere with natural illumination from windows. Visual problems such as blackboard and writing surface glare were observed in the majority of classrooms. These problems are basically related to orientation of classroom windows and improper shading conditions. Simple solutions include installation of external shading devices, painting of interior surfaces with light reflecting colors and improvements of the artificial lighting installations. Energy efficient fixtures should be evenly distributed over ceilings to ensure uniform light levels. In relation to the interaction of people with their physical surroundings, again, observation data shows the importance of environmental awareness. Participation of users, acting upon uncomfortable conditions, is of importance to attain a minimum level of comfort. Regulating light levels and eliminating glare through manipulation of curtains is a simple improvement measure.

All classrooms had serious acoustic problems, with high noise reverberation and speech recognition problems. In some cases, urban traffic noise was a problem in individual rooms. Interference of varying school activities, as well, was frequently observed. Finding solutions for acoustic problems is no simple matter. Since all buildings depend on natural ventilation for summer thermal comfort, windows and doors cannot be closed. In most cases, building renovation is necessary. Installation of acoustic material on ceilings to reduce noise reverberation can be considered a minimum improvement measure. The study results show that the main problems stem from noisy activities of users, such as rehearsals of dances and sports activities. Classroom discipline problems cause shouting and furniture moving disturbances. Behavior changes are thus seen as an important means of creating improved acoustic conditions for educational activities. The school administration, as well, can act on changes in activity schedules, preventing noisy activities from taking place near classrooms with quiet work tasks.

All schools had overcrowding and in some cases lack of appropriate furniture for age grouping. Equipment distribution was poor. Bathroom facilities in all schools were also found to be under-dimensioned. Overcrowding limits the variety of activities able to take place in classrooms and restricts movement of users. These conditions hinder the full working life of a school, limiting important educational experiences for children. In relation to these functional problems of buildings, simple solutions are more difficult to be established. Problems are mainly related to a lack of space. Additions are necessary. Where schools tried to improve conditions, the additions or adaptations were piecemeal solutions, often causing new interference problems. The case of the school (EEPG Armelinda da Silva) in Figure 1 can be used as an example. The school expanded through additions of wings as shown in Section A of Figure 1. However, the additions always interfered with existing constructions. The roof was extended to create new access corridors and classrooms impairing lighting and ventilation conditions. Simple changes as shown in Section B of Figure 1 can minimize the effect of the additions. The inversion of one of the classroom wings and the removal of the central corridor are sufficient to give natural light and ventilation to all classrooms. Classrooms in this school face east and west and appropriate shading must be provided.

Figure 1. A: Plan of the Armelinda da Silva School in its present conditions and B: Proposal of corridor inversion to improve environmental comfort conditions

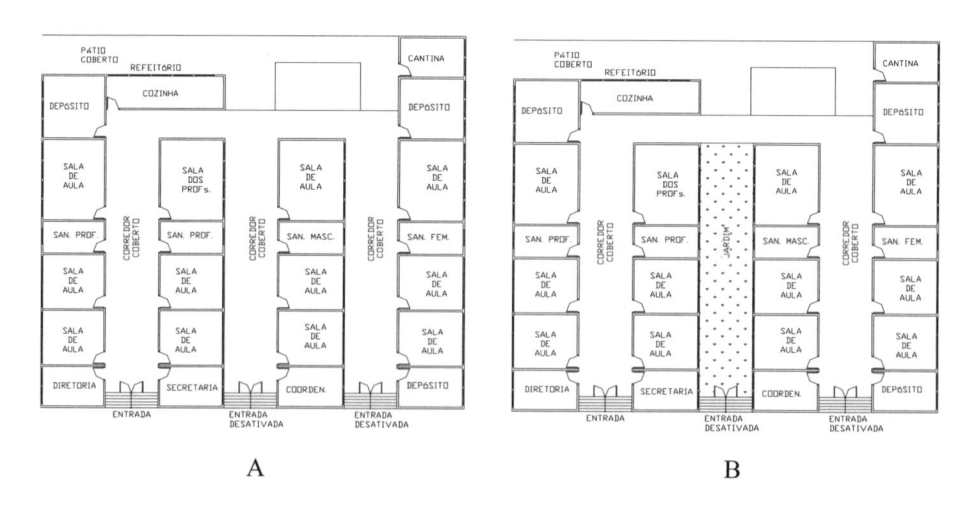

Satisfaction and desires

Despite the less than ideal comfort conditions, user satisfaction was found to be high. Problem perception is low and desires are often unrelated to the learning environment. Most students show major interests only in recreational facilities and activities not related to the school environment. When expressing their desires for improving the school environment, first grade children mainly expressed interest for items related to friendship and social and recreational activities as shown in Figure 2, Sections A–C. Some children demonstrated strong influences of family values in desiring a chapel or church in the public school as seen in Figure 2, Section D. Few students are aware of comfort detailing of classrooms. Desires of more learning material and equipment, even showing a ceiling fan, were expressed in rare cases (Figure 2, Sections E and F).

Figure 2. School children's desires expressed in drawings

The open questions on the preferred place in the school and on desires showed that the group spirit is the major force in expression. Thus with younger children the classroom is a major force. If one of the children expressed interest in the playing field, the majority of students in that group followed suit. Momentary desires are also strong in younger children and often context unrelated. Thus, many first-grade students showed in the drawings desires for items that are in fashion, unrelated to the school environment.

An overall analysis of the open questions on satisfaction and desires pointed out that most perceived problems are not related to the physical environment. Administrative and discipline issues, as well as the general spirit present in the user group, are reflected in the answers. Thus, personal relations and social activities play a major role in user evaluations of the school environment. When looking for specific physical aspects in satisfaction or problem expression, users are more aware of cleanliness or maintenance problems than serious environmental comfort issues. A tentative conclusion can be made. The actual non-ideal conditions diminish expectations and reduce the desire for a richer and more comfortable classroom setting.

School staff and teachers, in contrast to students, are more aware of comfort problems. The causes of uncomfortable conditions are, however, not always identified. Thus, these users are unaware of the overheating of classrooms by direct solar radiation. Curtains are drawn for reasons of glare and seldom to improve thermal conditions. Teachers are aware of the fact that they raise their voices for classroom communication. Resulting stress is attributed to discipline problems, not high noise reverberation conditions. The direction of schools has good intentions to improve the physical conditions of schools through spatial adaptations and piecemeal renovations. The unsatisfactory results are not linked to such interventions without planning, design, and technical support.

Conclusion

This paper presents, in part, the results of a comprehensive study of comfort conditions in 15 State Schools in the city of Campinas, Brazil. Users were involved in the study to express their satisfaction, perception, and desires in relation to the school environment. The study contributed with important data collection on local learning conditions and needs for improvement. The quantity of data collected was a stimulus to create an information system for design data for local school buildings. Thus the research data was organized specifically for local schools to guide their school building improvement plans. The study also stimulated the development of teaching material to increase user environmental awareness. The presence of the research team in the schools showed an increase in environmental perception by all users. The importance of user interaction with the environment was evident in comfort observations. Users can improve their conditions through simple interventions, such as increasing ventilation conditions by creating cross-ventilation.

The conclusion that can be brought forward is that a well functioning space, be it for work or learning, depends not only on the quality of its architecture, which supports the activity, but on the interaction of other factors. Thus the adequate disposition of equipment and furniture is important. Awareness and cooperation of users to act on adjustments of specific situations are essential elements in responding to the dynamics of actual space use. Participation is thus essential to create correct settings for momentary and specific activities.

To stimulate participation and increase awareness in schools, the teaching of environmental comfort concepts is recommended. More research is considered necessary to analyze user participation in depth. Preliminary findings indicate that every user group has a so-called leader (Bernardi, 2001). These individuals seem to have a sharper perception of environmental conditions and initiate participation, other acts of intervention follow, stimulated by example and peer pressure, present especially in the school environment. Investigations must detect the influence architectural elements can have on these leaders.

The question arises, if specific configuration and details can increase perception. Studies on the relationship of people and specific learning environments are thus being developed as they apply to local school building conditions.

References

Bernardi, N. (2001). *Avaliação da Interferência comportamental do Usuário para a Melhoria do conforto em ambientes Escolares: Estudo de caso em Campinas, SP [Evaluation of behavior interferences of users to improve environmental comfort in schools]*. Masters Thesis, School of Civil Engineering, State University of Campinas – UNICAMP. Campinas, Brazil.

INEP/MEC. (2000). Available online at *http://www.inep.gov.br/notícias/news_387.html*

Kowaltowski, D. C. C. K., Borges Filho, F., Labaki, L. C., Ruschel, R. C., Pina, S. A. M. G., & Bertoli, S. R. (2001). *Melhoria do Conforto Ambiental em Edificações Escolares Estaduais de Campinas- SP [Environmental Comfort Improvements in Public School Buildings in Campinas,SP.]*. Research Project Report, School of Civil Engineering, State University of Campinas – UNICAMP. Campinas, Brazil.

Gifford, R. (1997). *Environmental Psychology: Principles and Practice* (2nd ed.). Boston: Allyn & Bacon.

Yannas, S. (1995). *Educational buildings in Europe*. In proceedings of the III ENCONTRO NACIONAL: I ENCONTRO LATINO-AMERICANO DE CONFORTO NO AMBIENTE CONSTRUÍDO [Evaluation of behavior interferences of users to improve environmental comfort in schools], p. 49-69. Gramado, Brazil

FDE. (1990). *Especificações da Edificação Escolar de Primeiro Grau [Primary School building specifications]*. Fundação para o Desenvolvimento Escolar [Foundation for School Development]: São Paulo, FDE.

Ornstein, S. W., & Borelli, J. N. (1995). *O desempenho dos edifícios da rede estadual de ensino: O caso da Grande São Paulo-Avaliação técnica: primeiros resultados [Building performance of the state school system: the case of the Greater São Paulo, technical evaluation: first results]*. São Paulo: Laboratório de Programação Gráfica da Faculdade de Arquitetura e Urbanismo da Universidade de São Paulo [São Paulo: Graphic Programming Laboratory of the School of Architecture and Urbanism of the University of São Paulo].

Faccin, R., Kowaltowski, D. C. C. K., & Ruschel, R. C. (2001). *A Computer Tool for Environmental Comfort Management in Elementary Schools, Based on a Brazilian case Study*. Proceedings of the 18th International Conference on Passive and Low Energy Architecture, PLEA: Renewable Energy for a Sustainable Development on the Built Environment, vol. 2, pp. 997-1001, 7-9 November, 2001. Florianópolis, SC.

VI. The elderly and the environment

Satisfaction ratings and running costs of nursing homes
Advantages of smaller and more homelike units

Karin Høyland
SINTEF Civil and Environmental Engineering, Norway

Abstract. The new type of nursing home is divided into small living units (6–10 residents), each with their own kitchen and living room. The activities that earlier took place around the central kitchen and laundry room are now centred on the residents. Staff members are close by at all times, even when they are working with other tasks, and this increases safety for the residents. Also, relative's experience that it is easier to be familiar with the staff, and this increases nursing home visits. We have evaluated 3 new nursing homes; 2 in Bergen and 1 in Trondheim. These nursing homes were better in comparison to the old nursing homes in many ways. At the same time, the residents emphasised the importance of single rooms and the possibility of an increased private life. The residents feel safe being close to the staff and they can watch the activities and everyday life in the kitchen. The study also shows that having an easily accessible out-door recreation area increases the amount of time spent outside by the residents.

An important question to consider is if this modelling is less rational and more expensive to run than the traditional model. In comparison with figures from 14 other nursing homes, the new model is cheaper than the average to operate.

Despite the fact that the majority of reports from the involving parts were positive, there is room for improvement. One has to emphasise that a lot of groundwork has to be done in the units, and it has to be practical and easy to do this. Laundry rooms are too small and impractical. The storage spaces for aids and equipment are too far away and too small. Some places are too narrow for wheelchair users. Some of the

door thresholds are difficult to pass. Despite spacious bathrooms it was often crowded when a wheelchair user and assistant were present.
During the night, compared to daytime when rooms were side by side, it was more difficult for the small number of staff members to observe all the residents. Therefore, the challenge is to invest in technology that can assist in observation at nighttime.
There is little doubt that single rooms are a quality factor that residents relish. As one resident said: " I am very delighted to have a single room. I like to be alone, but also have the possibility to have visitors." The staff stated that smaller units and single rooms give more individual care, a fact that residents agreed with. "I think it's a grand place to live and the important thing is that everyone is allowed to be themselves."
The survey includes questionnaires answered by staff, residents, and relatives, in addition to interviews with the representative staffs and residents. Also operating economy and expenses for the three buildings have been analysed.

Key words: housing for the elderly, nursing home, long-term care

Introduction

During the last five years, a great deal of money has been used to rehabilitate and build new nursing homes in Norway. Between 1994 and May 2001, 19,700 lifetime care homes and nearly 11,300 units in nursing homes have been built. This large number of buildings gives an impression of many different solutions. It is therefore important to focus on how the different models function and what they cost. Investment costs are only a small fraction of the Municipalities' total expenses. The great expenses over time are related to care and nursing of the residents, not to the buildings themselves. This is why it is so important to consider all different aspects as a whole, and evaluate both user satisfaction and economy from a general point of view.

This paper presents a research project, which aims to contribute increased knowledge about how buildings influence service, use, expenses, and how the residents experience the housing and care.

Background, problem statement, and objectives

The aim of this project was to evaluate a new type of nursing home. How well did these new buildings work? Did residents find the new homes different? Was it cheaper or more expensive to run them this way? In summary, the new type of nursing home involves building a number of small living units that are connected to each other by a "central spine" where administration, daytime department, and other common functions are placed. The building and operation are organised in a way that makes all activities associated to the corresponding living unit. All deliveries of consumer goods, food, and textiles are made directly to the living unit, eliminating the need for a temporary storage room. Residents' laundry and cleaning take place inside the living unit, and are carried out by the available staff in the living unit.

Methods

Three new nursing homes were chosen. The intention of the project was to attain a general impression of both the operation and the buildings. It was important to get an idea of how satisfied the staff, residents, and their relatives were with the buildings. Even though the goal was to get a general impression, the main point of attention would be on the physical conditions; if, and possibly how, buildings and interior affect operation expenses and quality.

A case study method was used, in which data were collected in quantitative and qualitative forms.

Evaluation of the buildings

A guided survey was carried out in the three nursing homes. We evaluated the buildings based on a checklist (prepared in advance) explaining how nursing homes should be built (based on existing competence). Calculations and comparisons of areas were made. Finally, questionnaires were used to evaluate staff and relatives' impressions of the buildings.

Description of operation of the service

Interviews were held with the director and 3 staff members at each nursing home. This gave an impression of the ideology, objectives, division of labour, and use of resources in the building. The method used was evaluation of observations made by the participants. A nurse participated in care and daily life 2 days at each nursing home.

Step-counting registration, a small counter (pedometer), registered the number of steps and the length of steps made by staff members during the course of a work shift. This gave an estimate of how far staff members had to walk during a workday (night). A total of 136 step-counts were made.

User satisfaction: staff members

Four staff members at each nursing home were interviewed. Inquiries using questionnaires were further carried out at all nursing homes.

User satisfaction: relatives

Questionnaires were sent to the person listed as the resident's next of kin. This could be a child, spouse, or other relative.

User satisfaction: residents

Information about the residents' contentedness was obtained in several ways. Staff and relatives answered questionnaires, and 8 clear-minded residents at each nursing home were interviewed.

Operation expenses

Account analyses from October, November, and December 1999 were carried out at all three nursing homes. In addition, an account analysis for the whole year at one of the nursing homes was carried out.

Building expenses

Information about building expenses was obtained.

Figure 1: Ground floor plan (a) and exterior view (b) Skjoldtunet of nursing home

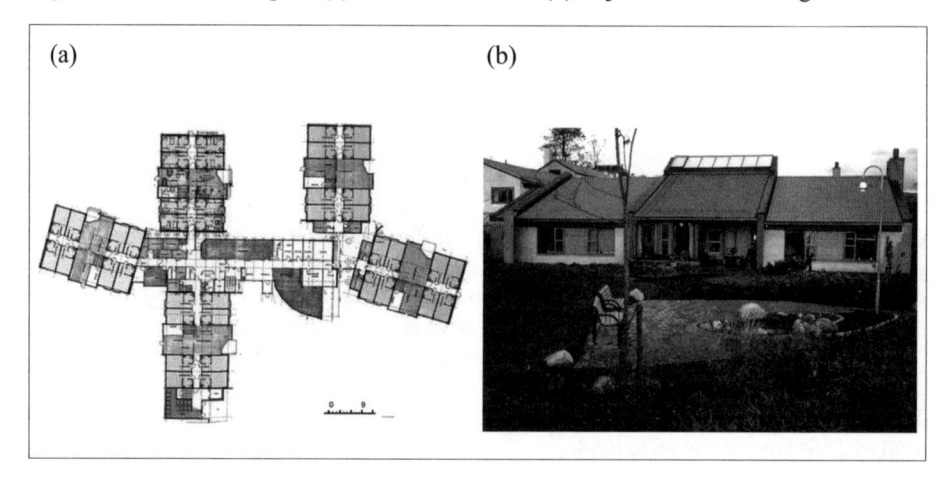

Experience with buildings, use, and operation expenses

Organization of the building and operation expenses

One of the objectives of the project was to evaluate whether this new model, in which the buildings are divided into smaller units, is more expensive to operate or not. During the planning time the one representing the more traditional organisation was changed. Our comparison of expenses is therefore based on three nursing homes with an approximately equal organizational model. In addition, we have used numbers from a report published by Ressurssenteret in Stjørdal (Nygård, 1999), which compares operation expenses for 14 nursing homes.

The survey shows that the three nursing homes are almost on average when considering operation expenses. If, however, one takes into consideration that Nygård's survey does not include cleaning expenses, operation expenses for our three nursing homes are less than the average cost for the 14 nursing homes analysed by Lars Nygård.

Figure 2. Distribution of expenses per user per month. (10, 000 Nkr = 1 316 EUR. NB: Cleaning expenses are not included in Lars Nygård's average of 14 more traditional nursing homes)

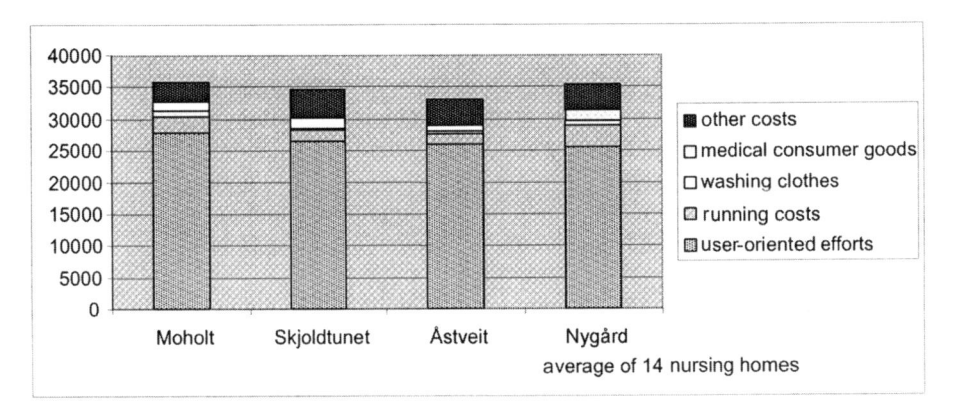

The numbers also show that a significant proportion of resources are used for user-oriented efforts at our three nursing homes, but this is a reality with certain modifications. The staffs (which are referred to, as "user oriented effort") are given a number of new tasks to do while helping the residents.

We see that the proportion of administrative services also vary somewhat. The smallest nursing home is lowest with only 1 man-labour year per 24 residents.

Experience with dividing the nursing home into smaller units

There is generally a very positive response to dividing the nursing home into smaller units. Since all three nursing homes are operated according to this model, we had to evaluate experiences from staff, relatives, and residents from other nursing homes. Seventy percent of the residents had been in another nursing home previously, and more than 90% of the staff members have previously worked in other, more traditional nursing homes. Relatives say that it has become more relaxed and pleasant, and it is easier to get to know the staff members.

Staff members' view of the division of the nursing home into units

Staff members, like relatives, agree that the division into units is advantageous for the residents. Approximately 91% think it is a good solution for the residents that the nursing home is divided into smaller units. Staff members say it has become more relaxed, it is easier to take care of individuals, and they emphasise meals times as positive situations. One could, however, imagine that this solution has negative implications for the staff's needs. They have to work more individually, there is more household work for nursing staff, and possibly less professional contact with staff members that work in other units. However, 93% of the staff members say that the division into units is very convenient for their work situation. It is obvious, though, that staff member have other requirements and needs during the night shift. The desire for control and overview is not fulfilled with the division into units. Especially in one of the units, which can only be accessed only from outside, this is a problem for night shift workers. A video camera was installed in the corridor to compensate for this, but the camera does not record sounds, and so this only partly solves the problem.

"We are always close by, even when we are doing something else. This makes people feel safe. I have less of a bad conscience when I leave work now than I had when I worked at the nursing home where I worked before. Even though I think we were just as many staff members there." (nurse/staff member, Skjoldtunet)

Organisation of services

All three nursing homes are operated according to a relatively similar organisation model. Most of the tasks are distributed among the units. In two of the nursing homes, the division of labour is aimed at making everybody take part in everything. This includes daily nursing and taking care of the residents in the unit, as well as taking care of food, clothes, and the house. There are certain leaders for each group, and they are responsible for shopping, the roster, budget, and organisation of activities. There are specific guidelines regarding medical responsibility. In one of the nursing homes, 1 out of 4 positions during the day is filled by an assistant that is responsible for cleaning.

Attitudes of the staff members

Generally, staff members respond positively to the model of organisation. Some of them seem dissatisfied with cleaning, while most others are satisfied with the organisation. In one of the nursing homes, it must be emphasised that staff members who were hired knew about this model when they started. They would most probably have chosen to work somewhere else if they did not like the new model. At the other two nursing homes, an old nursing home was closed down, and the staff members were offered a job at the new one. Nurses at one of the homes have a hard time finding their new role and are generally dissatisfied. They were against the new model prior to the reorganization, and this probably has an effect on their attitudes.

Experiences of relatives and residents

Relatives clearly state that quality factors for the residents are greatly concerned with safety and close contact with staff members, as well as the way in which staff members treat the residents. It may seem as though this way of organizing the home stimulates more contact between staff members and residents. Relatives say that it is easier for them to get to know the staff. This makes visiting the home more pleasant. Some also feel that there are more staff members at work than there was before.

No relatives comment on the staff's competence or lack of competence. We have chosen not to draw conclusions from this, as we did not have direct questions to relatives or residents about the competence of the staff.

The relationship between building and organisation model

The Government gives guidelines for dividing nursing homes into smaller units. As we can see from the experiences above, this has almost exclusively led to positive responses. It is, however, important to emphasise that some of the success is due to a combination of dividing the home into units and the organisation model. Without an altered organisation model, where also many jobs, and thus more personnel resources, are placed in the units, the staff would probably have to work more individually in the units. This is seen as a problem. There is a great difference in being alone or being two people sharing the responsibility. The fact that a person has other tasks in addition is not of much importance when a crisis or special situations take place. The most important thing is that there are more people to help out in such situations.

Staff members were asked about the extent to which organisation of the buildings contributed to the feeling of safety associated with knowing that they could get assistance from other colleagues if needed. Approximately 80% of the staff at two of the nursing homes answered that this was the case. In the nursing home with most distinctly divided units, only 55% answered the same. This is probably due to the fact that staff members spend time alone in the units during the evening (1.5 staff members per unit), and that some of the units are completely separated from the rest.

Safety of the residents is mainly dependent on contact with the staff. As deduced from the survey involving staff members and relatives, most residents are not able to understand the technical resources very well and thus are not rewarded with the safety these appliances can provide. A living unit without additional functions for laundry and cooking will have fewer personnel. The resident may again perceive this as less safe. With units of small sizes as those discussed here, during the nighttime the question is whether to have one person or two (possibly 1.5, with 1 person helping out in two units):

"It is not important for mother to be able to use an alarm system – from the living room there is always an open door leading into the kitchen/security room, so the staff can keep an eye" (relative, Skjoldtunet)

The living unit

The significance of single rooms with a bathroom

Both residents and relatives clearly state that having a single room is an important quality factor. Being able to have a private area seems to be important; it helps maintain a feeling of dignity and the residents are able to make decisions themselves. Residents express this clearly, but it can also be concluded from the other surveys. Many people seem satisfied with being able to spend time alone in their own room.

A number of people think of their room as a private domain where they can decide things for themselves. They also feel that their nursing home room is their "real" home.

Many people describe the ability to vary between being more private in their room and meeting others in the common living room as a quality factor, especially when they have visitors. All residents seem to like the idea of being able to have privacy in their room with their visitors. The size of private living units in our three cases varies from 24 to 26 m^2 (the bathroom included). A number of staff members think the bathroom is too small when they want to bring a wheelchair in. The number of m^2 alone is not of vital importance; the shape of the room and placement of installations are also important. Residents care most about the size of the room. If space is needed to maneuver a wheelchair, there is not really room for more than a bed, bedside table, armchair, and a small table. Many residents and relatives would have preferred a slightly larger room, into which they could bring more furniture and things.

About getting outside and design of outdoor areas

The questionnaires given to staff members show that, when comparing the three nursing homes, there is a great difference in how often the residents get to go outside. Most probably, this is directly related to accessibility and design of the outdoor areas.

From interviews with the director at one of the nursing homes, it seems that, during the planning phase, they were especially concerned with the importance of an outdoor area for the demented residents. This works well today. Later, however, one has realized that the importance of getting outside for the other residents has not been given enough attention. Today, people wish it were easier to access the garden from the other living units. There is no reason why the garden should only be used by the demented. If it is closed off so that the demented people do not wander off, it could still be well used by others.

Figure 3. Opportunities to go outside are important for residents of nursing homes

From the survey among relatives, however, it seems that accessibility and design of the outdoor areas do not really affect their bringing residents outside. Between 50% and 60% often or once in a while take their relatives for a walk when they visit. Relatives, however, have certain opinions about whether the outdoor areas are pleasant places to be. While 70% at one of the nursing homes say that the outdoor areas are pleasant, only 10% answer the same in another nursing home.

From interviews with the residents, many seem to want to spend more time outside. Ninety percent say they like being outside in the garden very much; with 83% saying they get in a good mood by being outside. The importance of going outside is most evident in the interviews with residents. While 25% of the residents in one nursing home say they spend time outside every day during summer, approximately 75% say the same in the two nursing homes with well-organised outdoor areas.

Organisation model, participation, and activity

The activities moved from corridors to living room/kitchen

One of the intentions with this organisation model is that it will lead to more participation and activity centred on the residents. Since all three nursing homes are relatively similarly organised, this can be evaluated based on references from earlier experiences at other nursing homes. As mentioned earlier, many of the staff members and relatives have experienced other nursing homes.

From a number of interviews, and also from the questionnaires, it seems that the activities to a certain extent have moved from corridors to the kitchen and living room, which have become the residents' meeting places. Staff members say that residents no longer sit in the corridors to see what is going on. The kitchen and living room are rooms where people like to spend time. This is confirmed by interviews with residents who say that they prefer sitting either in their rooms or in the living room/kitchen.

Meals are important and positive activities

It seems that simple activities, like making their own sandwiches, are not considered activities. This, however, is not necessarily the case in other operation concepts. In all three nursing homes, staff members think the meals are pleasant times that they organize well. They are able to stimulate residents to do things themselves, and they also have positive conversations with them.

Residents do not take part in activities in the kitchen to any significant degree. Besides helping themselves during meals, it does not seem like the residents are activated much in daily duties. Nursing home residents are very reduced, and staff members think most residents to a very small degree are able to participate in daily duties.

One resident, however, says that she helps out in the kitchen occasionally. She had, for example, explained how they should make fruit soup. Here, the resident is actually participating, while staff members probably would have called it activating. Therefore, facilitating activities, with today's manning situation and residents' level of function, mainly means that residents can sit and watch what is going on. One condition for this, however, is that there is enough space in the kitchen to sit in the wheelchair and watch.

More time and peace during meal times presumably stimulates more self-help and individual activity. The organisation model brings more activities into the kitchen. The living room/kitchen has therefore become the most attractive place to sit; earlier they would often sit in the corridor.

Is there too little activity, or do the residents prefer peace and quiet?

Relatives generally think there is too little activity in all three nursing homes. Approximately 60% of the relatives think days are too long for the residents. From interviews with residents, it also seems like they think there is too little activity in the nursing homes. They would like more entertainment, especially more singing and music. In one nursing home, they have organised bingo, but this was cancelled the day we were there for observations because there was too little interest for it. It may seem a dilemma that a lot of people want more events, but when it gets down to it, not many of them have the energy to participate anyway. This may reflect the fact that many residents wish to have a more active life even though they do not really have enough energy for it.

When residents are asked to mention events that they remember as pleasant, many of them mention changing of routines during the weekend as something positive. It may be minor things like dressing up a little bit or preparing something extra nice to eat. Another thing mentioned is that there are minor activities arranged by the staff themselves.

> Very pleasant during Christmas preparations; staff member's children came and made Christmas cookies. (resident, Skjoldtunet)

> Once, staff members and residents sang and played music together; that should happen more often. (resident, Åstveit)

> I think it is very nice that someone bakes bread once in a while. (resident, Åstveit)

One may also say that a number of the residents generally express satisfaction over having peace and time for themselves. This is emphasised more by the residents than both staff members and relatives.

Production and storage of food

Storage and preparation of breakfast and noon in the living units functions well, gives greater individual adaptation, and stimulates self-help (preparing sandwiches). Central production of dinner, and heating it in the living units, is a cost efficient food system, which quality-wise seems to function satisfactorily. This solution requires a large kitchen in the living unit, with a steamer and approx. 9m-kitchen inventory.

How many people in each group?

It is difficult to draw unambiguous conclusions from such a limited selection. The most important finding from this inquiry, however, is that all three group sizes are experienced as small and pleasant. If they are reduced down to groups of 6, one should, as at Moholt, facilitate co-operation of two groups during evenings. When expanding to groups of 10, it may seem as though there is more unrest, especially for the demented residents.

The physical organisation of the living group

The groups generally function as closed units (day and night) and staff members primarily want the secondary rooms placed inside the living unit. They also want the storage function and laundry room to be placed centrally in the living unit. The daytime step-counting measurements show that there is the least walking at Skjoldtunet (approx. 1 km during the day). At Moholt, even though there are smaller groups, there is more walking on the whole (2.8 km during the day). Experiences from these three nursing homes suggest that a small office should be established inside the living unit. Computer equipment and small meetings could be held here, close to the residents.

Since a lot of housework takes place in the living groups, there is a significant potential for improvement with respect to facilitating practical design of the side rooms. The study shows for example that the design of the laundry rooms and the lack of storage space for equipment make everyday life more bothersome than necessary.

Large or small nursing homes

One of the questions we wanted to elucidate here is to what extent large or small nursing homes are rational to operate, and to what extent these are quality factors that affect staff members, residents, and relatives. Regarding operation, it is only nightshifts that affect how efficiently it is to operate the nursing home with respect to size. This means that (as at Moholt) the director and administrative positions must be adjusted to the number of residents and, possibly, co-operation between several nursing homes. Both relatives and staff members prefer small nursing homes, even though these also have disadvantages. There is a lot indicating that the size is detrimental for what people perceive as homely and pleasant.

During programming and design it is important to search for and define functional solutions based on the different operation situations during daytime, evening, and night.

The relationship between operation expenses and investment expenses

Based on today's financing arrangements, loan expenses per year are approx. 11,357 Euro per unit. If we consider costs for operation and service, these amount to approx. 56,800 Euro per year, of which staff resources constitute the major part.

An important question to ask is whether it is possible to make technical improvements of the buildings, which will reduce operation expenses. As seen from the experiences, most factors perceived as quality by the residents are concerned with contact with the staff. Operation expenses are first and foremost concerned with staff resources. Technical resources may compensate contact with staff only to a very small degree. Therefore, we see no great potential in improving technical and physical solutions to achieve a goal that will reduce operation expenses for staff during the day. Better physical facilitation may, however, help spare time which staff members could use to organize activities, which both relatives and residents demand. We believe that technical resources may relieve the work situation during nights, and perhaps have an affect on staff resources.

Use of area

We have gone through plans for all three nursing homes and calculated user area (area inside surrounding walls), and divided this according to type of area. Based on this, we show different types of area in comparison.

If we compare the numbers with other nursing homes, we can see that these three have a small use of area per person. They represent a standard of area, which still appears relatively functional. It shows that this nursing home model, with a decentralized office, living room, etc., does not necessarily demand more space than more traditional solutions. Some of the rooms appear to be dimensioned too sparsely. In summary, we may say that this is area around dining room, bathroom, and cleaning rooms, plus space for storing equipment.

It also appears that some areas, such as common bathrooms, cafe/cafeteria, and ceremony room/chapel, are used relatively seldom. The survey shows that it is possible to find good functional solutions with approximately 65–70 m² per resident. Areas in the living units should be given priority.

Figure 4. How the available space is used is an important consideration

Conclusion

Judging from the project, there need not be any differences between these developmental models. "Concerns for low investment frames, concerns for efficient operation, and concerns for user friendliness." Organisation of smaller units clearly leads to greater user satisfaction, a fact that was especially emphasised by relatives. This leads to more peaceful surroundings, and it is easier for both relatives and residents to get to know the staff members. Being able to accommodate individual needs is an emphasised quality factor. Nursing homes operated in this way are not more expensive to run either. Compared with 14 other nursing homes, this form of organisation is somewhat cheaper than the average to run. Organisation of the buildings, however, requires new models to accommodate the services. Special functions, like the kitchen and cleaning of buildings and residents' clothes, are decentralized to staff members working in the units.

References

Dowling, J. (1995c). *Keeping busy: A handbook of activities for persons with dementia.* Johns Hopkins University Press.
Gottschalk G. (1995). *Boligstandarden i plejehjem og andre institutioner SBI-rapport*; 249 Trykt: Hørsholm: Statens Byggeforskningsinstitut.
Gottschalk, Bent Foerlev: *Boligforhold pleie og omsorg i ti kjøbehavnske pleiehjem.*
Hauge, S. (1998). Forskningsetiske spørsmål i samband med kvalitativ forskning i sjukeheim. Essey av Notat / Høgskolen i Vestfold; 4/98.
Hovdenes, G. H. (1998). *Et meningsfylt liv i sykehjemmet – vennlighet først og sist.* Hovedfagsoppgave i helsefag: Universitetet i Bergen.
Johan Ottoson Patrik Grahn. *Utemiljøns betydelse før eldre med stort vårdbehov.*
Johan Ottoson Patrik Grahn. Utemiljøns betydelse før eldre med stort vårdbehov.
Karr, K. (1991). *Promises to keep: The family's role in nursing home care.* Buffalo, New York: Prometheus Books.
Lindhardt, T. (1995). *Den opplevde erfaring med å anbringe en æktefelle på pleiehjem.* (Hovedoppgave sykepleiervidenskap.) Oslo: Universitetet i Oslo, Institutt for sykepleievitenskap.
Moland, L. (1999). *Suksess og nederlag i pleie og omsorgstjenestene.* FAFO rapport.
Nygård, L. (2000). Sammenligning av driftsutgifter ved 14 sykehjem. Ressursenteret for Omsorgstjenester.
Pedersen Brit, & Jensen Thor Øivind (1998). *Brukererfaring fra Bergenske sykehjem.*
Slagsvold, B. (1989). Knapphet som ramme for forandring: Omorganisering av sykehjemsposter. Oslo: Norsk gerontologisk institutt.
Slagsvold, B. (1995). Mål eller mening. Om å måle kvalitet i aldersinstitusjoner. Rapport: Norsk Gerontologisk Institutt.
Solbu, H. (1997). Kvalitet i Sykehjem. Kartlegging av sykehjemskvaliteten i Rapport III, November, 1997. Trondheim: Trondheim Kommune.
Sosial og helsedepartementets sammendragsserie. (1995). Nr. 11.

New Urbanism as a factor in the mobility of the elderly

Michael J. Greenwald
University of Wisconsin, Milwaukee, USA

Abstract. Proponents of "New Urbanism" have made several claims about altering individual behavior patterns by manipulating elements of the surrounding physical environment; neighborhoods with higher densities, narrower streets, greater mixes of retail and residential uses, and dedications of open space are all intended to simultaneously attack societal problems such as traffic congestion, urban sprawl, shortages of affordable housing, and class segregation (Calthorpe, 1993). Recent investigations into New Urbanist claims about urban form inspiring more walking relative to driving suggest there is at least some merit to the idea (Boarnet & Greenwald, 2000; Greenwald & Boarnet, 2001; Greenwald, 2003). Given this inducement to walking, we turn to the idea that new urbanist designs might be appropriate instruments in creating more manageable environments for the elderly. The first question this paper addresses is "To what degree is increased age associated with prevalence for living in new urbanist style communities?" From those findings, discussion is broadened to a second issue, the impact of urban form on travel decisions made by the elderly. The datasets used in this investigation include the 1994 Household Activity and Travel Diary Survey (hereafter referred to simply as the 1994 Travel Diaries), and the Regional Land Information System (RLIS) for the Portland, Oregon region.

Keywords: New Urbanism, residential location choice, elderly travel behavior

Introduction

Proponents of new urbanism have made several claims about altering individual behavior patterns by manipulating elements of the neighborhood. Specifically, the argument is physical environments with narrower streets aligned at right angles, combined with high density residential uses mixed with shopping and service activities, located within a quarter mile (approximately 402.3 meters) of a centralized public transit stop, act to simultaneously address multiple societal problems such as traffic congestion, urban sprawl, shortages of affordable housing, and class and age segregation by increasing the available housing stock and reducing dependence on the personal automobile (Calthorpe, 1993; Duany & Plater-Zyberk, 1991). In addressing these problems, new urbanists seek to enhance quality of life by increasing satisfaction of residents with their neighborhood and community, reduce expenditures for public services, provide more equitable distribution of resources, and provide multiple means for achieving one's daily tasks.

Because of the variety of social ills New Urbanism addresses, and because of the greater prevalence of private automobile use in the United States compared to the rest of the world, New Urbanism has received a great deal of interest from American scholars in testing it's effectiveness. Two key assumptions underlying the new urbanist argument are: 1) that the proposed benefits are linked as either a cause or effect of altering individual travel patterns. New Urbanist proponents argue individuals are forced into a sub-optimal behavior patterns based on the incongruity of current urban form with a preferred travel mode alternative. 2) That land use designs can have a measurable effects on individual travel behavior. The emerging literature testing new urbanist design influences on travel behaviors supports both assumptions.

This paper adds a new aspect to the travel behavior research by testing the potential of new urbanist designs to attract elderly individuals into environments more conducive to walking, and subsequently whether or not this induces a change in travel behavior in older residents. This requires asking three sequential questions. First, is there variation, based on age, in people choosing to live in environments consistent with new urbanist principles? Second, assuming there is difference in choice of residential environments based on age, what role does age play in selecting into new urbanist consistent environments? The answer to this second question could inform architects, planners and policy makers on how to use that relationship in addressing the housing needs of elderly residents at a neighborhood or city wide scale. Finally, how does new urbanist design impact travel mode choice behavior, when controlling for age? Answering this last question serves three purposes. First, it further validates the idea that urban design plays an important part in the travel decision-making process. Second, it helps familiarize urban designers and environmental psychologists with the research approaches adopted by travel behavior investigators when handling standard research issues (e.g., establishing causality; validity of measurement instrument, appropriateness of analytic method, etc.). Third, given the assumption that alterations of travel behavior are the manner in which the benefits of changes related to urban form are achieved, the approach applied here helps demonstrate how new urbanist design benefits accrue to older individuals.

Previous research

Levine (1998) helped clarify the connection between land use impacts on travel behavior by examining the impact of residential selection strategies on work commutes. Levine argues that many local land use policies in the United States result in urban designs which lead households to select sub-optimal residential locations with respect to their place of work (e.g., large residential lots, single use, segregation of employment and living areas). This forces individuals residing in those households into travel behaviors related to their daily tasks that are less efficient than they might otherwise choose. Levine's point speaks to a larger quality of life issue. Since travel involves the use of scarce resources (most notably travel time), any sub-optimal (i.e., unnecessary) travel represents a waste of that resource to the individual traveler.

Since Levine identifies land use practices contrary to new urbanist principles, the next logical step is to investigate whether or not elements of urban form consistent with new urbanist thinking help establish a more optimal set of travel behaviors. New Urbanists often argue this more optimal set involves reduced automobile travel concurrent with greater use of other travel modes, particularly walking (Calthorpe, 1993). The research thus far suggests that while urban design consistent with new urbanist ideals might induce a greater number of walking trips, urban design alone does not reduce the number of automobile trips made or the distance driven. The relationship appears to be that residential and employment densities, street grid connectivity and lot sizes all act to alter travel times for different modes of travel (e.g., walking and driving), which in turn changes the likelihood of adopting pedestrian or automobile travel modes for completing trips (Boarnet & Crane, 2001; Boarnet & Greenwald, 2000; Boarnet & Sarmiento, 1998; Greenwald & Boarnet, 2001). Using this research as a springboard, recent investigations have demonstrated how smaller lot sizes, increased street intersection density, greater mix of employment and housing land uses, and better access to transit all serve to support individual decisions to substitute walking for driving (Greenwald,2003).

This research still begs the further question as to whether or not this willingness to substitute only applies to additional trip making behavior that new urbanist (also known as neo-traditional) environments might induce, or if new urbanist urban form can actively encourage the abandonment of automobiles for increased walking behavior. The point that these "neo-traditional" design elements were never abandoned in Europe is not lost on those who compare travel mode usage in the United States and Europe. Pucher and Dijkstra (2002) have made specific note of the difference in the use of walking and cycling as modes of travel for Dutch, Germans and Americans, particularly among their respective elderly populations. They attribute the higher rates of pedestrian and cycling activities in the northern Europe in part to urban form elements which promote the use of non-automobile travel modes; narrower streets with dedicated traffic lanes for bicycles, traffic calming through raised intersections, artificial cul-de-sacs with cuts through for pedestrians and cyclists, and orientation of buildings frontages near the street, as opposed to a vast expanse of parking along the building front. Though Pucher and Dijkstra do not specifically address New Urbanism by name, the elements described are either directly associated or at least not inconsistent with new urbanist design concepts.

With regards to new urbanist designs holding special value for the elderly, Pollak discusses how new urbanist urban form could be used to improve the ability of the elderly to remain in familiar surroundings by addressing issues of housing costs. Starting from the premise that the elderly constitute a significant portion of the American population in need of affordable housing, Pollak identifies potential benefits for existing communities in considering the possibility of developing secondary residential units associated with single family homes, including additional affordable housing stock, an improved tax base, more efficient use and better care of existing stock, and greater social stability based on the interaction between age and sociodemographic groups (Pollak, 1989; Pollack, 1994). These ideas of increased affordability and greater communal engagement speak directly to enhancing the quality of life for the elderly.

Adding Pollak's arguments about elderly as a group in special need of affordable housing with Levine's ideas about sub-optimal residential location leading to sub-optimal travel behaviors, two points are raised about the impacts of New Urbanism on the elderly. First, how effective will new urbanist designs be in minimizing obstacles to neighborhood navigation faced by elderly residents? Second what motivates selection into new urbanist environments, and how do those factors change with age? It is to those questions we now turn.

Methodology

The data for this research comes from the 1994 Household Activity and Travel Behavior Survey conducted by the Portland Metropolitan Services District. The choice of this data set is guided by two principles. First, the survey was conducted with the explicit intent of incorporating new urbanists elements of urban design as explanatory factors in the travel decision-making process for individuals. Second, this data set contains enough observations at both the individual and household level (10,0048 persons in 4,451 households, respectively) to provide a sufficiently powerful test case so certain statistically valid conclusions might be drawn.

The first step in the process of identifying differential effects of urban form on travel behavior, based on age group, is to describe an age classification system. Table 1 shows how seven different age groups were applied to the dataset.

Table 1. Age group classification

Group	Age Range	Number of Observations
1	0–15	1,592
2	16–21	415
3	22–35	1,603
4	36–45	1,637
5	46–55	1,208
6	56–65	629
7	66 and Older	813

The choice of age classifications is guided by identifying significant points of change within a person's lifetime. Until the age of 16 most American youth do not possess a driver's license even if they do have access to a vehicle; hence, they have a limited set of mobility options available. The time frame between ages of 16 and 21 usually represents the first time an individual leaves home, and hence must personally face the trade offs involved in selecting residential locations. Between ages 22 and 35 the individual first enters the workforce, establishes a family, and consequently begins to attain their productive role in American society and becomes responsible for the welfare of others. The range between age 36 and age 64 represents the bulk of the individual's employment tenure, and in combination with family obligations, the individual acts to establish the primary activity patterns which define their travel behavior. Age 65 is the threshold for the final category because it represents the earliest opportunity most individuals have in the United States to retire from the active workforce. This is partly because age 65 is the earliest most Americans are eligible for benefits such as government pension (i.e., Social Security) assistance with medical costs and dependent living arrangements.[1]

After establishing the age categories, the next step was identifying elements of the individual's living arrangement, independent of age, which impacted travel decision-making. Given the new urbanist contention that urban form guides travel behavior, anything that guides selection into a particular urban environment needs to be accounted for as a possible factor in the travel behavior models. Examples of variables included under this idea include did the person have a handicap affecting travel or selection of residential environment? Does the person own or rent their home? Weisbrod et al. (1980) noted that different travel patterns exist for those who own their residence as opposed to renting. Someone not familiar with the language and/or culture might be less likely to engage it for discretionary activity, so there is a variable asking whether or not English is spoken in the home. Travel needs of children are also factors in determining travel arrangements, so the presence of children in the home is also considered. Income is included as a traditional sociodemographic variable; as income increases, the ability to engage in discretionary trip making is altered.[2]

One of the greatest difficulties in addressing the connection between urban form and travel behavior is the issue of incorporating elements of environmental design, which are fundamentally qualitative in nature, into modeling strategies, which are heavily reliant on quantitative data. As an example, the relationship between various densities (residential, commercial, employment, etc.) are often taken as a proxy of mix of land uses; although convenient for statistical modeling, the numbers alone say nothing about the aesthetics effects of a location. Addressing this issue of qualitative perception is important if the goal is to explain the role that urban design plays in altering observed behaviors. In this case, two measurements of new urbanist urban form included in the travel behavior models.

The first is a composite index known as the Pedestrian Environmental Factor (PEF). The PEF score for each transportation analysis zone (TAZ) ranges from four to twelve, a composite of points given for each of four criteria; ease of street crossing, sidewalk continuity, street connectivity (grid vs. cul-de-sac) and topography.[3] Scores for each crite-

---

[1] A grouping for individuals over age 76 was considered for this inquiry, to account for the potential of new urbanist environments to affect individuals with age related mobility impairments. This grouping exhibited fairly consistent travel behavior patterns (i.e., engaged in very little walking activity), so most of the travel behavior models failed to converge to a usable solution. Therefore, individuals aged 76 and older had to be included in the 65 and over age group.

[2] Although the income variable is used in the models for this research, the data loses a great deal of variation on this subject due to dummy coding. This additionally hampers the research by restricting the ability to test for non-linear effects of income on travel behavior. This problem has been partially mitigated through the use of linear regression and integral calculus to find an "average" value of income for each dummy coded category. The benefit of this procedure is that it converts a discrete variable into an approximation of a continuous one. The drawback is that it assigns the same value to all members of the same group, suppressing the distinctions which contribute to the income variable's statistical significance (e.g., for two households at either extreme of the income block $15,000 to $20,000 both were assigned a value of $17,500, regardless of the true values for the specific household). For a technical description of the process, see Sarmiento (1995).

[3] A transportation analysis zone (TAZ) is a geographic unit used in transportation modeling. It is used to forecast travel demand between sub-areas within a region. The transportation models assume all trips begin from the activity center of the zone. While the specific boundaries of a TAZ vary depending on the region for which the model is being run, in the United States TAZ's are based on census block groups as defined by the Federal Bureau of the Census.

rion can range from one to three points.[4] Because the PEF score criteria lend themselves to subjective interpretation, and due to their inherently constrained values, the PEF score is used as an organizing criterion for urban form. Scores of four through six classify the urban form of the neighborhood as not facilitating pedestrian activity; scores of seven through nine indicated the person lives in a neighborhood moderately conducive walking and/or transit use, compared to a low PEF neighborhood; scores of 10 or higher suggest the neighborhood is highly accommodating to non-automotive forms of travel, compared to low PEF neighborhoods.

Second, a continuous measure of intensity of mixed land uses, comparing jobs to housing available in the area, was incorporated into the travel decision-making model. Cervero (1989; 1996) has argued that areas which are not saturated with either jobs or housing facilitate travel by means other than private vehicle. Consistent with Cervero's explanation, the mixed use index is calculated by the Portland Metropolitan Services District as follows:

$$\text{Mixed Use Index} = \frac{((HH * (Emp * Factor))}{((HH + (Emp * Factor))}$$

where:
HH = Number of households within a half mile of the center of the specific TAZ
Emp = Total employment within a half mile of the center of the specific TAZ
Factor = Average of all HH values/Average of all Emp values (a constant used to weight TAZ's by the number of jobs they have relative to the regional average)

The result is a continuous index which is highest when the density of both housing and employment are high within a half mile of the center of a particular transportation analysis zone.[5]

Lastly, the degree to which interaction between intensity of mixed use affects the potential of PEF to alter particular travel behavior is included. Similar interactions are included between age and PEF category, to test for the possibility that improved pedestrian environment may have differential value to travelers as they get older.

Results and analysis

Table 2 addresses the first question about whether or not there is a significant underlying effect worth analyzing. Using the age categories established in Table 1, each individual was identified as living in an environment which had low, medium or high PEF scores. Using a two-way chi-square distribution, the analysis indicates there is a statistically significant relationship between the age of the resident and the PEF score of the home neighborhood. The results in Table 2 do not specify the nature of that relationship, although comparing the expected and observed cell counts gives some indication of the how the relationship might be tested.

---

[4] The PEF score was originally developed by a non-profit public interest group known as the 1,000 Friends of Oregon, for the purpose of indicating the degree to which specific transportation analysis zones supported travel via modes other than individual automobile usage For a complete description of the Pedestrian Environmental Factor score, see Cambridge Systems Inc. et al. *Making the Land Use Transportation Air Quality Connection, Vol. 4: Model Modifications*. 1000 Friends of Oregon, Portland, 1996.

[5] Mr. W. Stein, Portland Metropolitan Services District (personal communication, May 8, 2000).

Table 2. Age group counts × pedestrian environmental factor classification of home neighborhood crosstabulation

		PEF Category			Total
		Low (4-6)	Medium (7-9)	High (10-12)	
Age Group					
1 (0 -15)	Count	579	261	282	1122
	Expected Count	522.20	282.50	317.30	
2 (16-21)	Count	140	71	61	272
	Expected Count	126.60	68.50	76.90	
3 (22-35)	Count	396	249	363	1008
	Expected Count	469.10	253.80	285.10	
4 (36-45)	Count	499	276	338	1113
	Expected Count	518.00	280.30	314.80	
5 (46-55)	Count	400	208	204	812
	Expected Count	377.90	204.50	229.60	
6 (56-65)	Count	185	116	83	384
	Expected Count	178.80	96.70	108.60	
7 (66 & Older)	Count	187	110	119	416
	Expected Count	193.60	104,.80	117.70	
Total		2386	1291	1450	5127

Chi Squared Score, df = 12   66.660
Level of Significance $p > .001$

Counts for younger individuals (i.e., categories 1 and 2) living in low PEF score neighborhoods were substantially higher than expected, while counts for the same age brackets living in high PEF score neighborhoods were lower. However, in the middle age brackets, (categories 3 and 4), is a divergence. Individuals group into higher PEF neighborhoods at higher than expected rates and select low PEF neighborhoods at lower than anticipated rates. The underselection into low PEF neighborhoods appears to be totally offset by the overselection into higher PEF areas. In contrast, in categories 5 and 6 there is substantial under-selection into high PEF areas. In category 7, the oldest group, the observed selections between PEF categories appear more consistent with expected counts.

Tables 3 and 4 clarify the situation by examining the importance of age in the process of determining residential environment using a multinomial logit model. The multinomial logit is a statistical method employed to analyze discrete choice; rather than analyzing the choice making behavior directly, this model measures the impact of independent variables on the probability that a particular outcome will be chosen from the set, using one of the options as a frame of reference.[6] In the present case, the question involves what level of conformity to new urbanist principles will an individual choose given their age, command of the English language in the family environment, physical ability to navigate the environ-

---

[6] For a full technical description of the multinomial logit modeling processes used in this paper, see Johnston & DiNardo (1997), and the Stata Reference Manual, version 6.0 (1999).

Table 3. Model of predictors for selecting into new urbanist consistent environments (using low PEF environments as reference group)

	Moderate PEF		High PEF	
Variable	Coefficient	Z	Coefficient	Z
Age	0.0002	0.09	-0.0028	-0.95
Disabled (1 = Yes)	-0.1334	-0.41	0.0179	0.06
Own vs. Rent (1 = Own)	**-0.3611**	**-3.24**	**-0.4020**	**-3.54**
Driver's License (1 = Yes)	-0.2323	-1.27	-0.1367	-0.71
English Spoken in Home (1 = Yes)	0.2461	1.51	0.1606	0.91
Kids in Home (1 = Yes)	**-0.3893**	**-4.37**	**-0.3269**	**-3.53**
Household Income	0.0000	-0.62	**0.0000**	**-3.42**
Household Income. Squared	0.0000	0.52	**0.0000**	**3.87**
Median Rent in Census Block Group	**-0.0010**	**-3.45**	**-0.0029**	**-7.86**
Median Home Value in Census Block Group	**0.0000**	**-8.65**	**0.0000**	**-18.91**
Constant	**1.6991**	**5.39**	**4.8276**	**14.28**
Number of Observations				4052
Psuedo R^2				0.1106
Log Likelihood				-3848.31

All coefficients are rounded to four places to the right of the decimal
Coefficients in **bold** are significant at the five percent level
Coefficients in *bold italic* are significant at the ten percent level

ment, possession of a driver's license, presence of children in the household, household income, home ownership status, and costs of housing options in the area.

Table 3 suggests that once these other factors are considered, age does not play a statistically significant role in selecting residential environments; this stands in contrast to the results in Table 2. According to Table 3, the more important factors appear to be presence of children, cost of housing options, home ownership status, and household income, all of which reduce the likelihood of choosing neighborhoods with high levels of pedestrian accommodation.[7]

Table 4 investigates some of the discrepancy by conducting the model in Table 3 for separate age groups. While many of the same patterns as in Table 3 are observed, the decomposition by age group in Table 4 provides three interesting results. First, increased age may lower the willingness to live in neighborhoods with low compared to moderate PEF scores for individuals aged 22 through 35; presence of children and higher residential costs have similar (and stronger) effects of reducing the likelihood of selecting into neighborhoods with moderate compared to low PEF scores. Second, the only other group where increased age is associated with lower probability of choosing to reside in high compared to low PEF neighborhoods is for persons aged 46 through 55. Interestingly, increased age may have the opposite effect for persons aged 16 through 22. Not surprisingly, where possession of a driver's license is insignificant in the general model described in Table 3, it acts as a significant deterrent to selecting into high PEF scoring neighborhoods for persons in the oldest age group (66 and Older) and in the youngest age group (16 through 21 years old).

---

[7] The effect of increased income eventually maximizes at approximately $51,000.

Table 4. Multinomial logit probability model for selecting into new urbanist consistent environments (using low PEF environments as reference group), by age group

Age Group 2 (16–21)

	Moderate PEF		High PEF	
Variable	Coefficient	Z	Coefficient	Z
Age	-0.1041	-0.97	*0.2084*	*1.80*
Disabled (1 = Yes)	-0.1934	-0.15	-0.1683	-0.11
Own vs. Rent (1 = Own)	-0.1887	-0.42	0.5549	0.95
Driver's License (1 = Yes)	-0.4750	-1.43	**-1.0956**	**-2.60**
English Spoken in Home (1 = Yes)	0.4284	0.72	-37.5587	0.00
Kids in Home (1 = Yes)	-0.3549	-1.12	**-0.8791**	**-2.22**
Household Income	0.0000	-0.13	*-0.0001*	*-1.88*
Household Income. Squared	0.0000	0.01	**0.0000**	**2.20**
Median Rent in Census Block Group	-0.0002	-0.17	**-0.0034**	**-1.99**
Median Home Value in Census Block Group	0.0000	-0.43	**0.0000**	**-3.52**
Constant	2.2657	1.06	1.2162	0.52
Number of Observations				272
Psuedo R^2				0.1267
Log Likelihood				-244.1073

Age Group 3 (22–35)

	Moderate PEF		High PEF	
Variable	Coefficient	Z	Coefficient	Z
Age	*-0.0420*	*-1.83*	0.0092	0.40
Disabled (1 = Yes)	-1.4761	-1.25	-0.5627	-0.67
Own vs. Rent (1 = Own)	*-0.3379*	*-1.78*	-0.1968	-1.06
Driver's License (1 = Yes)	-0.3989	-0.68	-0.8083	-1.43
English Spoken in Home (1 = Yes)	0.1557	0.50	0.2210	0.67
Kids in Home (1 = Yes)	**-0.5001**	**-2.78**	**-0.6418**	**-3.62**
Household Income	0.0000	-1.15	**0.0000**	**-2.81**
Household Income. Squared	0.0000	1.04	**0.0000**	**2.77**
Median Rent in Census Block Group	**-0.0020**	**-3.09**	**-0.0042**	**-5.35**
Median Home Value in Census Block Group	**0.0000**	**-3.40**	**0.0000**	**-8.93**
Constant	**3.7531**	**4.05**	**6.0541**	**6.42**
Number of Observations				1008
Psuedo R^2				0.1369
Log Likelihood				-939.77727

All coefficients are rounded to four places to the right of the decimal
Coefficients in **bold** are significant at the five percent level
Coefficients in *bold italic* are significant at the ten percent level

Table 4. continued

| | Age Group 4 (36–45) | | | |
| | Moderate PEF | | High PEF | |
Variable	Coefficient	Z	Coefficient	Z
Age	-0.0308	-1.10	-0.0006	-0.02
Disabled (1 = Yes)	0.4074	0.44	0.0321	0.03
Own vs. Rent (1 = Own)	-0.0781	-0.34	-0.1683	-0.74
Driver's License (1 = Yes)	0.1907	0.30	0.3793	0.57
English Spoken in Home (1 = Yes)	0.2485	0.74	0.2928	0.84
Kids in Home (1 = Yes)	**-0.6627**	**-4.05**	**-0.4250**	**-2.54**
Household Income	0.0000	-0.85	**0.0000**	**-1.97**
Household Income. Squared	0.0000	0.67	**0.0000**	**2.19**
Median Rent in Census Block Group	-0.0004	-0.66	**-0.0036**	**-4.94**
Median Home Value in Census Block Group	**0.0000**	**-5.18**	**0.0000**	**-10.14**
Constant	**2.9670**	**2.17**	**5.1029**	**3.64**
Number of Observations				1113
Psuedo R^2				0.1275
Log Likelihood				-1036.5469

| | Age Group 5 (46–55) | | | |
| | Moderate PEF | | High PEF | |
Variable	Coefficient	Z	Coefficient	Z
Age	-0.0228	-0.71	**-0.0751**	**-2.12**
Disabled (1 = Yes)	-0.9354	-1.06	0.2709	0.40
Own vs. Rent (1 = Own)	-0.2600	-0.83	-0.3008	-0.93
Driver's License (1 = Yes)	-0.5821	-0.71	-0.3438	-0.38
English Spoken in Home (1 = Yes)	0.1360	0.38	-0.3329	-0.81
Kids in Home (1 = Yes)	0.0126	0.06	-0.0227	-0.1
Household Income	0.0000	0.92	0.0000	0.21
Household Income. Squared	0.0000	-1.05	0.0000	0.04
Median Rent in Census Block Group	-0.0001	-0.15	-0.0010	-1.27
Median Home Value in Census Block Group	**0.0000**	**-3.28**	**0.0000**	**-8.77**
Constant	1.6977	0.88	**6.8753**	**3.13**
Number of Observations				812
Psuedo R^2				0.095
Log Likelihood				-767.7102

All coefficients are rounded to four places to the right of the decimal
Coefficients in **bold** are significant at the five percent level
Coefficients in *bold italic* are significant at the ten percent level

Table 4. continued

	Age Group 6 (56–65)			
	Moderate PEF		High PEF	
Variable	Coefficient	Z	Coefficient	Z
Age	0.0570	1.14	-0.0096	-0.15
Disabled (1 = Yes)	0.4508	0.48	-0.3704	-0.34
Own vs. Rent (1 = Own)	**-1.1700**	**-2.04**	**-2.0638**	**-3.39**
Driver's License (1 = Yes)	-0.1563	-0.24	-0.7553	-1.07
English Spoken in Home (1 = Yes)	-0.1402	-0.22	-0.1289	-0.17
Kids in Home (1 = Yes)	-0.0542	-0.12	*-1.9892*	*-1.85*
Household Income	0.0000	-0.08	0.0000	1.12
Household Income. Squared	0.0000	0.29	0.0000	-0.91
Median Rent in Census Block Group	**-0.0018**	**-2.02**	**-0.0044**	**-3.24**
Median Home Value in Census Block Group	**0.0000**	**-4.24**	**0.0000**	**-6.09**
Constant	-0.0671	-0.02	*6.6792*	*1.66*
Number of Observations				384
Psuedo R^2				0.1597
Log Likelihood				-337.0601

	Age Group 7 (66 and older)			
	Moderate PEF		High PEF	
Variable	Coefficient	Z	Coefficient	Z
Age	-0.0285	-1.07	0.0376	1.50
Disabled (1 = Yes)	0.2465	0.43	-0.4719	-0.79
Own vs. Rent (1 = Own)	-0.2236	-0.42	-0.1177	-0.22
Driver's License (1 = Yes)	-0.4705	-0.86	*-1.0479*	*-1.84*
English Spoken in Home (1 = Yes)	0.5607	1.08	0.7628	1.43
Kids in Home (1 = Yes)	-1.1711	-1.41	-0.4263	-0.60
Household Income	0.0000	0.83	0.0000	-1.57
Household Income. Squared	0.0000	-0.28	0.0000	1.58
Median Rent in Census Block Group	*-0.0020*	*-1.72*	0.0000	-0.04
Median Home Value in Census Block Group	**0.0000**	**-4.66**	**0.0000**	**-6.38**
Constant	*4.4363*	*1.95*	1.4784	0.67
Number of Observations				416
Psuedo R^2				0.1344
Log Likelihood				-384.9792

All coefficients are rounded to four places to the right of the decimal
Coefficients in **bold** are significant at the five percent level
Coefficients in *bold italic* are significant at the ten percent level

Table 5. Tobit models for impact of age and residential environment on aspects of pedestrian travel behavior

Variable	Number of Walking Trips Completed Coefficient	T	Number of Driving Trips Completed Coefficient	T	Walking Mode Split Coefficient	T	Total Distance Walked Coefficient	T
Age	0.0001	0.02	*-0.0113*	*-1.89*	0.0000	0.01	-0.0071	-0.44
Disabled (1 = Yes)	**-1.1558**	**-1.76**	-0.0012	0.00	-0.1767	-0.81	-1.4885	-1.27
Own vs. Rent (1 = Own)	***-0.3915***	***-1.72***	0.2074	1.22	-0.1180	-1.56	-0.3888	-0.95
Driver's License (1 = Yes)	**-1.1560**	**-3.25**	**2.6031**	**9.27**	**-0.2935**	**-2.25**	**-2.1469**	**-3.39**
English Spoken in Home (1 = Yes)	0.5713	1.74	-0.4018	-1.61	0.1045	0.92	***0.9769***	***1.66***
Employed (1 = Yes)	**-0.4357**	**-2.08**	-0.0171	-0.11	-0.0957	-1.36	***-0.7248***	***-1.93***
Kids in Home (1 = Yes)	***-0.3433***	***-1.80***	**0.7049**	**5.26**	**-0.1361**	**-2.13**	***-0.5785***	***-1.68***
Household Income	0.0000	-1.22	0.0000	0.23	0.0000	-1.07	0.0000	-0.62
Household Income. Squared	0.0000	1.27	0.0000	0.20	0.0000	1.15	0.0000	0.51
Mixed Use Index	**0.0117**	**5.11**	0.0026	1.48	**0.0036**	**4.64**	**0.0190**	**4.62**
Medium PEF (1 = Medium. 0 = Low)	**1.6624**	**2.65**	0.1112	0.25	**0.5293**	**2.50**	***2.0541***	***1.82***
Medium PEF × Age	-0.0163	-1.26	0.0079	0.86	-0.0045	-1.02	-0.0217	-0.93
Medium PEF × Mixed Use Index	**-0.0085**	**-3.68**	**-0.0045**	**-2.50**	**-0.0028**	**-3.56**	**-0.0129**	**-3.10**
High PEF (1 = High. 0 = Low)	**2.5308**	**4.39**	0.2209	0.52	**0.7760**	**3.93**	**3.4697**	**3.36**
High PEF × Age	-0.0149	-1.23	0.0006	0.07	-0.0037	-0.89	-0.0085	-0.39
High PEF × Mixed Use Index	**-0.0092**	**-3.97**	**-0.0050**	**-2.74**	**-0.0028**	**-3.54**	**-0.0161**	**-3.86**
Constant	**-2.0525**	**-3.00**	**3.0794**	**6.21**	**-0.8913**	**-3.69**	**-4.1363**	**-3.38**
Number of Observations (uncensored observations in brackets)	4052 (873)		4052 (3866)		3866 (748)		4052 (873)	
Psuedo R-Squared	0.0412		0.0108		0.0485		0.0301	
Log Likelihood	-3593.3361		-10842.3080		-2243.8854		-3978.6990	

All coefficients are rounded to four places to the right of the decimal
Coefficients in **bold** are significant at the five percent level
Coefficients in ***bold italic*** are significant at the ten percent level

Table 6. Tobit models for impact of age and residential environment on aspects of pedestrian travel behavior – Elderly sub group

Variable	Number of Walking Trips Completed Coefficient	T	Number of Driving Trips Completed Coefficient	T	Walking Mode Split Coefficient	T	Total Distance Walked Coefficient	T
Age	*-0.1365*	*-1.88*	-0.0975	-2.08	**-0.0480**	**-2.36**	-0.2365	-1.54
Disabled (1 = Yes)	0.4549	0.47	0.3277	0.45	0.2047	0.83	-0.6860	-0.33
Own vs. Rent (1 = Own)	*-1.2972*	*-1.70*	-0.3197	-0.52	-0.2373	-1.18	-1.1855	-0.74
Driver's License (1 = Yes)	*-1.5152*	*-1.76*	**1.7383**	**2.76**	-0.3462	-1.46	**-4.6265**	**-2.54**
English Spoken in Home (1 = Yes)	-0.6484	-0.68	-0.3647	-0.57	-0.2203	-0.84	1.5755	0.87
Employed (1 = Yes)	-0.5287	-0.83	-0.1421	-0.31	-0.1236	-0.76	1.1211	0.89
Kids in Home (1 = Yes)	0.2801	0.25	-0.2482	-0.28	-0.1020	-0.33	**4.7070**	**2.17**
Household Income	0.0000	0.16	0.0000	0.46	0.0000	-0.24	0.0001	0.99
Household Income. Squared	0.0000	-0.17	0.0000	-0.12	0.0000	0.20	0.0000	-1.04
Mixed Use Index	**0.0187**	**3.18**	0.0038	0.77	**0.0046**	**3.04**	**0.0375**	**3.08**
Medium PEF (1 = Medium. 0 = Low)	-10.9322	-1.40	-5.7011	-1.03	*-3.5312*	*-1.72*	-19.6487	-1.19
Medium PEF × Age	0.1617	1.47	0.0849	1.10	*0.0519*	*1.79*	0.2914	1.26
Medium PEF × Mixed Use Index	**-0.0187**	**-2.01**	-0.0017	-0.23	**-0.0046**	**-1.92**	*-0.0330*	*-1.71*
High PEF (1 = High. 0 = Low)	-3.1411	-0.46	-1.3130	-0.26	-0.8398	-0.43	-9.8884	-0.69
High PEF × Age	0.0597	0.62	0.0253	0.37	0.0163	0.59	0.1780	0.88
High PEF × Mixed Use Index	**-0.0182**	**-3.07**	-0.0062	-1.22	**-0.0047**	**-3.04**	**-0.0369**	**-3.01**
Constant	**8.9528**	**1.66**	**10.1636**	**2.81**	**3.1589**	**2.11**	12.4285	1.09
Number of Observations (uncensored observations in brackets)	416 (85)		416 (402)		402 (77)		416 (85)	
Psuedo R-Squared	0.0379		0.0137		0.0572		0.0410	
Log Likelihood	-339.4493		-1059.2003		-200.9687		-385.8705	

All coefficients are rounded to four places to the right of the decimal
Coefficients in **bold** are significant at the five percent level
Coefficients in ***bold italic*** are significant at the ten percent level

Tables 5 and 6 tie together the relationship between urban form and transportation behavior by testing the impact of new urbanist environment elements on the willingness to choose pedestrian and automobile travel modes. Looking at Table 5, increased age is related to reductions in the number of driving trips completed, although the mode split model suggests the proportion of trips made by walking as compared to automobile is not affected by age. In contrast, the PEF classification and increased mixed use index both act to promote walking behaviors, both in absolute and relative terms (as described in the Number of Walking Trips Completed and Total Walking Distance, and the Walking Mode Split columns, respectively). Higher PEF classifications did not, however, induce more automobile trips.

While moderate and high PEF classification and greater intensity of mixed use separately encouraged walking, the interaction between these variables acts to suppress pedestrian activity. This pattern may be explained by dual impacts that mixing residential, retail and employment uses can have on trip making behavior in general. Higher conformity to new urbanist principles (as indicated by moderate or high PEF classification) may induce more walking behavior, arguably because the environmental design is more accommodating of this behavior. However, the clustering of residential and non-residential (e.g., retail and employment) uses facilitates the completion of many tasks at the same location; to the extent this centralized location is close to home, this would facilitate walking over driving. However, a centralized activity location also reduces the total number of trips needed to complete a set of daily tasks. By combining new urbanist design with intense mixed use, the centralization effect of reducing trips appears be stronger than the inducement to greater walking behavior.

In contrast, the results in Table 6 suggest the relationships in Table 5 are not maintained as one gets older. Looking at the travel behavior model for the oldest sub-group, the importance of PEF classification of the neighborhood disappears, while increased age becomes a significant impediment to travel by both automobile and walking. Age further translates into a lower proportion of walking trips made compared to car trips, as noted by the walking mode split model. Interestingly, while the mixed use index still acts to encourage walking behaviors relative to driving, the interaction between PEF classification and higher mixed uses indices act to reduce walking activity.

Taken as a set, the findings in Table 6 suggest that new urbanist urban form can have a beneficial impact on the mobility of the elderly, although not in the form of increased travel activity. The interactions between the mixed use index and the higher PEF classified neighborhoods, between age and higher PEF classified neighborhoods, and the suppressing impact of age on walking behavior for older residents, all suggest that rather than promoting pedestrian travel, New Urbanism acts to consolidate activities for older residents at centralized locations, so that aggregate trip making can be minimized regardless of travel mode. Further, although there is variation in choosing to reside in new urbanist environments based on age, the results from Table 4 suggest age is a proxy for other lifestyle elements, such as family status (e.g., children in the household), income, and ability to opt for different modes of travel (i.e., possession of a driver's license).

Conclusion

New Urbanist design does hold benefits for the elderly, although the results presented here give some insight on what New Urbanism can and cannot be expected to achieve. New

Urbanist design appears to have great promise in getting people in general to walk in greater proportion to driving, although for the elderly the main benefit appears to be the reduction in travel by consolidating activities (as evidenced by the impact of the mixed use index). This finding would be consistent with what other travel behavior researchers looking at New Urbanism have found. Further, with respect to residential location selection, the effects for elderly are only visible when contrasting extremely low and extremely high conformity to New Urbanism; small changes of urban form don't seem to have much effect.

The reduction in travel by the elderly can be a double-edged sword. Reducing travel implies more time can be devoted to longer participation in or a greater number of more enjoyable or socially engaging activities. It also means that the elderly need not travel as often to accomplish the same number of tasks. In the United States, where the elderly sometimes become unable or unwilling to leave their homes, these findings should serve as a warning to new urbanists that designs must be accompanied by or facilitate creating social support networks. Otherwise the designs may result in social withdrawal rather than promote social connectivity.

This still begs the question of how to get people to select into new urbanist environments in the first place. The results from Tables 2 through 4 suggest the most effective way to achieve that goal is to attract residents to new urbanist neighborhoods early in their life cycle. At these earlier stages, children, housing costs and income appear to be the strongest factors for not selecting into new urbanist areas. To the extent new urbanist calls for higher density can translate into lower housing costs, the free market alone is sufficient to promote integration of socio-economic and age groups. But, if new urbanists fail to address the social needs of younger residents, they are not likely to get the opportunity to address the needs of older ones.

References

Boarnet, M. G., & Greenwald, M. J. (2000). Land use, urban design, and non-work travel: Reproducing for Portland, Oregon empirical tests from other urban areas. *Transportation Research Record*, 1722, 27-37.

Boarnet, M. G., & Sarmiento, S. (1998). Can land-use policy really affect travel behaviour? A study of the link between non-work travel and land-use characteristics. *Urban Studies* 35, 7, 1155-1169.

Calthorpe, P. (1993). *The next American metropolis: Ecology, community and the American Dream*. New York, NY: Princeton Architectural Press.

Cambridge Systematics. (1996). *Data collection in the Portland, Oregon Metropolitan Area: Travel Model Improvement Program, Track D. Data Research Program*. Oakland, CA: Author.

Cervero, R. (1989). Jobs-housing balancing and regional mobility. *Journal of the American Planning Association,* 55, 2, 136-150.

Cervero, R. (1996). Jobs-housing balance revisited: Trends and impacts in the San Francisco Bay area. *Journal of the American Planning Association*, 62, 4, 492-511.

Duany, A., & Plater-Zyberk, E. (1991). *Towns and town making principles*. New York, NY: Rizzoli International Publications, Inc.

Greenwald, M. J. (2003). The Road Less Traveled: New Urbanist Inducements to Travel Mode Substitution for Non-Work Trips. *Journal of Planning Education and Research*, 23(1), 39-57.

Greenwald, M. J., & Boarnet, M. G. (2001). The built environment as a determinant of walking behavior: Analyzing non-work pedestrian travel in Portland, Oregon. *Transportation Research Record,* 1780, 33-47.

Johnston, J., & DiNardo, J. (1997). *Econometric Methods* (4th ed.). New York, NY: McGraw-Hill, Inc.

Levine, J. (1998). Rethinking Accessibility and Jobs-Housing Balance. *Journal of the American Planning Association,* 64, 2, 133-149.

Pollak, P. B. (1989). Community-Based Housing for the Elderly, Planning Advisory Service Report No. 420. Chicago, IL. American Planning Association.

Pollack, P. B. (1994). Rethinking Zoning to Accommodate the Elderly in Single Family Housing, *Journal of the American Planning Association,* 60, 4, 521-531.

Pucher, J., & Dijkstra, L. (2002). Making walking and cycling safer: Lessons from Europe, *Transportation Quarterly,* 54, 3, 25-50.

Sarmiento, S. (1995). *Studies in Transportation and Residential Mobility.* Unpublished Ph.D. Dissertation, University of California, Irvine.

StataCorp. (1999). *Stata Reference Manual: Release 6.0, vol. 3.* College Station, TX: Stata Corporation.

Weisbrod, G. E., Lerman, S. R., & Ben-Akiva, M. (1980). Tradeoffs in residential location decisions: Transportation versus other factors. *Transportation Policy and Decision Making,* 1, 13-26.

Founding houses for the elderly
On housing needs or dwelling needs?

Wim J. M. Heijs

Eindhoven University of Technology, The Netherlands

Abstract. Present research for housing policy, as well as a number of social scientific studies on housing, concentrate on *housing needs*, preferences, wants and demands of the elderly. These perspectives are somewhat hazardous, because they may produce results that are based on a limited choice of alternatives, inadequate knowledge of the possibilities or the apparent inescapability of "homes" for the elderly. To prolong independence in one's own house, it is essential to have an understanding of the *dwelling needs* of elderly, i.e., needs on the user side that follow from the process of dwelling, and of the changes in those needs during the dwelling and ageing processes. Having an overview of these needs might prevent the taking of decisions in moving or design that promote certain needs while obstructing others, thereby possibly inducing an unnecessary future house move. Some studies in the social sciences do assess dwelling needs, primarily by using qualitative methods. These methods are suitable and valid in the circumstances in which they are used. However, for reasons discussed below, a quantitative method is preferred in the present case.

This method takes person-environment fit theory as a starting point. Using functional problems that people experience if needs or abilities are not in line with resources or demands from the environment, statistical procedures can be applied to uncover dwelling needs. Tangible problems with these resources and demands will reveal related dwelling features and possibilities to promote independence. The method was tested in a study among tenants of a housing association in Eindhoven, The Netherlands. Although the procedure was somewhat hindered by the sample size, the results indicate that the method has potential. Dwelling needs of three age groups were established and compared. The relevant dwelling features of single family houses were classified and, on those grounds, recommendations were formulated.

Keywords: elderly residents, needs analysis, housing needs, dwelling needs, person-environment fit, environmental psychology

Introduction

Research on the housing needs of the elderly is increasingly important because it occupies a position at the cross-roads of two developments. First, the housing market is changing from being dominated by the supply side ("These are the houses from which you may choose") to being more demand oriented ("What kind of house do you want?"). Second, the elderly population is growing and suitable housing has to be provided for them. Government policy as well as elderly dwellers themselves state a clear preference for prolonging the duration of independent living in one's own house (Timmermans et al., 1997). To achieve this goal, various initiatives are taken to compensate for physical impairments in old age, such as supplying additional care and adapting housing circumstances. Planning these initiatives requires an adequate knowledge of the housing needs of the elderly. This paper describes a method to uncover those needs and to establish related dwelling features, which may suggest possibilities for the promotion of independence. It also includes a study to assess the potential of the method among tenants of a local housing association. The later sections contain an outline of the background, the theoretical and methodological framework, a summary of the findings, and conclusions with respect to the analysis of the needs of elderly dwellers.

Background

Because of the vast amount of studies on housing for the elderly, many policy makers believe that this knowledge is an adequate basis for the initiatives mentioned above. These studies largely consist of quantitative market research conducted by order of local governments or housing associations for management or planning purposes. The starting point is the housing market (the supply side) and the housing needs at issue are usually preferences for types of houses, locations and services. There is also a lesser amount of policy-oriented studies that look at housing needs from the perspective of the user (the demand side). These are mostly qualitative by nature and conducted on smaller scales. In-depth interviews or focus groups of elderly sometimes produce goals or activities (such as safety, comfort or social interaction) and in other cases wants or demands regarding the design (which are aspects of the supply side, e.g., two bedrooms or a large kitchen).

Taking the market as the point of departure instead of the user may be a cause of several problems that face policy-oriented research (e.g., Sommer, 1983; Heijs, 2001; Houben, 1985; Lawton, 1990). Satisfaction, wants and demands are known to reveal inadequate information regarding actual well-being. The elderly, in particular, often express a higher satisfaction with their present situation than might be expected, probably because they do not want to be a burden or because they don't expect to find an alternative. They are inclined to name preferences they do not actually have but which agree with societal norms (the reputed "inevitability" of homes for the elderly). A confrontation with physiological impairments will often prompt a choice (a small house) that promotes certain needs (easy cleaning) while hindering others (giving up valued belongings that would otherwise enhance identification). Sometimes they are not aware of the effects because the one need overrides the others or other needs are not recognised (until afterwards). Furthermore, prompting for demands or wants is prone to yield known solutions, thereby possibly preventing alternatives. These problems are confirmed by the fact that actual relocation is not in line with predictions from this type of research, even if the desired housing situations are made available.

Apparently, there is some confusion regarding the concept of housing needs, which is used to denote needs based on the supply side (preferences, wants and demands) as well as on the dwelling process (like goals and activities). This confusion might cause the amount and significance of existing data, which are mainly of the first type, to be overestimated. The goal being a prolongation of independence in one's own house, the source for knowledge about housing needs should be the dwelling process, because this is the basis of the needs and changes in those needs that occur as a result of physiological, social and psychological aspects of ageing (a temporal aspect; see e.g., Lawrence (1987) and Saegert (1985)). Insights in the dwelling process and in these changes might offer alternative solutions that prevent (further) troublesome relocation of residence. Therefore, it is useful to introduce *dwelling needs* as the term indicating needs that relate to the dwelling process, apart from *housing needs* as those associated with the supply side.

Dwelling needs of the elderly (or housing needs according to the existing terminology) are being examined in several ways in environmental psychology. Firstly, qualitative methods (interviews, focus groups) and design participation are recommended to attain performance criteria (e.g., Dieckman, 1998; several authors in Preiser, Vischer & White, 1991; Sanoff, 2000). In some cases dwelling needs are established as an intermediate step. A second means is the transformation of general psychological needs, mostly from Maslow's need hierarchy, into dwelling needs (illustrations are provided by among others Flade, 1993, and Lawrence, 1987). A third way is to obtain needs from reviews of the literature, which is done in research to test models of competence, congruence and person-environment fit in relation to the dwelling process of elderly (Carp & Carp, 1984; Cvitkovich & Wister, 2001; Kahana, 1982; Flade, 1998; Lawton & Nahemow, 1973; Linneweber, 1993).[1]

These methods suit their particular application (such as designing for specific situations or model testing) and they also have significance for understanding the dwelling process by tapping this process, providing insight in the nature of dwelling needs and offering ways to use them in design. However, comprehension of the dwelling process in relation to ageing, to reach the goal of prolonged independence, probably requires an approach that generates theoretical assertions that are more widely applicable. The following requirements are gathered from various sources (Cvitkovich & Wister, 2001; Heijs, 2001; Kahana, 1982; Lawrence, 1987; Sommer, 1983). (1) Research must use a definition of dwelling needs that allows a distinct operationalisation. (2) Quantitative data and large enough samples are needed to check reliability, validity and generalisability. (3) Using standardised procedures, needs should be established across target groups and episodes in the dwelling and ageing processes to allow theoretical conjectures regarding dependencies between both processes. (4) The method should strive for an adequate overview of needs to prevent decisions that favour certain needs while obstructing others because they are less known. (5) Priorities must be established because it is likely that dwelling needs are not equally important all the time to all groups. (6) Explanations of differences in content and priorities must be made possible by using existing theoretical frameworks. (7) And there should be a way to connect needs to performance criteria that can be used in design.

---

1 This brief description does not do justice to all of the efforts, and their different characteristics, made to uncover dwelling needs in general and those of the elderly in particular. It is, however, not intended to be a review, which is beyond the scope of this paper, but only to substantiate the arguments that follow.

Theoretical and methodological framework

The present project is not an ideal one according to the previous demands. The goal was to take a small step in this direction in order to explore prospects and obstacles one might encounter. First, a definition of dwelling needs was formulated. In general, a need can be thought of as a vital goal or condition one is trying to reach. Thus, a dwelling need is: *"a physiological or psychological condition related to the dwelling process, or activities that are necessary to reach this condition"*. Activities are mentioned separately because some conditions are attainable if the environment allows them (e.g., safety if certain measures are taken) while others are dependent upon actions that inhabitants should be able to carry out (e.g., inviting people for social interaction) and also because abstract conditions (health) can be expressed more practically as required activities (washing, cleaning).

It is not feasible to ask directly about dwelling needs because people generally do not think of their relation with a dwelling in those terms. Instead, they usually phrase their needs as shortcomings or wants (i.e., assumed solutions for these needs: "Our house is too large." or "We want a smaller dwelling." and not "We need to be able to clean it easily and we want a solution."). In line with the accepted position of human needs in the behavioural sciences, it is assumed, however, that people do have dwelling needs which are shared to some extent. The method to derive these needs is an indirect one, based on person-environment fit theories.

These theories suggest two key aspects of person-environment interaction: (1) the degree to which the environment, through its resources, can meet the needs of people and (2) the way in which people, because of their abilities, can meet the demands posed by the environment (Caplan, 1983; Conway, Vickers & French, 1992; Cvitkovich & Wister, 2001). Here, needs and abilities are joined to represent the user side of the dwelling process, whereas resources and demands depict the supply (or housing) side (see figure 1). To trace dwelling needs, the model is reversed by taking misfit as the starting point. If a need is not met or an ability proves to be insufficient, this will cause *functional problems* on the user side: the intended condition cannot be reached (e.g., a lack of hygiene) or necessary activities cannot be carried out (e.g., not being able to clean). At the same time this means that environmental resources and/or demands are at odds with those needs and/or abilities, thereby representing *tangible problems* on the supply side (e.g., spots that cannot be reached easily). The ensuing functional and tangible problems and their connections are the basis of the method.

Functional problems serve to construct dwelling needs. Dwellers will respond more easily, and probably more reliably, to questions regarding everyday problems than to questions about needs or satisfaction because the former problems are real. From existing research literature a comprehensive list of functional problems must be compiled (e.g., difficulties with safety, cleaning, washing, thermal comfort, having family over for the night, carrying out hobbies, hearing the doorbell and so on).

The analysis consists of three steps. First, one problem may correspond to one need (e.g., a cleaning problem – a need of hygiene), several problems may jointly indicate the same need (e.g., cleaning and bathing problems – a need of hygiene) and one problem may be a part of several needs simultaneously (e.g., a cleaning problem – needs of hygiene and safety). These structures can be uncovered by factor analysis. The factors are sets of functional problems which, after checking internal consistency (Cronbach's a), may be interpreted as needs.

Figure 1. Mechanisms of fit and problems of misfit

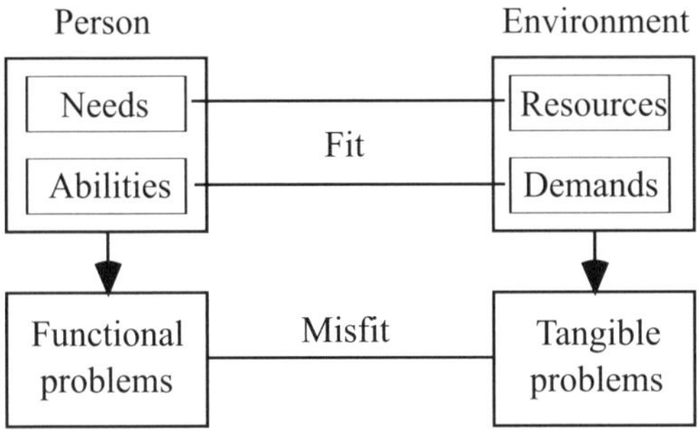

```
      Person                          Environment
    ┌─────────┐                      ┌───────────┐
    │  Needs  │────────┐     ┌───────│ Resources │
    ├─────────┤        Fit           ├───────────┤
    │Abilities│────────┘     └───────│  Demands  │
    └─────────┘                      └───────────┘
         │                                 │
         ▼                                 ▼
    ┌─────────┐        Misfit       ┌───────────┐
    │Functional│───────────────────│  Tangible │
    │ problems │                    │  problems │
    └─────────┘                     └───────────┘
```

Second, the relative weights of these needs must be determined to establish priorities. This is accomplished using correlation analyses and multiple regression with a measure of overall residential satisfaction as the dependent variable. A low weight does not imply that a need is meaningless. It might have a significant role that is currently less crucial but may be more substantial in the future (like goals, needs loose their significance once they are fulfilled and they will regain interest if problems arise).

And third, an actual value is determined for each need to reflect its concern for present policy decisions. This value is composed of the indications of its existence (from the factor analysis) and its relative importance (from the correlation and regression analyses), ranging from *very actual* (a strong cohesion in a factor, a strong relation with satisfaction) to *not actual* (a moderate cohesion and an insignificant relation with satisfaction). Again, a non-actual designation does no imply that a need can be ignored: it might just temporarily have a subordinate role. When designing solutions *all* needs have to be considered, though some are more important than others. The method has one complication however: because the origins of needs are problems, needs that are met in almost every case will not be found (i.e., there are no problems). How then can all needs be considered? This complication is common and manageable. Because needs will (re)appear in case of problems, the method will uncover those that are of interest at a given point in time. The dynamic qualities of the dwelling process, of society and of technology urge repetitive research. If this is carried out, needs that are now apparently non-existent will show up when they are meaningful.

The tangible problems on the other side of Figure 1 serve to study the material counterparts of the needs, to find causes and to suggest improvements if needs are not satisfied. Dwellers are not always cognisant of the causes. A lack of privacy might, for instance, be attributed to a family member while the real cause is a lack of space. Hence, an indirect, statistical way is preferred here, too. This is accomplished by examining the relations between the needs on the one hand and problems that are experienced with parts and properties of the dwelling on the other (using correlation analyses, *t* tests and analyses of variance). The result is a list of dwelling features for each need, indicating the association with that need and the direction in which to search for solutions if a need is not met (e.g., the list for a need of hygiene will show rooms and other characteristics that are important in that respect because they are easy or difficult to clean, thus contributing in a positive or negative way). The method does not offer final solutions. These must be found by designers and inhabitants, e.g., in design participation meetings, where the data can be used as input.

So far, the method observes most of the requirements that were mentioned in the Background section. The third one was not yet mentioned: the necessity to study the development of dwelling needs across time in relation to the ageing process (which is crucial for finding ways to prolong independence). One could use a longitudinal design in a single user group to probe the needs continuously, but this is not practicable because of the efforts involved and also because the findings will be biased on account of learning effects. An alternative is to compare the needs of different age groups in similar dwelling circumstances at the same time and to repeat this procedure for each dwelling type that is of interest. This, too, is not an easy task, since it requires large samples and a multitude of analyses.

A form of the latter strategy has been employed in a study among elderly tenants of a housing association in Eindhoven (aged 55 or over).[2] The goal was twofold: to test the method and to establish dwelling needs and recommendations to prolong independent living. All of these households received the questionnaire and approximately 60% returned it in a satisfactory form ($n = 3902$). The questionnaire consisted largely of five-point scales for measuring 50 functional and 88 tangible problems accumulated from research literature (no problem, hardly, rather small, rather big, big problem), and satisfaction with the dwelling and its components (10 items). Other questions concerned characteristics of the household, the house and a number of other subjects that are not of interest for the present purpose.

Single family dwellings and younger households were over-represented and the number of households in the highest age group in any single dwelling type was insufficient to perform the analyses. Thus, it was impossible to use the preferred strategy (comparing age groups in similar dwellings). A decision was taken to perform two sets of analyses. In the first, dwelling needs of three age groups were compared in all dwelling types simultaneously: aged 55 to 68 years ($n = 1693$), 69 to 76 years ($n = 650$), and 77 years and above ($n = 368$). In the second, dwelling needs of tenants in single family houses were examined without a division in age groups. Relating tangible dwelling features to needs in a meaningful fashion, however, requires the dwellings to be of the same type. Hence, this part of the method could only be used in the second set. The section below contains the results of the needs analyses in the age groups and of the analysis of dwelling features in single family dwellings. Although the study does not conform to all of the requirements mentioned, it does offer an illustration of the possibilities and also usable data (provided they are considered with some caution).

Findings

Dwelling needs

Classical factor analyses were carried out using varimax rotation on 43 of the 50 functional problems as the remaining items showed no variation (e.g., driving a wheel chair) or were irrelevant for most houses (e.g., controlling a ventilation unit that is absent in single family dwellings). Interpretability was used as the primary criterion to determine

---

[2] The survey was organised by Stichting Interface, an institute in the Department of Architecture, Building and Planning of Eindhoven University of Technology, specialised in conducting applied research.

the correct amount of factors (starting with the number suggested by the Kaiser criterion of eigenvalues = 1). Some of the functional problems had very little in common with others: they showed a low communality (or a high degree of unicity) in the factor analyses. These problems can be considered to represent separate dwelling needs. In the regression analyses used to establish the relative weights of the needs, the independent variables were factor scores for the needs that result from the factors and standardised scores for the separate, unique problems. The dependent variable was the overall satisfaction with the dwelling.

55 to 68 years of age

Factor analysis revealed 13 dwelling needs in this group (9 factors and 4 unique problems). Table 1 contains an overview in which these needs are arranged according to their relative weights (starting with the most important ones). Behind every need is its actual value (Av.: ++ very actual, + fairly actual, 0 neutral and – not actual) and the last column lists the main functional problems in that particular factor (or – if the need is similar to a unique problem).

The interpretation of the first need as *controlled social interaction* is obvious because the associated functional problems relate to both sides of the interaction continuum (contact and privacy). The possibility to stow away possessions fits this description: having guests or carrying out hobbies requires that one should be able to create room and order. Second in row is *thermal comfort*, which is also easy to identify from the set of problems. The third factor presents a collection of problems that share the degree to which inhabitants are able to operate a variety of controls. This need of *easily operating equipment and controls* is clearly significant for the elderly in view of diminishing physical capacities (e.g., eyesight and diseases like rheumatic arthritis). The problems in the fourth factor evidently imply a need of *avoiding nuisance from inside the dwelling*. In fifth place is the need to *carry out activities of daily living (ADL)*. It is based on a larger collection of problems that are bound up with doing the household as well as taking care of oneself. This substantiates notions in the person-environment fit literature on the existence of life-maintenance needs (ADL) next to higher order needs (Carp & Carp, 1984; Cvitkovich & Wister, 2001). A noticeable item is home-care: presumably, this is a relatively normal element of ADL. The sixth factor resembles the fourth, but here it is a need of *avoiding nuisance from outside the dwelling*, including the protection of visual and auditory privacy. The need of *safety of movement* follows directly from the related problems, pertaining to the increase of mobility risks while getting older. All of these needs are very actual.

The next four needs are based on single, unique problems and they are fairly actual: *physical warmth*, *a feeling of social safety inside the dwelling*, the possibility of *hearing sounds from equipment* (door bell, telephone, radio, TV, etc.) and a *comfortable air humidity* (elderly are more sensitive to a dry atmosphere). In twelfth place is the need to *keep the dwelling in order*, by adequate furnishing and normal upkeep. And last is the need of *stowing away possessions* for shorter or longer periods of time. This provides an example of a problem belonging to more than one need because there are several aspects involved: stowing away things is required for interaction but it also implies having storage space. The latter two do exist but are not actual, presumably because single family dwellings that form the greater part of the sample (78%) provide sufficient opportunities to meet them.

Table 1. Needs, actual values and related functional problems

Dwelling need	Av	Functional problems with
55 to 68 years		
1. controlled social interaction	++	– letting guests stay the night, receiving visitors, carrying out hobbies, having a private spot, stowing away things
2. thermal comfort	++	– avoiding cold, avoiding draught, keeping a comfortable temperature, avoiding a moist atmosphere
3. easily operating equipment/ controls	++	– operating taps, switches, radiator knobs, locks, doors and windows, lights for special activities
4. avoiding nuisance from inside	++	– ventilating, avoiding smells, avoiding noise
5. carrying out act. of daily living (ADL)	++	– cleaning, washing, ironing, cooking, getting up, going to bed, bathing/ showering, stowing away things, home-care
6. avoiding nuisance from outside	++	– protecting privacy, avoiding smells, avoiding noise
7. safety of movement	++	– walking, risks of injury, accessing the dwelling, stepping in and out of bathtub or shower, going to toilet
8. physical warmth	+	–
9. feeling of social safety inside	+	–
10. hearing sounds from equipment	+	–
11. comfortable air humidity	+	–
12. keeping the dwelling in order	0	– furnishing the dwelling, upkeep
13. stowing away possessions	–	– stowing away for short periods, for longer periods
69 to 76 years		
1. avoiding nuisance from in- and outside	++	– avoiding smells and noise from outside and inside, protecting privacy from outside
2. keeping the dwelling in order/ home-care	++	– furnishing and decorating the dwelling, upkeep, cleaning, home-care, warning others when ill
3. thermal comfort	++	– avoiding cold, avoiding draught, keeping a comfortable temperature, avoiding a moist atmosphere
4. healthy atmosphere	++	– avoiding smells from inside, ventilating
5. safety of movement/carrying out ADL	++	– bathing/showering, getting up, going to bed, walking, going to toilet, risks of injury, cooking, accessing the dwelling
6. controlled social interaction (contact)	++	– letting guests stay the night, receiving visitors, hearing sounds from equipment

Table 1. continued

Dwelling need	Av	Functional problems with
7. controlled social interaction (privacy)	++	– carrying out hobbies, having a private spot
8. feeling of social safety inside	+	–
9. physical warmth	+	–
10. comfortable air humidity	+	–
11. easily operating equipment/ controls	+	– operating taps, switches, radiator knobs, locks, doors and windows, decorating the dwelling
12. carrying out household activities (HA)	–	– cleaning, washing, ironing, cooking
13. stowing away possessions	–	– stowing away for short periods, for longer periods
77 years and above		
1. carrying out ADL	++	– washing, ironing, cooking, getting up, going to bed, bathing/showering, stowing away things, cleaning, upkeep, decorating the dwelling, home-care
2. shelter inside	++	– avoiding smells, ventilating, protecting privacy from outside, stepping in/out bathtub/shower, going to toilet, social safety inside, operating locks/doors/windows
3. controlled social interaction	++	– receiving visitors, carrying out hobbies, letting guests stay the night, stowing away things, decorating the dwelling
4. easily operating equipment/ controls	++	– operating switches and radiator knobs, lights for special activities, outside view, warning others when ill, upkeep
5. comfortable air humidity	++	–
6. safety of movement	++	– accessing the dwelling, risks of injury, walking, stepping in/out bathtub/shower, going to toilet, bathing/showering
7. privacy	++	– having a private spot, protecting privacy from outside
8. avoiding nuisance from in- and outside	+	– protecting privacy from outside, avoiding noise from inside and outside, avoiding smells from outside
9. keeping the dwelling in order	+	– furnishing and decorating the dwelling, upkeep, cleaning
10. thermal comfort	–	– avoiding cold, avoiding draught, keeping a comfortable temperature, avoiding a moist atmosphere
11. physical warmth	–	–
12. hearing sounds from equipment	–	–

69 to 76 years of age

The factor analysis yielded 13 needs (10 factors and 3 unique problems), which illustrate that their content and priority are partially dependent on age. The first one is a combination of needs four and six in the previous group: *avoiding nuisance from inside the dwelling and from outside*. One possible explanation for this blend and its slightly increased weight is the fact that a number of tenants have moved to smaller, more compact dwellings, which is a fairly normal course of events (the amount of larger dwellings in this group is 12% less). The proximity of family members and neighbours may cause nuisance from in- and outside the dwelling to be subjectively merged. Another reason (not restricted to smaller dwellings) might be that inhabitants are more often at home after retirement, so there is an increased possibility of being disturbed by smells or noises and of an invasion of privacy. There is still a separate need of avoiding smells from inside and of an adequate ventilation (in fourth position): a need of *a healthy atmosphere*. The former two explanations demonstrate one of the major complications in dwelling research: it is an interactive process, made up of mutual influences of the inhabitants and their physical environment, which are not easily untangled, even if the preferred method would have been possible. The best option is to study events from both sides and to perform additional research in order to establish true effects.

The second factor is also different: a need of *keeping the dwelling in order and receiving home-care,* including the ability to warn others in case of illness. Naturally, possibilities to call for help and to get home-care are valued more strongly when getting older, but the likely reason for the jump from a neutral to a very actual level is the addition of cleaning activities. These are important in case of home-care because hygiene is required and someone else has to carry them out.

A third difference is in the fifth need, which is a combination of *safety of movement and carrying out activities of daily life (ADL)*. Presumably, activities of daily life that imply a higher risk of injury (like stepping in and out of a bathtub) have a stronger bond with safe movement than before (e.g., walking) because of a decreasing mobility and a greater concern for safety. Other ADL items make up a separate need of *carrying out household activities (HA)*. The latter need has a low actual value, possibly because these activities are less risky in general and less strenuous in smaller dwellings.

The fourth change concerns social interaction. This need is divided into two components: a need of *contact* (including hearing of sounds from equipment e.g., a telephone or a doorbell) and a need of *privacy*. The principal explanation is the retirement of family members. They will spend more time at home and might experience a longing for a personal territory and opportunities to perform personal activities (hobbies). Thus, the privacy aspect is assigned a role of its own. Alternatively, tenants that have moved to a smaller house might encounter the various effects of reduced space on contact and privacy.

Fifth, in this group the need of *easily operating equipment and controls* is fairly, instead of very, actual. This can be attributed to the same two causes. After retirement there are more family members present to help solve problems. And smaller (often also newer) dwellings may offer more comfort in this respect. The content and priorities of other needs (*thermal comfort, feeling of social safety inside, physical warmth, comfortable air humidity,* and *stowing away possessions*) have not changed.

77 years of age and above

The factor analysis in this last age group showed 12 needs (9 factors, 3 unique problems). As in the youngest group, there are separate needs of *ADL* and of *safety of movement*. The ADL need, which includes home-care, is the most important one, and safety of movement has about the same actual value as in the other groups. The prime position of ADL and home-care is not surprising. The disconnection of safety of movement is probably due to the increase of the number of smaller dwellings in this group (from 26% to 40%). These dwellings may allow household activities to be carried out more easily so that all these activities are again clustered and mobility issues are set apart.

There is a distinction between a need of *avoiding nuisance from inside and outside* on the one hand and a healthy atmosphere on the other, similar to the second age group. The latter need, however, has a broader scope, encompassing not only avoiding smells, but also a feeling of social safety, the ease of operating locks, doors and windows, and a set of items that seem to involve a sense of security in sanitation (bathing, showering, going to the toilet and privacy). Hence, this is termed a need of *shelter*. Its second place is understandable in view of the fact that people this age often consider themselves more vulnerable. The former need occupies a lower position with a fairly actual value.

Needs involving social interaction also resemble those in the second group. There is one apparent difference: hobbies belong to the need of contact. This seems to indicate a change in the nature of those hobbies. While they probably required a separate space in the second group, they are now part of the interaction with others (perhaps suited to the living room). This can be the result of impairments that prevent hobbies demanding physical effort or of a reduction of the available space in smaller dwellings. Consequently, the first factor is a need of *controlled social interaction* and the second one, containing only privacy items, a *need of privacy*.

In the former age group, the need of *easily operating equipment and controls* had a fairly actual value and one of the alleged reasons was that after retirement family members may help to solve problems. The same need has now a very actual value that could originate from physical ailments (e.g., arthritis) and from widowhood. The physical condition of the very high aged also causes the need of *comfortable air humidity* to rise to a very actual level. In contrast, the needs of *keeping the dwelling in order*, *thermal comfort*, *physical warmth* and *hearing sounds of equipment* have lower priorities. In the first case this is caused by the fact that home-care is covered by the ADL need. In the others, the conditions in smaller, compactly built (and newer) dwellings may be held responsible (a better thermal climate and smaller distances that mitigate hearing problems). Finally, the need of stowing away possessions does not exist. The related problems are part of the ADL and social interaction needs, and inhabitants that have moved to smaller houses will already have taken account of the reduced space by parting with some of their belongings. Stowing away possessions will primarily concern objects for daily use and cleaning up when receiving visitors.

Dwelling features and recommendations

As stated in the Theoretical and Methodological Framework Section, this analysis concerns the material counterparts of dwelling needs in single family houses. These needs are similar to those in the younger age group (which is understandable because they are a majority in the sample as a whole and in these dwellings in particular). There are 12 needs (9 factors and 3 unique problems): (1) controlled social interaction; (2) avoiding nuisance from inside; (3) thermal comfort; (4) easily operating equipment and controls; (5) carrying out of activities of daily life; (6) avoiding nuisance from outside; (7) safety of movement; (8) keeping the dwelling in order; (9) feeling of social safety inside; (10) comfortable air humidity; (11) hearing sounds from equipment; and (12) stowing away possessions.

Correlations between tangible problems with dwelling features and dwelling needs are the main source. Additionally, several exact measures were used to establish cut-off points by means of t tests and analysis of variance (i.e., boundaries between a satisfactory and less satisfactory size of a kitchen or the number of rooms that are required). In addition, scatter plots and analysis of variance were used to check for curvilinear relations. The first part of this section contains a limited outline of the principal correlates (see Heijs, 2001 for a more comprehensive list). It has to be kept in mind that the values of the cut-off points are based on single family dwellings; for other types they will probably be smaller. In the second part, recommendations are given for prolonging independence in these dwellings.

Dwelling features

Controlled social interaction is primarily related to the number of bedrooms (minimum 3 for letting guests stay the night, home-care and hobbies) and their size, to the sizes of the living room (minimum 25 m^2) and other spaces that are normally used to receive visitors and to prepare and consume meals. Other features include the location of equipment (e.g., of the washing machine), a closed kitchen, two toilets, a relative position of various rooms and spaces that guarantees visual and auditory privacy, and a spacious entrance.

Avoiding nuisance from inside requires among other things adequate ventilation facilities and isolation within the dwelling, a closed kitchen and a suitable relative position of rooms, a proper finishing of walls and floor, and no moisture.

Thermal comfort is connected to the thermal insulation of the dwelling, the quality of the heating equipment, the finishing of walls and floor, ventilation facilities, the quality of doors and windows and the position of the dwelling relative to other houses.

Easily operating equipment and controls demands that the controls are reachable and in safe places (no sharp edges or corners nearby, not underneath a window-sill), that there is enough light and that one is able to find switches in the dark. Locks, doors and windows should open and close effortlessly. And the kitchen size is important for comfortably operating equipment.

The carrying out of ADLs is, of course, related to many dwelling features. The most important ones are the number of bedrooms (minimum 3) and the sizes of various rooms and spaces (minima are 25 m^2 for the living room, 7.5 m^2 freely usable space for the kitchen, 12 m^2 for the main bedroom and 3.5 m^2 freely usable space for the bathroom; and maximal requirements for outdoor space are: 75 m^2 for the backyard and 23 m^2 for the front yard). The relative position of rooms and spaces is also of interest (a horizontal floor-plan and the absence of stairs, direct connections between the living room and the main bedroom and between the main bedroom and the bathroom). Other features are: special appliances for elderly people (in the bathroom, the toilet and the kitchen), no doorsteps, ample storage space, adequate lighting, enough electric outlets, adequate space for waste disposal, no sharp edges or corners and suitable location of doors and windows (in view of cleaning, decorating and furnishing).

Avoiding nuisance from outside requires a proper insulation to prevent smells and noise, and a position of the rooms and the garden that foregoes an invasion of privacy.

Safety of movement is ameliorated by the absence of doorsteps and stairs (horizontal floor-plan), by an adequate relative position of rooms and spaces (direct connections between the living room and the main bedroom, and between the main bedroom, the bathroom and the toilet), special appliances for elderly people in the bathroom and the toilet, no sharp edges or corners, a comfortable finishing of the floors, adequate lighting and a satisfactory location of radiators, windows and doors.

The demands for keeping the dwelling in order are similar to those that are required for the needs of ADL and safety of movement. A horizontal floor-plan and the absence of stairs and doorsteps are at the top of the list, followed by adequate sizes (see above) and a good relative positioning of rooms and spaces, enough light to carry out these activities, no sharp edges or corners and a backyard that is less then 75 m^2.

Feelings of social safety are promoted by the quality of locks, windows and doors, outside lighting at the front- and the backdoor, and a proper finishing of the outside of the dwelling.

The stowing away of possessions is, of course, primarily dependent upon available space. The preferred location of this space for shorter periods of time is in the living room and for longer periods of time in the bedrooms. Other features are: a horizontal floor-plan, enough lighting in the relevant rooms and spaces, a usable height of cupboards in the kitchen, and proper facilities for waste disposal.

Material correlates for the needs of a comfortable air humidity and the hearing of sounds from equipment could not be established because these needs relate to features that are too specific to be identified in the questionnaire.

One unexpected result is that nearly none of the respondents mentioned tangible problems regarding too much space or too large rooms. This is notable because the prevailing opinion is that the elderly want a smaller dwelling. In fact, many tenants in the younger as well as in the older age groups stated a preference for more space and larger rooms.

Most of the other correlates that were mentioned are obvious. And this is also one of the reasons that overviews of needs and related features are necessary. If they are apparent, they may play a lesser role in design or be overlooked (assuming that they are taken care of). Only by acquiring an overview, can the (dis)advantages of design decisions be made manifest so that satisfactory trade-offs are possible.

Recommendations for prolonged independence in single family dwellings

Tangible problems concerning dwelling features can be more or less serious, and in many cases these problems are related to more than one need. To arrive at recommendations, the dwelling features were categorised as performance criteria according to the number of needs involved, their actual values, the seriousness of the tangible problems and the strength of the relation between seriousness and ageing (if a problem occurs sooner it must be dealt with at an earlier stage). Table 2 shows the most important criteria arranged by relevance.[3]

---

[3] This table only serves the present purpose; it is not a program of demands. The demands in such a program must contain links between needs and performance criteria in order to take proper design decisions.

Table 2. Most important performance criteria for prolonged independence

- a horizontal floor-plan and no stairs
- direct connections between the living room and the main bedroom and between the main bedroom, the bathroom and the toilet
- no doorsteps
- special appliances for elderly in the bathroom and the toilet
- a backyard of less than 75 m^2
- social safety measures like lighting outdoors
- a closed kitchen
- sufficient storage space
- a relatively large toilet, bathroom, kitchen and entrance
- a usable height of cupboards
- a proper waste disposal
- comfortable pavement
- enough space for equipment in kitchen and bathroom
- ease to open and close windows and doors
- taps, switches and radiator knobs that are easy to operate
- the prevention of a dry atmosphere
- no sharp edges or corners
- adequate lighting
- a good finish of the floor and the walls
- sufficient and large enough bedrooms

Because of the floor-plan, single family dwellings are less suitable than other types. In most cases, it is impossible or too costly to correct sizes or the relative positions of rooms. The sole option is to find some kind of compensation. Because various needs, as well as many of the tangible problems, indicate that a horizontal floor plan is essential, one could try to locate the principal functions downstairs (installing additional bathing, toilet, sleeping and storage facilities), to add equipment (an elevator) or to provide service.

In comparison, other measures are easy to take. Some are rather costly, such as the removal of doorsteps and the installation of special appliances. But several helpful modifications are certainly cost efficient: outdoor lighting and suitable locks for social safety; an adjustment of the height of cupboards; the addition of storage space where it is needed; providing a practical waste disposal, comfortable pavement, good doors and windows, lighting and taps, switches and radiator knobs that are appropriate for the elderly; the installation of equipment to prevent a dry atmosphere; rounding sharp edges or corners and applying a good finish to the floor and the walls for isolation and safety. The inhabitants might be assisted in looking for appliances that take up less space and in replacing the old ones.

Finally, the recommended reduction of outdoor space (mainly the backyard) is related to maintenance, which is a serious burden. It may be possible for housing associations to help design gardens that require less physical effort or to provide gardening services. Another possibility is to use the extra space in the garden to accommodate the downstairs facilities mentioned above (by adding a bathroom, a bedroom or even an external elevator).

Conclusions

The goal of this project was twofold: (1) to test a quantitative method for the uncovering of dwelling needs, for tracing changes in needs during the process of ageing and for classifying related dwelling features; (2) to establish dwelling needs in the population of elderly tenants of a housing association and derive recommendations for prolonging independent living.

The method has proven to be functional. It allowed the construction and ranking of dwelling needs for four selected samples (three age groups and one type of dwelling), the verification of changes in needs for the age groups and the categorisation of relevant dwelling features for single family dwellings. Though the method is somewhat laborious and large samples are required, the needs seem to be valid (as witnessed by their interpretability, by confirmation from their relation with dwelling features and by the agreement with some of the needs that are mentioned in the literature). At the same time it offers new insights in the nature and the relative importance of needs. For assembling general knowledge of the dwelling process and for understanding the needs of larger groups, it may be the preferable choice, while other, qualitative methods may be of service in a smaller and more specific context. A combination is recommended by using the outcomes of the first as input for the second. Methods like design participation may profit from an overview of needs and features. A consideration of all known needs (also the less actual ones) and features (including the obvious ones) may prevent decisions that further certain needs while obstructing others, thus improving trade-offs between solutions. With minor changes (to make its application easier) the method could also be used as a measure of congruence in research on person-environment fit.

A complication of this type of research is that mutual influences of users and environment are not easily untangled. Needs are altered in an interactive process that involves changes in both inhabitants and the dwelling. To gain insights in this process and in the dynamics of needs in relation to ageing, a longitudinal study is indicated. But since this type of research meets with objections such as learning effects, it would be better to examine the needs of different age groups in similar dwelling circumstances and to perform the analyses for each of the relevant dwelling types. Developments have to be looked at from both the user and the dwelling perspectives and in order to verify the effects it may be useful to formulate hypotheses for testing in separate projects. Reliability and generalisability (e.g., for home owners) are also key issues to be scrutinised in additional research. This study is only a first step, limited to one measurement, to tenants and to a sample size that did not allow a full implementation of the method.

Another reason for additional and continuous research is the ambition to create a complete set of needs. Existing methods cannot guarantee this, which is perfectly understandable because needs (like goals) tend to disappear when they are satisfied and turn up again if problems arise. Hence, a measurement at any given moment will necessarily be incomplete. However, quantitative methods using larger samples will probably provide a more complete overview. Furthermore, because of changes in society and technology, needs will alter. If continuous research is implemented, they will be detected when surfacing (again).

Keeping in mind the limitations mentioned above, the second goal has largely been reached. A first finding is that, while current initiatives for housing the elderly generally emphasise aspects of a practical nature (safety, useability, accessibility and adaptability), controlled social interaction is more important and probably undervalued as a need. Other significant needs that are usually not mentioned as such are: avoiding nuisance from inside and from outside the dwelling, easily operating equipment and controls, and keeping the dwelling in order. The remaining needs bear more resemblance to those in existing literature although there are also some differences with respect to content.

Second, it is obvious that a set of general needs will not suffice. Some needs show a fairly constant composition and weight (e.g., thermal comfort, the carrying out activities of daily living, easily operating equipment and controls, a feeling of social safety inside and stowing away possessions). However, the nature and relevance of other needs differ from one group to another and they change over time as a result of ageing and modifications in housing.

And finally, relations between needs and tangible problems with dwelling features clearly indicate that the popular belief that elderly prefer small(er) dwellings is at odds with the facts. On the contrary, the dwelling is often perceived as being not large enough. The main problem is not size but the absence of a horizontal floor plan, a variety of aspects involving safe movement (e.g., no doorsteps, usable height of cupboards, no sharp corners), sanitation and the maintenance of the garden. Attempts to improve single family dwellings with a view to prolonged independence will inevitably run up against the absence of a horizontal floor plan. This can be partly compensated for by (re)locating main functions downstairs (a complex and costly solution). Other measures are relatively easy to take, although they are sometimes expensive too. If the elderly indicate a desire to move to a smaller house, a housing association must ascertain that this is a well considered wish and not the result of choosing according to societal norms or insufficient knowledge of alternatives, the fact of the matter being that a number of important dwelling needs (e.g., social interaction, carrying out ADL and stowing away possessions) are not easily satisfied in a small dwelling.

References

Caplan, R. (1983). Person-environment fit: Past, present and future. In C. Cooper (Ed.), *Stress research* (pp. 35–76). New York: Wiley.

Carp, F. & Carp, A. (1984). A complementary/congruence model of well-being or mental health for the community elderly. In I. Altman, M. Lawton, & J. Wohlwill (Eds.), *Human behavior and environment: the elderly and the physical environment* (pp. 279–336). New York: Plenum.

Conway, T., Vickers, R., & French, J. (1992). An application of person-environment fit theory: perceived versus desired control. *Journal of Social Issues, 48(2)*, 95-107.

Cvitkovich, Y., & Wister, A. (2001). A comparison of four person-environment fit models applied to older adults. *Journal of Housing for the Elderly*, 14(1), 1-25.

Dieckman, F. (1998). Nutzorientierte Programmentwicklung [Use-oriented program development]. In F. Dieckmann, A. Flade, R. Schuemer, G. Ströhlein, & R. Walden (Eds.), *Psychologie und gebaute Umwelt [Psychology and the Built Environment]* (pp. 117-144). Darmstadt: Institut Wohnen und Umwelt.

Flade, A. (1993). Wohnen und Wohnbedürfnisse im Blickpunkt [Dwelling and dwelling needs in focus]. In H. Harloff (Ed.) *Psychologie des Wohnungs- und Siedlungsbaus [Psychology of Housing and Settlement]* (pp. 45-55). Göttingen: Verlag für Angewandte Psychologie.

Flade, A. (1998). Einleitung [Introduction]. In F. Dieckmann, A. Flade, R. Schuemer, G. Ströhlein, & R. Walden (Eds.). *Psychologie und gebaute Umwelt [Psychology and the Built Environmant]* (pp. 3-43). Darmstadt: Institut Wohnen und Umwelt.

Heijs, W. (2001). *Woonbehoelften van ouderen: een onderzoele onder oudeere huurders von de Woningstichting Hhvl [Dwelling needs of the elderly: an investigation among elderly tenants of housing association Hhvl].* Executive and Research Reports. Eindhoven: TUE, Interface.

Houben, P. (1985). *Maatschappelijke participaatie van ouderen en volkshuisvesting [Participation in society of the elderly and public housing].* Dissertation. Groningen: RUG.

Kahana, E. (1982). A congruence model of person-environment interaction. In M. Lawton, P. Windley, & Th. Byerts (Eds.), *Aging and the environment. Theoretical approaches* (pp. 97-121). New York: Springer

Lawrence, R. (1987). *Housing, dwellings and homes. Design theory, research and practice.* Chichester: Wiley.

Lawton, M. (1990). Ageing and performance of home tasks. *Human Factors, 32(5)*, 527-536.

Lawton, M., & Nahemow, L. (1973). An ecological theory of adaptive behavior and ageing. In C. Eisdorfer & M. Lawton (Eds.), *The psychology of adult development and ageing* (pp. 657-667). Washington: APA.

Linneweber, V. (1993). Wer sind die Experten? „User need analysis" (UNA), „Post occupancy evaluation" (POE) und Städtebau aus sozial- und umweltpsychologischer Perspektive [Who are the experts? "User need analysis" (UNA), "Post occupancy evaluation" (POA) and urban planning from the perspective of social and environmental psychology]. In H. Harloff (Ed.), *Psychologie des Wohnungs- und Siedlungsbaus [Psychology of Housing an Settlement]* (pp. 75-85). Göttingen: Verlag für Angewandte Psychologie.

Preiser, W., Vischer, J., & White, E. (1991). *Design intervention: towards a more humane architecture.* New York, Van Nostrand Reinholt.

Sanoff, H. (2000). *Community participation methods in design and planning.* New York: Wiley.

Saegert, S. (1985). The role of housing in the experience of dwelling. In I. Altman & C. Werner (Eds.), *Home environments.* New York: Plenum Press.

Sommer, R. (1983). *Social design.* New Jersey: Prentice Hall

Timmermans, J., Heide, F., de Klerk, M., Kooiker, S., Ras, M., & van Dugteren, F. (1997). Vraagevekenning wonen en zorg voor ouderen *[Exploration of the demands of the elderly concerning housing and care].* Rijswijk: SCP.

Older Spanish adults' involvement in the education of youngsters

Vicente Lázaro & Alfonso Gil

University of La Rioja, Spain

Abstract. The purpose of this study was to show older adults' activities carried out in the educational environment of children and young persons at different ages. A total of 179 participants between 60 and 80 years of age were included in the study and divided into four age groups: under 65, 65 to 69, 70 to 74 and over 75 years of age. Questionnaire results show older adults' active participation in educational and raising activities, which are mainly focused on children and young persons from their immediate family in various circumstances. The research also shows that the older the children are, the more they cut back on caregiving hours, and it should be also noted that as older adults age their participation is less significant. These educational and raising activities give them a feeling of satisfaction as they aid their own children doing so. Besides, they allege that they don't interfere with the educational guidance that parents offer their children and, in general, they think that significant changes have taken place in educational contexts from the past and present.

Keywords: old age, educational environments, educational contexts, family microsystems, intergenerational relationships

Introduction

Up to now, the relevance of older people has never been so considerable in society. According to Report 2000 from the IMSERSO (2001, p. 5) in 1999 Spain had a population of 40,202,160, of whom 16.2% were 65 or over 65, about six million and a half people. This population has increased their life expectancy. The INE (1993) set life expectancy around 73.40 years for men and 80.49 years for women. It is expected that by the year 2026 life expectancy will increase in Spain, about 77.65 years for men and 85.5 years for women.

Apart from a life expectancy increase, there are many other factors which show older adults' high quality of life, such as, better health conditions, improved social services, higher incomes, etc. This quality of life allows them to live in a more independent way until very old (Musitu, 2000).

Judging by this, it can be stated that up to now there have never been so many old people – and so healthy – participating actively in society. As research by Del Vado indicates (2001, p. 8): "they have retired from work but, fortunately, not from life. They are older not only due to their old age, but also in their spirit of sacrifice to the rest of people."

Older adults' active participation in society is channelled in very different ways. Their functional autonomy allows them to carry out a wide variety of tasks. Some of these tasks have been called *activities of daily living* (ADL; Fitzgerald et al., 1993), which, at the same time, have been divided into *basic activities* (feeding themselves, getting dressed, combing themselves, cooking, going to bed, bathing, etc.) and *instrumental activities* (travelling, shopping, cooking, house chores, medication, handling money, etc.). But, in addition to this, older people interact with other people at different levels and take on other types of *recreational activities*.

These findings also provide knowledge about the performance of these activities. Thus, for instance, in a study related to older people in the province of Cuenca by Martinez and Lozano (1998, p. 61), they found that "the participants in the study managed to do them without help, and with the following percentages, tasks such as: travelling (46.6%), shopping (72.6%), cooking (68.5%), house chores (56.4%), medication (80.2%) and handling money (74.0%)." Or, to mention another example, which is the most complete study ever done in our country on older adults, the report "La soledad en las personas mayors" ("Solitude among elderly people") from the CIS-IMSERSO (1998), reveals some activities done by older persons and how often they do them. Results show some leisure activities such as: mass media following – television (96.9%) and radio (71.4%), going for walks (70%), shopping and short trips (70%), going to a club, home-related or recreational activities (around 20%)

A recent study from the UK (National Statistics, 2001) shows a wide range of older people's ADL, and some significant results show that the older they are, the more difficulties they find in doing certain activities. For example, 13% of participants between 65–74 years of age found difficulties concerning PAs (gardening, decorating or doing household repairs) as opposed to 31% of participants over 75.

In spite of all the data accumulated, little quantified knowledge of the activities related to educational environment has still been obtained. Focusing on children and young people's educational contexts it's known that older people have played a role of great importance in their care and education, and that this role has changed, evolved and adapted to social transformations. Nowadays, Servais points out (2001, p. 98):

> "One of the most important factors that affect grandparents position in modern societies is the change brought about in family authority patterns. Today, respect

gains priority over authority in intergenerational relationships and confidence is ahead of respect as well; this is a principle accepted and proclaimed by parents and children."

But older people have not only changed the conceptions of their behaviour patterns, they are also caring more for children and young people of different ages than ever before. In fact, Vila affirms (1998, p. 157):

"we have observed that regarding many children between 0 to 12 years of age, their family is not only made up of mother and father, but it also extends to grandparents whose caregiving tasks play a very important role."

Life expectancy increases and improvements in economic and social conditions mean that there are more and more older adults who can care for children's needs and, at the same time, can play an active role in family and social support. The latest findings confirm this very same direction. Hernandez (2001, p. 139) indicates, when commenting on FOESSA study (1994), that "the most important weight of older adults' well-being support clearly benefit their sons and daughters (75% of all tasks.)." When analysing the content of these tasks we observe that they are distributed as follows: taking care of children (69%), house chores (42%), shopping (26%), making clothes, etc. (12%), bureaucratic managements (9%), and other tasks (11%).

Therefore, taking care of children represents the greatest older adults' support of families. The family becomes, as stated by Palacios and Rodrigo (1998, p. 35), an intergenerational meeting scene where third age people widen their vital horizons, laying a bridge towards the past (grandparents' generation) and towards the future (children's generation) ... in a network of social support for diverse vital transitions... "

However, we may wonder, before dealing with these types of tasks performed in family microsystems, in which contexts older adults' educational activities take place? Are they confined to what we know as the extended family or to other types of wider environments and other informal groups, like neighbourhoods, groups of friends, needy children, etc.?

Finally, what are the main activities that older people carry out for children's and young people's education, whatever the context is? Arago (1999, p. 313), in reference to the caregiving role grandparents play, comments: " grandparents can play very positive roles in assistance, help and encouragement; but, on the other hand, too much affection or strict discipline or interventionism, which annuls or decreases their parents figure, are clearly negative." In this sense, Serra, Gomez, Perez-Blasco, and Zacares (1998, p. 160) express that:

"... an essential contribution from grandparents is the transmission of that wisdom with respect to non-regulated, vital events or circumstantial transmissions of life and the way of facing it through their experience. Grandparents, nevertheless, should be aware of contexts and the historical-generational change that affects individual and family development."

In any case, it's a parents' decision to adopt a certain educational style (Ceballos & Rodrigo, 1998).

In reference to what has been said so far, some of the objectives set for this study are:
(1) to describe functions and activities that older adults regularly do, especially, functions and activities related to children and young people's education, care and raising in their affective environment
(2) to show which children and young people are the beneficiaries of these educational and child-raising activities
(3) to find out how often and in which circumstances older adults' interventions in the education and raising of children and young people take place.
(4) to verify how many hours older people dedicate to each of these three age groups: first childhood, childhood, adolescence and youth.
(5) to detect the circumstances which make older adults carry out educational and raising activities.
(6) to see which the interrelations between older adults and younger adults (children and young people's parents) related to educational functions are
(7) to find out older adults' opinions about the changes brought about in modern educational fields

Method

Individuals

A total of 179 individuals were included in the study, 124 older women and 55 older men between 60 and 80 years of age, arranged in four age groups: 24 participants were pre-retired under 65 (13.5%), 63 were between 65 and 69 (34.4%), 59 between 70 and 75 (31.1%), and 32 were over 75 (18.0%).

According to the objectives of the study, only older adults who could look after themselves were selected and those who were close to children or young people. The percentage of old people from the sample who had children or young people in their affective environment were: with children under 3 (41.9%), with children between 3 and 7 (42.5%), with children between 8 and 11 (34.6%), with adolescents between 12 and 14 (21.2%), and with young people between 15 and 18 years of age (26.3%).

Most of older adults belonged to a homogeneous social group that corresponds with a middle-low social status. Overall, participants' level of studies was: complete primary education (75.3%) or non-concluded primary education (14.0%). Their professions were "housewife" for women (71.7%) and "salary earner" for men (69.1%).

The immense majority came from the north of Spain (La Rioja, Navarre and the Basque Country) and they lived in localities of less than 200,000 inhabitants: in a locality of less than 500 inhabitants (2.8% of older adults), between 500 and 2,000 (8.9%), between 2,001 and 5,000 (12.3%), between 5,001 and 10,000 (10,1%), between 10,001 and 50,000 (106%), between 50,001 and 200,000 inhabitants (37.7%), and in a locality of more than 200,000 inhabitants (17.9%).

A great percentage of them (90.5%) had had children and, when the data were collected, some of their children were still living in the same locality as them (77.8%). Older adults who were married or living together (64.0%), widowers (28.7%, with 37.1% of widows as opposed to 9.3% of widowers), single (5.6%) and separated or divorced (1.7%). According to their living arrangements, individuals who lived alone (21.2%), with their spouse or partner (41.3%), in their home with their spouse and with some children (18.4%), with their spouse in their children's house (1.7%), in their house with some children and without spouse (5%), with other relatives (1.1%), in older adults' homes (10%), and in other circumstances (0.6%).

Procedure

Data were collected through a questionnaire. First, all the interviewees were asked personal questions. They were asked about age, marital status, place of residence, life style, number of children and grandchildren, grandchildren's ages, older adults' qualifications, and their occupations.

Having identified the individuals from the objective sample (participants who had children and young people in their affective environment regardless of their relationship) they were asked if they devoted their time, at present or in the past, to the education and raising of children and young people, and how often they did these activities nowadays.

Now the study focuses on the percentage of older people who perform a series of daily life tasks affecting children and young people in their family or their affective environments, frequency of these tasks and on which characteristic group of children and young people they are focused.

Children and young people were divided into three groups: children between 0 to 3 years of age, not attending school; children between 3 and 11, attending school and until childhood age limit because, of "complete neuronal development" (Rodrigo & Acuña, 1998, p. 263.); and adolescents and young people between 12 and 18 years of age, end of legal minority.

Results

Table 1 shows the percentage of randomized older adults who devoted their time to five suggested tasks in different contexts.

To begin with, 69.7% of participants spent time doing household chores (ironing, cleaning, washing etc.), although the frequency of these activities decreases as time passes. Basically, all participants in the under 65 age group are involved (95.8%), as opposed to the over 75 age group (38.7%). Gender highly conditions who performs these types of tasks: they are carried out by 87.8% of older women and 29.1% of older men.

Second type: House-related tasks such as doing small house repairs and decorating, are performed by 88.8% of older adults. As in the previous case, the frequency of these tasks diminishes as they grow older, from 100% in the under 65 and 65 to 69 age group to 51.6% in the over 75 age group. In this type of tasks, gender, as opposed to findings in the previous case, does not have any influence: they are done by 90.9% of men and 87.8% of women.

Practically all of them (97.2%) devoted time to sporting and recreational activities and socializing with other people. The percentages are very high in all age groups regardless of sex and age.

Tasks like "giving a hand to their children or other relatives" were performed by 75.3% of older adults; from 91.7% in the youngest group to 25.8% in the oldest. Men and women show similar percentages.

Grandchildren's care and education activities were carried out by 67.0% of them. Regarding age, the older their age, the less participation older adults show. Gender conditions to a lesser extent the performance of these types of tasks.

Table 1. Percentages of older adults who devote time to the performance of diverse child raising and educational activities

	TOTAL	GENDER		AGE			
		Man	Woman	under 65	65-69	70-75	over 75
Household chores (washing, ironing...)							
Yes	69.7	29.1	87.8	95.8	85.7	57.6	38.7
No	30.3	70.9	12.2	4.2	14.3	42.4	61.3
House-related activities (small house repairs, decorating...)							
Yes	88.8	90.9	87.8	100.0	100.0	91.5	51.6
No	11.2	9.1	12.2	.0	.0	8.5	48.4
Recreational and social activities							
Yes	97.2	98.2	96.7	100.0	96.8	96.6	96.8
No	1.7	.0	2.4	.0	1.6	1.7	3.2
No Know	1.1	1.8	0.8	.0	1.6	1.7	.0
Aid to their children or other relatives							
Yes	75.3	76.4	74.8	91.7	92.1	78.0	25.8
No	24.7	23.6	25.2	8.3	7.9	22.0	74.2
(*N*)	(178)	(55)	(123)	(24)	(63)	(59)	(31)
Care and education of their grandchildren or other young people and children							
Yes	67.0	61.8	69.4	79.2	77.8	69.5	31.3
No	33.0	38.2	30.6	20.8	22.2	30.5	68.8
(*N*)	(179)	(55)	(124)	(24)	(63)	(59)	(31)

Table 2 collects older adults' answers to the question to which children and young people they devoted their time. Findings show that practically all of them carry out this type of activities with their grandchildren. This was suspected from the beginning of the study, so we decided to consider other reasonable options apart from grandchildren. Thus, the greater beneficiaries are grandchildren (75.8% of the cases), as opposed to 7.8% of cases in which the beneficiaries are relatives' children, 1.7% friends' children, and 1.7% needy children and young people. Remarkable differences in gender can be observed: the percentage of women who devote their time to grandchildren is greater than that of men (79.7% as opposed to 67.3%), whereas the percentage of men who devote time to friends' children is greater than that of women (3.6% as opposed to 0.8%). Age seems not to have an outstanding influence.

Table 2. Beneficiaries of older adults' child-raising and educational activities

	TOTAL	GENDER		AGE			
		Man	Woman	under 65	65-69	70-75	over 75
Grandchildren							
Yes	75.8	67.3	79.7	70.8	79.4	74.6	74.2
No	24.2	32.7	20.3	29.2	20.6	25.4	25.8
Relatives' children							
Yes	7.8	7.3	8.1	8.3	7.9	6.8	9.4
No	92.2	92.7	91.9	91.7	92.1	93.2	90.6
Friends' children							
Yes	1.7	3.6	.8	.0	1.6	3.4	.0
No	98.3	96.4	99.2	100.0	98.4	96.6	100.0
Needy children and young people							
Yes	1.7	1.8	1.6	4.2	.0	.0	6.3
No	97.8	92.2	97.6	95.8	100.0	100.0	90.6
No answer	.6	.0	.8	.0	.0	.0	3.1
(N)	(179)	(55)	(124)	(24)	(63)	(59)	(32)

Table 3 gathers the results of frequency and circumstances of older adults' interventions in the care and education of young children.

Of the sample, 25.9% perform educational or caregiving tasks quite regularly; constantly, 29.3% when they are asked to, and 12.1% of them only when they themselves consider their help necessary. On the whole, the frequency of performance in each of the circumstances diminishes as participants grow older.

It is obvious that gender is very significant: 30.0% of women as opposed to 16.7% of men, and a great percentage of men (18.5%), compared to women (9.2%), spend time on this type of tasks only when they consider it necessary.

Table 4 shows the number of weekly hours they devoted to educational and child-raising activities in each of the three children and young people's age groups. The older the children are, the less hours they spend. To the smallest children, they devote an average of 14.58 hours per week, 13.98 hours to children attending school, and 5.40 hours to adolescents and young people.

As far as gender is concerned, it can be observed that men devote more time to the youngest children (0 to 3 years of age) than women; and women are more concerned about small children (3–11 years of age) and young people than men. Age is a relevant factor, the older the participants are, the less number of hours they employ.

Table 3. Frequency and circumstances of older adults interventions in the education and raising of children and young people

	TOTAL	GENDER		AGE			
		Man	Woman	under 65	65-69	70-75	over 75
Very frequently (constantly)	25.9	16.7	30.0	34.8	35.5	23.7	3.4
When my children, relatives, friends or some institution ask me to	29.3	24.1	31.7	34.8	25.8	32.2	24.1
When I consider my help necessary . . .	12.1	18.5	9.2	8.7	16.1	11.9	6.9
No answer	32.8	40.7	29.2	21.7	22.6	32.2	65.5
(N)	**(174)**	**(54)**	**(120)**	**(23)**	**(62)**	**(59)**	**(29)**

Table 4. Weekly hours devoted by older adults to educational and child-raising activities focused on children and young people

	TOTAL	GENDER		AGE			
(Media)		Man	Woman	under 65	65-69	70-75	over 75
To youngest children. .	14.58	15.35	14.26	17.08	12.95	16.15	7.50
(N)	(67)	(20)	(47)	(13)	(22)	(27)	(4)
To small children	13.98	12.71	14.53	10.75	15.57	13.73	2.33
(N).	(92)	(28)	(64)	(8)	(44)	(37)	(3)
To young people	5.40	4.25	5.72	12.75	7.63	4.69	1.63
(N	(55)	(12)	(43)	(4)	(19)	(16)	(16)

A total of 119 older adults (66.5%) out of 179 were involved in the raising and education of children and young people when the survey was done. These 119 participants were asked which the two most important characteristics of taking an active part in the education and care of their grandchildren and young people were. The answers are shown in Table 5, presented in descending order, were: 1) they felt they were helping their children and relatives (84.0%), 2) just a way of entertainment (52.9%), 3) they had a feeling of accomplishment doing these tasks (31.9%), 4) considered as an obligation (25.2%), 5) considered as a heavy load (0.8%), and 6) 3.4% of them answered otherwise.

With respect to age, there are some striking differences in the data. The percentage of older adults who consider it to be a great help for their children increases from 80.0% in the under 65 age group to 89.7% in the 70 to 75 age group, but descends to 60.0% in the over 75 age group. The three youngest groups of older adults consider that the most important aspect of taking care of their children and young people is that it helps to keep them entertained, a percentage of about 50.0%, but the percentage increases to 70.0% in those who are over 75.

It is also striking the percentage obtained to the answer "it was a feeling of accomplishment" descends as age increases: 50.0%, 34.7%, 20.5%, and 20.0% respectively in the different age groups. On the other hand, the consideration of these tasks as a moral obligation increases as they age: from 20% in the under 65 age group to 40% in the over 75 age group.

The differences due to gender are reflected especially in the percentage of older women (36.0%) who consider these tasks as a way of feeling accomplished as opposed to 21.2% of older men.

Table 5. Reasons for older adults' interventions in the education and child-raising of children and young people

Multipleanswer Question	TOTAL	GENDER Man	GENDER Woman	AGE under 65	AGE 65-69	AGE 70-75	AGE over 75
Aid to children	84.0	87.9	82.6	80.0	85.7	89.7	60.0
A moral obligation	25.2	27.3	24.4	20.0	24.5	25.6	40.0
A way of keeping entertained	52.9	54.5	52.3	50.0	51.0	53.8	70.0
A way of feeling fulfilled	31.9	21.2	36.0	50.0	34.7	20.5	20.0
A load	.8	3.0	.0	.0	2.0	.0	.0
Other answers	3.4	3.0	3.5	.0	2.0	7.7	.0
(N)	(119)	(33)	(86)	(20)	(49)	(39)	(10)

Table 6 shows the habitual forms of interaction between older persons and children and young people in educational activities; These are the percentages obtained when they were asked about what they advised their children and young people's parents to do to care for and raise their children in a proper way

A large percentage of respondents (38.8%) answered: "I try to advise them when my children or other relatives or friends ask me to," 16.3% of participants answered "I don't interfere in my children and other relatives' lives when they perform these tasks," 12.4% state that "I give my children or other relatives some advice when I think they need it," and 1.7% of them said, "I try to have things my way if I consider it necessary." A substantial percentage of them, 30.9%, did not answer this question.

Gender differences condition the way in which older adults interact with parents. In general, according to their answers, older women seem to interact more than older men when it comes to educational tasks, (72.4%) as opposed to (61.8%). The same holds true in each type of answer; thus, for example, in the first answer: "when my children or other relatives or friends ask me to," women seem to be in more demand (40.7%) than men (34.5%).

Age differences also seem to condition the answers. Overall, as they grow older, the percentage of grandparents that carry out educational activities somewhat decreases. For example, when older adults were asked if they give advice to others when they are asked to, the percentages range from (58.3%) of respondents under 65 to (15.6%) of respondents over 75 years of age.

Table 6. *Older adults' ways of interacting with the education and child-raising of children and young people*

	TOTAL	GENDER		AGE			
		Man	Woman	under 65	65-69	70-75	over 75
I advise them when they ask me to	38.8	34.5	40.7	58.3	45.2	35.6	15.6
I advise them when I think they need some advice	12.4	10.9	13.0	8.3	16.1	16.9	.0
I do not interfere in their lives	16.3	14.5	17.1	16.7	17.7	15.3	15.6
I try to have things my way...	1.7	1.8	1.6	4.2	.0	3.4	.0
No answer	30.9	38.2	27.6	12.5	21.0	28.8	68.8
(N)	(178)	(55)	(123)	(24)	(62)	(59)	(32)

They were also asked what they thought about the changes that have taken place in modern educational contexts; that is, comparing present contexts to those they knew from the past. the answers are collected in Table 7.

Table 7. Older adults' opinion about changes brought about in educational contexts

	TOTAL	GENDER		AGE			
		Man	Woman	under 65	65-69	70-75	over 75
They have varied a lot	93.8	96.4	92.7	87.5	92.1	96.6	96.8
They have changed little	5.6	1.8	7.3	12.5	7.9	3.4	.0
No answer	.6	1.8	.0	.0	.0	.0	3.2
(N)	(178)	(55)	(123)	(24)	(63)	(59)	(31)

Of the sample, 93.8% said that nowadays education and taking care of children differ considerably from what they experienced as young parents. Only 5.6% say that educational contexts have varied little.

A greater percentage of men (96.4%), women (92.7%) think that education has changed considerably. Perhaps the most striking characteristic of the data regarding intergenerational changes is that age differences are highly relevant, because the percentage of people who think that the contexts have changed considerably increases proportionally with maturity: from 87.5% of respondents under 65 to 96.8% of respondents over 75 years of age.

When they were asked in which way they thought that care and education had changed, a great deal of older adults mentioned freedom. They talked about it in general terms saying that nowadays' there is more freedom, or they were more specific in some of its aspects: "more freedom, more dialogue," or in other expressions such as "nowadays there is more freedom; not long ago everything was prohibited," "nowadays there is more freedom, too much," "there is too much freedom in sexuality and couple issues," etc.

Also, they talked about other concepts, such as communication ("now there is more information, more communication"), discipline ("now there is less discipline," "there is a lack of discipline," "much less respect"), economy ("now life is better," "there is more comfort, more means"), and they also mentioned other more formal educational aspects ("nowadays, teachers are more cultured and better trained," "there is more education," "parents play a more meaningful role in education").

Discussion and conclusions

These findings show that older adults between 60 and 80 years can be considered to be active people. They perform a wide range of ADL, getting in touch with other people, and support their own children raise and educate their children.

In this study, it was considered quite relevant to ask the interviewees about five different types of tasks they are able to carry out, and we paid special attention to the fifth question. These are the tasks considered in the survey:

(1) household chores (washing, ironing, cleaning etc.)
(2) house-related activities (small house repairs, decorating, etc.)
(3) recreational and social activities
(4) supporting activities, giving a hand to their children, young people or other relatives
(5) raising and educational activities focused on their grandchildren and other children and young people

Before offering findings on each of these types of tasks, in general terms, it should be pointed out that age represents the most conditioning factor, which has a very similar influence on all types of activities.

It can be stated that when older adults age the frequency of these tasks is lower. Proof for this has been obtained as participants were divided into 4 groups: 1) under 65, 2) between 65 and 69, 3) between 70 and 75, and 4) over 75 years of age. Depending on the type of activities there is a decrease in the percentages, and in the over 75 age group this decrease is particularly significant compared to the other three age groups. Results are in line with the National Statistics study (2001), which highlights that "…as time passes, older persons find greater difficulties in performing a certain kind of activities of daily living." All these results seem to point out that a certain pattern of decline is somehow emerging: "…while in the psychological field, and especially in those dimensions of motivation and a person's self concept, the perception of stability might be the significant feature throughout their lives, the biological field might show clear subjective patterns of decline..." (Triadó & Villar, 1999, p.76), because there is a decrease in the percentage of older adults who perform certain physical tasks requiring effort as they grow older.

In the psychological field, some authors report that certain cognitive processes, such as emotion (Isaacowitz, Charles & Carstensen, 2000), experience no declining trends as age increases, and stability patterns seem to explain how these processes work in a better way.

The percentage of older adults who do house chores (washing, ironing, etc.) is relevant (69.7%), but it should be noted that marital rates by gender are distinctly different: 87.8% of women and 29.1% of men; that is to say, the total percentage might vary considerably if the number of individuals of both sexes increased.

As to house-related tasks (small house repairs, home modifications etc.), results show differences from the previous case. The percentage of older adults who do them is very high (88.8%), with responses from both genders being very similar, older men (90.9%) and older women (87.8%). Work distribution by gender is likely to account for these results, in this case, older men have a higher percentage compared to house chores data.

Previous task percentages seem to be in line with results from other studies. For example, Martinez and Lozano (1998) found that tasks like shopping were done by 72.6% or cooking by 68.5% of their sample; or in the studies from the INSERSO (1998), where 70% of them do the shopping or short trips.

As far as recreational and social activities are concerned, practically all of them carry them out (97.2%), regardless of sex and age.

Data shows that 75.3% of older adults care for their children or other relatives ("giving a hand"); older men's (76.4%) and older women's (74.8%) percentages are very similar. These data coincide with the report from FOESSA (1994), which indicates that these tasks represent 75% of all older adults' caregiving performances.

Finally, specific support activities related to care and education of their grandchildren or other children and young people were carried out by 67.0% of them. Gender, to a lesser extent, conditions performance and accomplishment of these types of task: 69.4% of older women and 61.8% of older men. These data are also coincide with the report from FOESSA (1994), which indicates that these tasks are done by 69%.

Older adults' educational and raising interventions occur mainly in family microsystems. According to the data, 75.8% of their interventions are focused on their grandchildren. For example, research by Villa (2001) indicates that today's life style, which implies work obligations for most Spanish couples, has forced them to have to care more for their grandchildren. Results show that only 7.8% of the interventions are focused on relatives' children, 1.7% on friends' children, and 1.7% on needy children and young people.

Analysing these data, the general impression is that older adults with grandchildren devote plenty of time to them, and when it is not with their grandchildren, these caregiving activities are focused on other groups of children. It would be highly interesting to observe if childless, unmarried older adults act as caregivers to friends' children or needy children. Probably results would show that some of them somehow would care for other people's children, similar to older adults' caregiving associations from the U.S.A. As Martin says (1994, p. 277) in reference to Jusenius (1987) commenting on the Retired Senior Volunteer Program (RSUP), "it is a program open to all older adults who want to participate as tutors at educational centres (registered: 300,000 older people, 80% of older women and two thirds are unmarried)".

A great percentage of older adults (67.2% of our sample) carry out educational or raising tasks with children and young people, although in different circumstances: one fourth of them (25.9%) perform them quite regularly, 29.3% when they are asked by their children, relatives, friends or some institution, and 12.1% of respondents when they consider their help necessary. Older women perform these tasks more frequently than older men. Progressive aging makes them abandon educational activities, except when their children, relatives, friends or institutions ask them to.

Regarding the number of hours per week older adults dedicate to different activities with children and young people, data indicates that there is a connection between children, young people and participants' age and the number of hours. The younger children and older adults are, the larger the number of hours is. Thus, for example, it should be noted that the highest average of dedication (17.08 hours) belongs to the youngest children's age group and to the under 65 age group.

In reference to older adults' opinion about which the most significant aspects are when devoting their time to the education and raising of their grandchildren or other children and young people, it can be stated that they volunteer to do these tasks which does not involve a sacrifice but a satisfaction. Social reasons may account for these behaviours. For example, Pruchno and Mckenney (2002) find certain similarities in well-being patterns among black and white caregiving grandmothers; Goodman and Silverstein (2002) detect greater well-being among Hispanic grandmothers than among white and Afro-American ones with grandchildren's custody gained after parents' incarceration. Glasser and Tomassini (2000) observe that cultural differences are quite significant: British grandmothers come closer to their children to be cared for whereas Italian grandmothers do so to care for their grandchildren's needs.

In our study, results show that the main reason older adults have for performing caregiving activities, is to "give a hand" to their children and relatives (84.0%). Additionally, these activities may represent a way of keeping them entertained (52.9%), a feeling of accomplishment (31.9%), or moral obligation. Few consider it to be a load (0.8%).

Among the data, some incidental percentages seem to reflect a consequence of the traditional separation between male and female roles. A greater percentage of older women (36.0%) consider raising and educational activities as a "personal accomplishment" compared to older men (21.2%)

Older adults' interactions with parents, children and young people, focused on what advice or help they give in children's education and care, do not interfere in parents' educational orientation. So, they seem to line up with Palacios, Hidalgo and Moreno (1998) study, who indicated that parents set the educational guidance of their children.

Therefore, support from older adults seems not to become an obstacle for the achievement of the objectives that the parents themselves set. Besides, there is some evidence that it may not be negative, since, in fact, some authors such as Solomon and Marx (1995) do not find any substantial differences between children raised solely by grandparents and those raised by parents.

A great percentage of older adults (38.8%) affirm, "I give them some advice when my children or other relatives or friends ask me to". As participants become older, less advice is demanded from them: from 58.3% of respondents under 65 to 15.6% of them over 75 years of age. Some feasible causes may be that, the younger the group of older adults is, the better the relationship between children-parent and parents-grandparent generations and it should be also noted that the oldest adults live more secluded, even in older adults' homes, which make intergenerational relationships more difficult. These relationships have a positive effect on each of the groups (Newman, Faux & Larimer, 1997) provided that there is no "over stimulation," which may lead to stressful situations in their own family (Silverstein & Angelelli, 1998) as well as in other intergenerational contexts (Salari, 2002)

It is practically unanimous, from older people's point of view, that substantial changes have been brought about in diverse educational contexts that they have known as children, parents, and, at present, as grandparents. Third generation people consider modern education and childcare to be very different from the past, both in formal and informal aspects.

The percentage of older adults who think that contexts have varied a lot increases proportionally as there is an increase in the age of the individuals: from 87.5% of respondents under 65 to 96.8% of them over 75.

To conclude, older adults think that nowadays there is more freedom in education, more communication, more means and facilities, better trained teachers greater parental participation, and less discipline than when they were young parents caring for their small children.

References

Aragó, J. M. (1999). Aspectos psicosociales de la senectud [Psychosocial aspects of old age]. In M. Carrero, J. Palacios, & A. Marchesi (Eds.), *Psicología evolutiva 3. Adolescencia, madurez y senectud [Evolutionary pschology 3. Adolescende, maturing and old age]*, pp. 289-325. Madrid: Alianza.

Ceballos, E., & Rodrigo, J.M. (1998). Las metas y estrategias de socialización entre padres e hijos [Socialisation goals and strategies between parents and children]. In M. J. Rodrigo, & J. Palacios (Eds.), *Familia y desarrollo Humano [Family and human development]*, pp. 225-243. Madrid: Alianza.

Cis-Imsero. (1998). *La soledad en las personas mayores [Solitude in older people]*. Estudio núm. 2.279 de febrero de 1998.

Del Vado, S. (2001). Mayores y solidarios [Old and supportive]. *Sesenta y más,* 193 (Mayo), 8-13.

Fitzgerald, J.F., Smith, D.M., Martin, D.K. et al. (1993). Replication of the multidimensionaly of activities of daily living. *Journal of Gerontology: Social Sciences,* 48, 28-91.

FOESSA. (1994). *V Informe Sociológico sobre la situación social de España [Fifth sociological report on Spain's social situation]*, pp.481-484. Madrid: Fundación FOESSA.

Glaser, K., & Tomassini, C. (2000), Proximity of Older Women to Their Children: A Comparison of Britain and Italy. *Gerontologist,* 40 (6), 729-737.

Goodman, C., & Silverstein, M. (2002). Grandmothers Raising Grandchildren: Family Structure and Well-Being in Culturally Diverse Families. *Gerontologist,* 42 (5), 676-689.

Hernández Rodríguez, G. (2001), Familia y ancianos [Family and older people]. *Revista de educación,* 325, 129-142.

IMSERSO. (2001). *Informe 2000: Las personas mayores en España [Report 2000: Older people in Spain]*, pp.5-47. Madrid: Ministerio de Trabajo y Asuntos Sociales, Instituto

de Migraciones y Servicios Sociales. <www.seg-social.es/imserso/mayores/docs/iO.mayobs.22.html>.
INE (1993): *Tablas de mortalidad de la población española 1990-91 [Mortality rates in the Spanih population]*. Madrid: INE.
Isaacowitz, D.M., Charles, S.T., & Carstensen, L.L. (2000). Emotion and Cognition. In F.I.M. Craik & T.A. Salthouse (Eds,), *The Handbook of Aging and Cognition*, pp. 593-631. London: LEA.
Jusenius, R.C. (1987). Retirement and Older Americans Participation in Volunteer Activities. In S.H. Sandell, *The Problem isn't age Word and Older Americans*. New York: Praeger Publisher.
Martín, A. (1994). Relaciones intergeneracionales y educación: el concepto de comunidad de generaciones [Integenerational relationships and education: The concept of a community of generations]. *Bordón,* 46 (3), 273-281.
Martínez Vizcaíno, V., & Lozano, A. (1998). *Calidad de vida de los ancianos [Older people's quality of life]*. Cuenca: Ediciones de la Universalidad de Castilla-La Mancha.
Musito, G. (2000). Las redes de apoyo social en la persona mayor [Older people's social support networks]. In M. Medina & M. Ruiz (Eds.), *Políticas sociales para las personas mayores en el próximo siglo [Social policies regarding older people in the next century]*, pp. 61-86. Murcia: Universidad de Murcia.
National Statistics (2001). Living in Britain 2001. Avaiable online at <http://www.statistics.gov.uk/lib2001/Section3745.html>.
Newman, S., Faux, R., & Larimer, B. (1997). Children's views on aging: their attitudes and values. *Gerontologist,* 37, 412-417.
Palacios, J., Hidalgo, M.V., & Moreno, M.C. (1998). Ideologías familiares sobre el desarrollo y la educación infantil. In M. J. Rodrigo & J. Palacios (Coords.): *Familia y desarrollo Humano [Family and human development]*, pp.181-200. Madrid: Alianza.
Palacios, J., & Rodrigo, J.M. (1998). La familia como contexto de desarrollo [Family as development context]. In M. J. Rodrigo & J. Palacios (Coords.), *Familia y desarrollo Humano [Family and human development]*, pp. 25-44. Madrid: Alianza.
Pruchno, R.A., & McKenney, D. (2002). Psychological Well-Being of Black and White Grandmothers Raising Grandchildren: Examination of a Two-Factor Model. *Journal of Gerontology: Social Sciences,* 57 (5), 444-452.
Rico Sapena, C., Serra Desfilis, E., & Viguer Seguí, P. (2001): *Abuelos y nietos. Abuelo favorito, abuelo útil [Grandparents and grandchildren. Favourite grandparent, helpful grandparent]*. Madrid: Pirámide.
Rodrigo, M.J., & Acuna, M. (1998). El Escenario y el curriculum educativo familiar [Family scenario and educational curriculum]. In M. J. Rodrigo & J. Palacios (Coords.), *Familia y desarrollo Humano [Family and human development]*, pp. 261-276. Madrid: Alianza.
Salari, S.M. (2002). Intergenerational Partnerships in Adult Day Centers: Importance of Age-Appropriate Environments and Behaviors". *Gerontologist, 42,* 321-333.
Serra, E, Gómez, L., Pérez-Blasco, J., & Zacarés, J. J. (1998). Hacerse adulto en familia: una oportunidad para la madurez [Growing up in an family: an opportunity for matury]. In M. J. Rodrigo & J. Palacios (Coords.), *Familia y desarrollo Humano [Family and human development]*, pp.140-160). Madrid: Alianza.
Servais, P. (2000). La perspectiva sociológica, histórica de intervención. Las personas mayores y la familia en Europa del siglo XVI al siglo XX [Sociological and historical

perspective of intervention. Older people and family in Europe from the 16th to the 20th century]. In S. Adroher (coord.): *Mayores y Familia [Older people and family],* pp. 37-53. IMSERSO y Universidad Pontificia Comillas: Madrid.

Silverstein, M., & Angelelli, J.J. (1998). Older parents' expectations of moving closer to their children. *Journal of Gerontology: Psychological Sciences and Social Sciences,* 53, 153-163.

Solomon, J.C., & Marx, J. (1995). 'To grandmother's house we go', Health and school adjustment of children raised solely by grandparents". *Gerontologist,* 35, 386-394

Triadó, C., & Villar, F. (1999), Teorías implícitas del cambio evolutivo en diferentes cohortes: representación de pérdidas y ganancias en la adultez [Implicit therories of evolutionary change in different cohorts: Representation of losses and gains in adulthood]. *Infancia y aprendizaje,* 86, 73-90.

Trilla, J. (1996). *La educación fuera de la escuela. Ámbitos no formales y educación social [Education outside school. Informal spheres and social education].* Barcelona: Ariel.

Vila, I. (1998). *Familia, escuela y comunidad [Family, school and community].* Barcelona: Horsori.

Vila, J. (2001). Actitudes sociales de los mayores [Older people's social attitudes]. *Sesenta y más,* 195-196 (Julio-Agosto), 64-69.

Authors

M. Teresa Anguera Argilaga
University of Barcelona
Faculty of Psychology
Campus Mundet
Passeig de la Vall d'Hebrón, 171
08035 Barcelona
Spain
tanguera@ub.edu

Marcello Balzani
University of Ferrara
Dept. of Architecture
Via Quartieri, 8
44100 Ferrara
Italy
bzm@unife.it

Jussara Basso
Faculty of Architecture/PROPUR
Federal University of Rio Grande do Sul
Rua Sarmento Leite, 320
Porto Alegre, RS
CEP 90050-170
Brazil
tarcisio@orion.ufrgs.br

Stelamaris R. Bertoli
State University of Campinas – UNICAMP
Department of Architecture and Construction
School of Civil Engineering
CP 6021
13084-971 Campinas, SP
Brazil
rolla@fec.unicamp.br

Miriam Billig
Dept. of Geography
Bar Ilan University
Ramat Gan
Israel
biligm@mail.biu.ac.il

Sarah Blandy
School of Environment and Development
Sheffield Hallam University
City Campus
Howard Street
Sheffield S1 1WB
United Kingdom
s.blandy@shu.ac.uk

Francisco Borges Filho
State University of Campinas – UNICAMP
Department of Architecture and Construction
School of Civil Engineering
CP 6021
13084-971 Campinas, SP
Brazil
borges@fec.unicamp.br

Antonio Borgogni
UISP Ferrara
Via Verga, 4
44100 Ferrara
Italy
aborgogni@tin.it

Arza Churchman
Faculty of Architecture and Town Planning
Technion, Israel Institute of Technology
Haifa
Israel
arzac@tx.technion.ac.il

Fitnat Cimsit
Istanbul Technical University
Faculty of Architecture
Taksim 80191
Istanbul
Turkey
fitnatc@beykent.edu.tr

Karen Cronick
Institute of Psychology
Central University of Venezuela
P.O. Box 47018
Caracas 1041-A
Venezuela
kcronick@reacciun.ve

Authors

Maria Cristina Dias Lay
Faculty of Architecture/PROPUR
Federal University of Rio Grande do Sul
Rua Sarmento Leite, 320
Porto Alegre, RS
CEP 90050-170
Brazil
tarcisio@orion.ufrgs.br

Erincik Edgü
Istanbul Technical University
Faculty of Architecture
Taksim 80191
Istanbul
Turkey
erincik@superonline.com

Nan Ellin
School of Architecture
College of Architecture and Environmental Design
Tempe, AZ 85287-1605
USA
nan.ellin@asu.edu

M. Celia Espinosa Arámburu
Universidad Nacional Autónoma de México
Facultad de Psicología
Av. Universidad 3004
Col. Copilco-Universidad
Delegación Coyoacán
México, 04510, D.F.
Mexico
celyesp@yahoo.com

Ricardo García Mira
People–Environment Research Unit
Social Psychology Laboratory, Dept. of Psychology
Faculty of Educational Sciences
University of Corunna
Campus of Elviña
15071 A Coruña
Spain
fargmira@udc.es

Alfonso Gil
Avda. de La Rioja, 25, 1º
26520-Cervera del Rio Alhama
La Rioja
Spain
rmmmmm@teleline.es

Michael J. Greenwald
Dept. of Urban Planning
School of Architecture and Urban Planning
University of Wisconsin – Milwaukee
2131 East Hartford Avenue
Milwaukee, WI 53211
USA
mgreenwa@uwm.edu

Wim J. M. Heijs
Department of Architecture, Building and Planning
Eindhoven University of Technology
PO Box 513
5600 MB Eindhoven
The Netherlands
w.j.m.heijs@bwk.tue.nl

Karin Høyland
SINTEF Civil and Environmental Engineering
Architecture and Building Technology
Alfred Getz vei 3
7465 Trondheim
Norway
Karin.hoyland@civil.sintef.no

Doris C.C.K. Kowaltowski
State University of Campinas – UNICAMP
Department of Architecture and Construction
School of Civil Engineering
CP 6021
13084-971 Campinas, SP
Brazil
doris@fec.unicamp.br

Chin-Chin (Gina) Kuo
Faculty of Architecture
University of Sydney
Room 506, Building G04
148 City Road
New South Wales 2006
Australia
kuo_c@arch.usyd.edu.au

Department of Architecture
Feng Chia University
100 Wen Hua Road
Taichung, Taiwan 427
cckuo@fcu.edu.tw

Lucila C. Labaki
State University of Campinas – UNICAMP
Department of Architecture and Construction
School of Civil Engineering
CP 6021
13084-971 Campinas, SP
lucila@fec.unicamp.br

Vicente Lázaro
Departamento de CC. Humanas y Sociales
Universidad de La Rioja
Edificio Vives. C/ Luis de Ulloa, s/n
26004 Logroño (La Rioja)
Spain
vicente.lazaro@dchs.unirioja.es

Yung-Jaan Lee
Graduate Institute of Architecture and Urban Planning
Chinese Culture University
55, Hwa-Kang Road, Yang-Ming-Shan
Taipei
Taiwan 111
yjlee@faculty.pccu.edu.tw

Alain Legendre
CNRS Laboratoire de Psychologie Environnementale UMR 8069
Institut de Psychologie
Université René Descartes-Paris V
71, avenue Edouard Vaillant
92774 Boulogne-Billancourt Cedex
France
alain.legendre@psycho.univ-paris5.fr

María Dolores Losada Otero
University of Corunna
Psychology Department – Social Psychology Laboratory
Faculty of Education Sciences
Campus of Elviña
15071 A Coruña
Spain
lolilo@udc.es

Laura Migliorini
University of Genoa
Department of Anthropological Sciences
Vico S.Antonio 5/7 Sc.B
16126 Genoa
Italy
migliori@nous.unige.it

Silvia A. Mikami G. Pina
State University of Campinas – UNICAMP
Department of Architecture and Construction
School of Civil Engineering
CP 6021
13084-971 Campinas, SP
Brazil
smikami@fec.unicamp.br

Gabriel Moser
CNRS Laboratoire de Psychologie Environnementale UMR 8069
Institut de Psychologie
Université René Descartes-Paris V
71, avenue Edouard Vaillant
92774 Boulogne-Billancourt Cedex
France
moser@psycho.univ-paris5.fr

Michael Ornetzeder
Centre for Social Innovation
Koppstraße 116/11
A-1160 Vienna
Austria
ornetzeder@zsi.at

David Parsons
School of Environment and Development
Sheffield Hallam University
City Campus
Howard Street
Sheffield S1 1WB
United Kingdom
d.e.parsons@shu.ac.uk

Antonella Piermari
University of Genoa
Department of Anthropological Sciences
Vico S.Antonio 5/7 Sc.B
16126 Genoa
Italy
anto_pi@libero.it

A. Terry Purcell
Faculty of Architecture
University of Sydney
Building G04
New South Wales 2006
Australia
terry@arch.usyd.edu.au

Eugénia Ratiu
CNRS Laboratoire de Psychologie Environnementale UMR 8069
Institut de Psychologie
Université René Descartes-Paris V
71, avenue Edouard Vaillant
92774 Boulogne-Billancourt Cedex
France
ratiu@psycho.univ-paris5.fr

Maaris Raudsepp
Environmental Psychology Research Unit
Tallinn Pedagogical University
Narva Road 25
10120 Tallinn
Estonia
maaris@tpu.ee

Antonella Rissotto
Institute of Cognitive Sciences and Technologies
National Research Council
Viale K. Marx 15
00137 Rome
Italy
rissotto@ip.rm.cnr.it

José Romay Martínez
University of Corunna, Spain
Social Psychology Laboratory, Dept. of Psychology
Faculty of Education Sciences
University of Corunna
Campus of Elviña
15071 A Coruña
Spain
romay@udc.es

Regina C. Ruschel
State University of Campinas – UNICAMP
Department of Architecture and Construction
School of Civil Engineering
CP 6021
13084-971 Campinas, SP
Brazil
regina@fec.unicamp.br

José Manuel Sabucedo Cameselle
Faculty of Psychology
University of Santiago de Compostela
Campus Sur
15706 A Coruña
Spain
sabucedo@usc.es

Euclides Sanchez
Institute of Psychology
Central University of Venezuela
P.O. Box 47018
Caracas 1041-A
Venezuela
eusanche@reacciun.ve

Carlos Santoyo Velasco
Universidad Nacional Autónoma de México
Facultad de Psicología
Av. Universidad 3004
Col. Copilco-Universidad
Delegación Coyoacán,
México, 04510, D.F.
México
carsan@servidor.unam.mx
celyesp@yahoo.com

Andrew D. Seidel
College of Architecture
Texas A&M University
College Station, TX 77843-3137
USA
a-seidel@tamu.edu

Quentin Stevens
School of Geography, Planning and Architecture
University of Queensland
Brisbane QLD 4072
Australia
q@uq.edu.au

Alper Ünlü
Istanbul Technical University
Faculty of Architecture
Taksim 80191
Istanbul
Turkey
aunlu@itu.edu.tr

David Uzzell
Department of Psychology
University of Surrey
Guildford
Surrey GU2 7XH
United Kingdom
d.uzzell@surrey.ac.uk

Lucia Venini
University of Genoa
Department of Anthropological Sciences
Vico S.Antonio 5/7 Sc.B
16126 Genoa
Italy
venini@nous.unige.it

Charles A. J. Vlek
Department of Psychology
University of Groningen
Grote Kruisstraat 2/1
9712 TS Groningen
The Netherlands
c.a.j.vlek@ppsw.rug.nl

Esther Wiesenfeld
Institute of Psychology
Central University of Venezuela
P.O. Box 47018
Caracas 1041-A
Venezuela
ewiesen@reacciun.ve

John Zacharias
Urban Studies Programme
Concordia University
1455 de Maisonneuve Boulevard West
Montreal, Quebec H3G 1M8
Canada
zachar@vax2.concordia.ca

Reviewers

Thanks are due to the following individuals who have served as reviewers on this project:

María Amérigo Cuervo-Arango, *University of Castilla La Mancha, Spain*
María Teresa Anguera Arguilaga, *University of Barcelona, Spain*
Juan I. Aragonés Tapia, *University of Madrid, Spain*
Constantino Arce, *University of Santiago de Compostela, Spain*
Mirilia Bonnes, *University of Rome 'La Sapienza', Italy*
Lineu Castello, *University of Porto Alegre, Brazil*
Arza Churchman, *Technion – Israel Institute of Technology, Israel*
Victor Corral, *University of Sonora, Mexico*
José Antonio Corraliza, *Autónoma University, Madrid, Spain*
Ricardo de Castro, *Junta de Andalucía, Spain*
Ricardo García Mira, *University of Corunna, Spain*
Maria Vittoria Giulliani, *National Research Council, Rome, Italy*
Terry Hartig, *University of Uppsala, Sweden*
Antonio Hernández Mendo, *University of Malaga, Spain*
Bernardo Hernández Ruiz, *University of La Laguna, Spain*
Liisa Horelli, *Helsinki University of Technology, Finland*
Florian Kaiser, *Eindhoven University of Technology, The Netherlands*
Satoshi Kose, *Tsukuba Science City, Ibaraki-ken, Japan*
Roderick Lawrence, *University of Geneva, Switzerland*
Alain Legendre, *University René Descartes – Paris V, France*
Bob Martens, *Vienna University of Technology, Austria*
Jeanne M. Moore, *University of Teesside, Middlesbrough, United Kingdom*
Gabriel Moser, *University René Descartes – Paris V, France*
Toomas Niit, *Tallinn Pedagogical University, Estonia*
Maria Nordström, *Stockholm University, Sweden*
Jose Q. Pinheiro, *Federal University Rio Grande Norte, Brazil*
Enric Pol Urrútia, *University of Barcelona, Spain*
Eulogio Real Deus, *University of Santiago de Compostela, Spain*
Mauro Rodriguez Casal, *University of Santiago de Compostela, Spain*
José Romay Martíne, *University of Corunna, Spain*
Ombretta Romice, *University of Strathclyde, Glasgow, United Kingdom*
José Manuel Sabucedo, *University of Santiago de Compostela, Spain*
César San Juan, *University of the Basque Country, Gipuzkoa , Spain*
Andrew Seidel, *Texas A&M University, USA*

Acknowledgements

The editors want to acknowledge the work carried out by Gemma Blanco Martínez, research fellow of the Social Psychology Laboratory of the University of A Coruña, who helped with the co-ordination of this work, specially during the review process of the contributions of this book.

iaps 2002

The editors express their gratitude to the following institutions and organizations for their cooperation in and support of this project:

Ayuntamiento de La Coruña
Concello de A Coruña

Fundación Biodiversidad

FUNDACIÓN CAIXAGALICIA

MINISTERIO DE CIENCIA Y TECNOLOGÍA

MINISTERIO DE MEDIO AMBIENTE

Coca-Cola
BEGANO, S.A.

REPSOL YPF

XUNTA DE GALICIA
CONSELLERÍA DE MEDIO AMBIENTE
Dirección Xeral do Centro de Información e Tecnoloxía Ambiental

Hogrefe & Huber's Online Journals Service

PsyJOURNALS

The online psychology information service

PsyJOURNALS is Hogrefe & Huber's online journals service. In PsyJOURNALS you can search online and access the full text (in most cases, from 1999 onwards) of the 25 psychology and psychiatry journals published by the Hogrefe & Huber group.

If you have a personal subscription to one or more of the Hogrefe & Huber journals, you are entitled to free online access to the full text of that journal. You can use the online activation to obtain password-protected access to the corresponding journal(s).

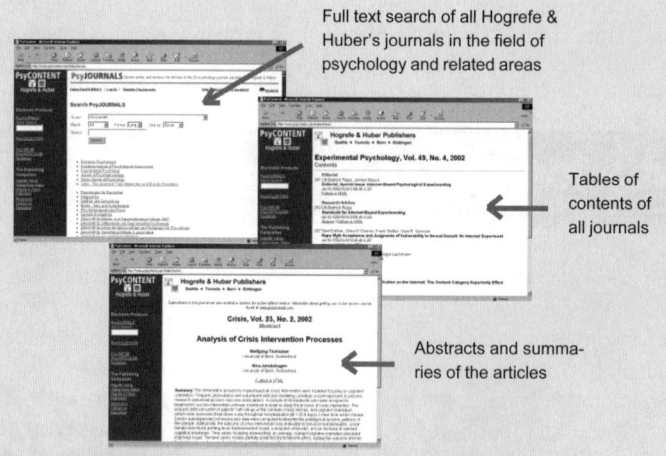

Full text search of all Hogrefe & Huber's journals in the field of psychology and related areas

Tables of contents of all journals

Abstracts and summaries of the articles

You can also obtain online access to the full text of all 25 Hogrefe & Huber journals if you have a personal subscription to at least one of them and register for PsyJOURNALS, for an annual access fee of only US $148.00 / € 148.00.
Online registration: www.psycontent.com

www.psycontent.com

People, Places, and Sustainability

Edited by *Gabriel Moser, Enric Pol, Yvonne Bernard, Mirilia Bonnes, José A. Corraliza, Vittoria Giuliani*

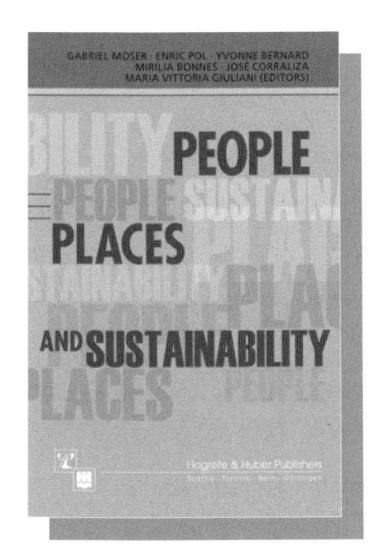

2003, 352 pages softcover
US $ / € 44.95, 0-88937-263-2

Sustainable development involves satisfying the needs of the present generation without compromising the chances for future generations. Quality of life thus plays an important part in determining how we can achieve sustainable development. What are the perspectives for the 21st century? *People, Places, and Sustainability* presents new approaches to traditional issues of people-environment studies and environmental psychology, looked at in the light of sustainability. The contributions brought together in this book cover the main issues addressed by the International Association of People-Environment Studies (IAPS), which includes psychologists, sociologists, architects, and designers. The book is divided into four main sections. Urban Change and Sustainability discusses the cultural and historical references as models for sustainable cities. Today's metropolises host increasingly culturally heterogeneous populations. Community, Attachment and Identity looks at the conditions for their sustainable development in the light of communities' participation, through processes of identification and place attachment. People's relations to their immediate residential surroundings, their workplaces, or learning environments, significantly influence their health and well-being. The contributions to Proximal and Specific Spaces concern requirements of environmental layout and design which enable them to become sustainable. Finally, Global Environment Issues points at ways to promote ecologically favorable behavior in order to achieve the conditions required for sustainable development.

List of Contributers

Florence Bordas-Astudillo, Annie Moch, Danièle Hermand, Djamel Boussaa, Edwin S. Brierley, Tim Brindley, Rocío Martín, Anthony Craig, Susan J. Drucker, Youhansen Y. Eid, Guido Francescato, Hisham S. Gabr, Ricardo García Mira, María del Mar Durán Rodríguez, José Romay Martínez, Eulogio Real Deus, M. Vittoria Giuliani, Fiorenza Ferrara, Silvia Barabott, Sengül Öymen Gür, Ayhan Bekleyen, Mia Heurlin-Norinder, Peter Kellett, Wendy Bishop, Alain Legendre, Dorothée Marchand, Annie Matheau-Police, Anna Maria Nenci, Annamaria S. de Rosa, Giuseppe Carrus, Giorgio Testa, Wouter Poortinga, Linda Steg, Charles Vlek, Teresa Ribeiro, Ombretta Romice, Valérie Rozec, Maruja Torres, Mary Joyce Hasell, John Scanzoni, Hülya Turgut, Arzu Ispalar Çahantimur, Hazem Ziada

Order at www.hhpub.com

Hogrefe & Huber Publishers

USA: 30 Amberwood Parkway, Ashland, OH 44805, Phone (800) 228-3749 or fax (419) 281-6883
Europe: Rohnsweg 25, D-37085 Göttingen, Phone +49 551 49609-0, Fax +49 551 49609-88